Topology Design for Large-Scale Interconnection Networks

Topology Design for Large-Scale Interconnection Networks

A Modern Approach

Ziyu Shao
ShanghaiTech University, Shanghai, China

Yicheng Deng
ShanghaiTech University, Shanghai, China

WILEY

Published by John Wiley & Sons, Inc., Hoboken, New Jersey.

Library of Congress Cataloging-in-Publication Data

Names: Shao, Ziyu author | Deng, Yicheng author
Title: Topology design for large-scale interconnection networks : a modern approach / Ziyu Shao, Yicheng Deng.
Description: Hoboken, New Jersey : John Wiley & Sons, [2026] | Includes bibliographical references and index.
Identifiers: LCCN 2026017383 (print) | LCCN 2026017384 (ebook) | ISBN 9781394362516 cloth | ISBN 9781394362530 adobe pdf | ISBN 9781394362523 epub | ISBN 9781394362547
Subjects: LCSH: Computer networks—Design and construction | Large scale systems—Design and construction | Computer network architectures | Electric network topology | High performance computing
Classification: LCC TK5105.5 .S525 2026 (print) | LCC TK5105.5 (ebook)
LC record available at https://lccn.loc.gov/2026017383
LC ebook record available at https://lccn.loc.gov/2026017384

Cover Design: Wiley
Cover Image: Generated with AI using Gemini 3

To Helen and Callie, for their loving support.

Contents

Preface

Ten years ago, I encountered a fundamental open problem in the field of network communication and computing: "How can we systematically design the topology of large-scale networks using a top-down approach to meet diverse performance requirements?"

At that time, the prevailing topology designs were predominantly bottom-up, relying heavily on the designer's engineering intuition and ingenious heuristic ideas. Taking the data center network in cloud computing as an example: research into its topology has a long history, giving rise to numerous topologies such as Fat-Tree, HyperX, DCell, and BCube. However, these bottom-up designs are difficult to extend or upgrade effectively to broader application fields, and several key questions remained unanswered: (1) Why do certain topologies, like Fat-tree, perform so well in real systems? (2) Under what conditions do these topologies perform poorly? (3) How can we systematically improve existing topology design and design new topologies?

In short, transforming the design of large-scale network topologies from *art* to *science* became my research focus.

My approach to solving this problem was to combine reverse engineering with forward engineering. The guiding principle of this approach is that well-performing engineering designs, born from experience and heuristics, should implicitly correspond to some underlying mathematical problem and its associated principles. The design can thus be viewed as a solution or approximate solution to that implicit mathematical problem.

Why should we uncover the hidden mathematical problems and principles when a solution already exists? Because only by doing so can we understand when such solutions work well, when they do not, and how to systematically improve them in a top-down manner. In other words, we seek to understand not only *how* but also *why*. In fact, many classic designs in network communication and computing – including TCP and BGP on the Internet, CSMA in wireless networks, and BitTorrent in peer-to-peer networks, have had their underlying mathematical problems and principles uncovered, serving as successful examples of the reverse-and-forward engineering approach.

My starting point at the time was to perform reverse engineering on the Fat-tree topology, the most commonly used topology in data center networks. This was a crucial step. I tried numerous approaches and tools, yet after one year I remained stuck without progress.

Fortunately, in the spring of 2017, I arrived at the University of California, Berkeley, as a Visiting Professor in the EECS Department. I printed out the Fat-tree topology diagram and hung it in my office, staring at it repeatedly every day. Simultaneously, I began voraciously attending various lectures, hoping to find inspiration. In one lecture, I learned about the *renormalization group* in physics. This is a powerful tool for studying multiscale complex systems. Its main idea is to focus on the macro while letting go of the micro, proceeding step-by-step: one "coarse-grains" the system, integrating microscopic patterns into effective macroscopic variables, and then observes how the coarse-grained configurations evolve. This implies that macroscopic laws are not merely a simple superposition of microscopic laws, but are emergent outcomes derived through scale transformation.

This was the first ray of light in my journey through the dark. Previously, I had always understood topology from a conventional graph perspective, seeing only node-to-node connections. This was a microscopic perspective, which naturally made it difficult to find macroscopic laws – namely, the universal rules of topology design. Inspired by the renormalization group idea, I began coarse-graining topologies, viewing them as composite graphs in an unconventional way, focusing on connections between subgraphs.

The next question was: "How to systematically decompose a topology into connections of multiple subgraphs?" Inspired by another lecture, I re-read the classic work of Claude Shannon (the father of information theory): "A Mathematical Theory of Communication." From this paper, I learned Shannon-style thinking on the decoupling and modularization of complex systems and a first-principle approach to understand the fundamental limit of the system. This was the second ray of light and my mind was completely opened.

After further attempts, I finally discovered the hidden mathematical problem behind the Fat-tree topology design: "The optimal trade-off design of a parameterized topology, constructed by connecting block incidence graphs as subgraphs via balanced incomplete block designs (BIBD), under given performance metrics and constraints." Generalizing this problem by extending subgraphs from block incidence graphs to general *Ramanujan graphs* and extending BIBD to general combinatorial designs, the mathematical principle of large-scale network topology design can be summarized in one sentence:

"Construct the topology through multi-round recursive iteration, where in each round Ramanujan graphs are adopted as subgraphs connected via combinatorial designs, ultimately achieving hierarchical load balance, high throughput in local regions, and low latency across global regions."

At this point, the fog cleared, and the path became distinct. On this foundation, I led my research group to distill a general topology design framework from the above mathematical principles and engineering structures. We found that almost all known important topologies in data centers and supercomputers can be systematically generated from this framework. Moreover, through this framework, we can systematically improve existing topologies and understand their performance limits. By combining this framework with a three-layer topology model in forward engineering, we systematically generated various new topologies tailored to different performance needs. Parts of these results were published at the IEEE INFOCOM 2020 conference and received enthusiastic responses from both academia and industry.

As I continued to refine this framework, I also initiated the writing of this book. Interestingly, over the past five years, with the rapid development of artificial intelligence, GPU-centric intelligent computing centers (AI clusters) have expanded explosively. Once again, academia and industry adopted a bottom-up, heuristic approach to design several topologies for these AI clusters. My student Yicheng and I were delighted to find that, with only slight adjustments to my proposed general topology design framework, existing topologies of AI cluster could be unified, their strengths and limitations understood, and they could be systematically improved or entirely new ones designed. The relevant details have become a part of this book.

Now the book writing is finished. To the best of my knowledge, this book is the first monograph dedicated to the design of modern large-scale network topologies. Readers are advised to follow the chapters in order.

- Chapter 1 first introduces the background of network topology design, then covers foundational mathematics including graph theory basics, expander graphs, and combinatorial design. Finally, it introduces two parameterized tools essential for topology design: the block incidence graph and the transversal design graph.
- Chapter 2 details a general framework for large-scale network topology design. In this framework, topology design is decoupled into three granularities and integrated into a unified design flow (including intra-module design, connection design, inter-module design, and finalization). Based on this framework, Chapter 2 reinterprets several representative topology design instances, providing a new way of understanding these classic designs and exemplifying the rationality and effectiveness of the framework.

- Chapter 3 focuses on systematic incremental improvements to existing classic topologies. It first details Tree-like and HyperX-like topologies in switch-centric designs, then briefly introduces DCell-like and BCube-like topologies in server-centric designs.
- Chapter 4 focuses on designing a completely new topology from scratch. It first summarizes 24 design patterns and their corresponding design parameters, then introduces performance metrics for the design models. Guided by these metrics, it presents principles for selecting design patterns and parameters, illustrated with detailed examples of representative topologies.
- Chapter 5 first briefly discusses potential application scenarios of the proposed general topology design framework, then details a unified scheme for topology design of AI cluster (intelligent computing center), extending from the original single-domain design to a dual-domain joint design of scale-up and scale-out. Finally, it looks ahead to future topology design, covering both theoretical exploration and practical extension.
- All proofs of lemmas and theorems in this book are provided in the Appendix for reference.

Reflecting on this scientific odyssey of the past decade, I am filled with mixed emotions. In today's restless and utilitarian era, where superficial exploration and quick success often capture the spotlight, few possess the patience to settle down and grapple with fundamental open problems.

Yet, 10 years ago, I made a bold decision to double down on the challenge of network topology design framework. I have always maintained that believing is seeing, rather than seeing is believing. It was this unwavering faith that sustained me through those early years of fruitless searching, eventually leading me to the light. I am fortunate that the outcome has been fruitful. This was truly *my road less traveled by*, and as the poet Robert Frost so timelessly wrote:

"Two roads diverged in a wood, and I—I took the one less traveled by, And that has made all the difference."

Looking back at this mental journey – from confronting an open problem to developing a universal design framework – I am reminded of what my academic *grand-grandfather*, MIT Professor Robert Gallager, once said: "A good theory should shrink rather than grow the knowledge tree." In this sense, I hope the work described in this book stands as a *good theory* – one that simplifies and clarifies our understanding of network topology design. Finally, I close with a saying that has long been my personal conviction: "There is nothing more practical than a good theory."

Shanghai, China
January 2026

Ziyu Shao

Acknowledgments

This book is the culmination of years of research, and it would not have been possible without the collective effort and support of many individuals.

First and foremost, I would like to express my deepest gratitude to my students: Yijia Chang, Xi Huang, Yinxu Tang, and Longxiulin Deng. Their dedication, brilliant insights, and tireless efforts were instrumental in developing and refining the general topology design framework presented in this book. It has been a privilege to work alongside such talented individuals.

Most importantly, I want to thank my family. Throughout the long years of dark exploration and the intense process of writing this book, their infinite support and tolerance were my constant sanctuary. Their belief in me, even when the path forward was unclear, made this achievement possible. This book is as much theirs as it is mine.

Ziyu Shao

Acronyms

BIBD Balanced Incomplete Block Design
BIG Block Incidence Graph
CBD Complete Block Design
CBG Complete Bipartite Graph
CG Complete Graph
CN Computing Node
COTS Commercial Off-the-shelf
CPU Central Processing Unit
DCN Data Center Network
GDD Group Divisible Design
GN Gateway Node
GPU Graphics Processing Unit
HN Hybrid Node
HPC High-performance Computer
IoT Internet of Things
ISN Intermediate Super-node
NIC Network Interface Card
PBD Pairwise Balanced Design
POPS Peta Operations Per Second
RN Routing Node
SLEM Second Largest Eigenvalue Modulus
SUHN Scale-up Hybrid Node
SURN Scale-up Routing Node
TD Transversal Design
TDG Transversal Design Graph
TOPS Teta Operations Per Second
TPU Tensor Processing Unit

1

Introduction

1.1 Motivation

The past decade has witnessed the rapid evolution of a wide range of large-scale computer networks, such as data center networks (DCNs) (Chen et al. 2016; Andreyev 2014), high-performance computers (HPCs) (Trobec et al. 2016), artificial intelligence infrastructures (Gershon et al. 2024), distributed machine learning systems (Yang et al. 2012), artificial intelligence of things (AIOT) systems (Siam et al. 2025), multi-agent networks (MANs) (Oroojlooy and Hajinezhad 2023), and low Earth orbit (LEO) satellite networks (Pan et al. 2023). Such systems are continuously serving as the cornerstones that underlie the proliferation of various network applications and services, unleashing the power of cutting-edge technologies such as artificial intelligence and driving the world toward a more digitalized and intelligent future.

To accommodate the ever-growing network traffic and provide more diversified services with stringent requirements, Cisco Corporation (2020), the crux lies in the design of the large-scale networks that underpin such systems. The design is overwhelmingly complicated as it involves concerns from various aspects, including but not limited to network topology (Xia et al. 2017), routing protocol design (Chen et al. 2016), and traffic engineering mechanism (Shao et al. 2013). Particularly, *network topology*, defined as the *interconnection* relationship among the devices (e.g., switches, servers) in a network, plays a fundamental role that imposes profound impacts on other aspects of the network.

For network topology design, the academia and the industry have devoted years of efforts in developing numerous solutions that are applied to various scenarios and systems. Overall, there are two veins that lead existing works.

Works along the first vein focus on designing concrete topologies for large-scale networks, especially for DCNs (Al-Fares et al. 2008; Walraed-Sullivan et al. 2013; Liu et al. 2013; Singla et al. 2012; Gyarmati and Trinh 2010; Shin et al. 2011; Yu and Qian 2016; Valadarsky et al. 2016) and HPCs (Kim et al. 2008; Ahn et al. 2009;

Shpiner et al. 2017). Such concrete topology designs can be roughly divided into *deterministic* topologies (Al-Fares et al. 2008; Walraed-Sullivan et al. 2013; Liu et al. 2013; Kim et al. 2008; Ahn et al. 2009; Shpiner et al. 2017) and *random* (or *pseudo-random*) topologies (Singla et al. 2012; Gyarmati and Trinh 2010; Shin et al. 2011; Yu and Qian 2016; Valadarsky et al. 2016). The deterministic topologies are mostly designed based on engineering heuristics. Owing to their decent performance and low complexities in deployment and management, such topologies have been widely acknowledged and commonly adopted in practice, with a wealth of follow-on works (e.g., the well-known Fat-Tree topology [Al-Fares et al. 2008] and its variants Facebook Fabric [Andreyev 2014] and Google Jupiter [Singh 2015]). In parallel, some other works adopt tools from graph theory (especially expander graphs [Hoory et al. 2006])[1] to devise topologies that are conducted in a randomized manner in pursuit of better performance (e.g., the average shortest path length and the number of paths between devices). Nonetheless, the applicability of such solutions is hindered by high complexities and prohibitive costs of deployment and management in practice.

Works along the second vein aim to design network topology in a *synthesized* fashion. The key idea is to search topologies by combining engineering insights based on existing topology designs with optimization-based or learning-based approaches. For example, Schlinker et al. (2015) present Condor, an efficient framework that generates candidate topologies subject to pre-specified design objectives and constraints in a declarative manner. Later, Wang et al. (2018) propose an online traffic-driven deep learning solution to generate high-performance network topologies for reconfigurable networks such as optical/wireless networks.

Despite the justified effectiveness of these works, today's network topology design remains an empirical discipline that heavily relies on engineering heuristics and experimental verification. Moreover, the evaluation is often coupled with other technical implementations (e.g., particular routing protocol, traffic engineering, and network administration strategies), making it hard to investigate the merits of the topologies themselves. Unlike some other disciplines in the field of network system optimization (e.g., layered network protocol design) with mature design frameworks (Chiang et al. 2007; Wei et al. 2006; Griffin et al. 2002; Lee et al. 2007), a *systematic* framework to understand, analyze, and evaluate different network topologies, especially for large-scale networks, is still missing. On its way toward being a science, there are two impeding factors centering around the *intrinsic* complexities of topology design and the lack of a *fundamental* understanding of its design space.

On one hand, topology design usually involves various performance metrics that are potentially conflicting with each other, such as scalability, cost, latency,

1 Intuitively, expander graph is a type of large-scale graph with dense connectivity properties.

throughput, and fault-tolerance. On the other hand, large-scale networks contain numerous devices with a massive number of possibilities in their interconnection relationship. Even a minor modification of the interconnections among devices can lead to a significant performance change (Liu et al. 2013). Without a fundamental understanding of the trade-offs among such metrics, topology design remains a costly trial-and-error-based procedure – more like an art than a science. Although existing optimization-based or learning-based approaches are promising to accelerate this procedure, the search space is limited by human heuristics. Besides, the resulting designs still resort to empirical evaluation that only explains *how* they work but not *why* they work, not to mention how they can be improved systematically toward better performances.

To fill such a gap, in this monograph, we present a unified design framework of topology design for large-scale networks, aiming to provide a *first-principled* perspective to understand, analyze, evaluate, and generate topology design in a *systematic* fashion. An overview of the design framework is given in the next section.

1.2 Overview of the Design Framework

In this section, we briefly introduce the three essential parts of the design framework.

1. **A unified design procedure:** Inspired by Shannon's style thinking of *modularization* and *decoupling*, at the core of the framework lies a general design procedure serving as a *unified* language to describe network topology design. Based on extensive reverse-engineering over existing solutions, the procedure provides a *novel* modularized perspective to decouple network topology. Following the procedure, a network topology can be generated step by step in a *top-down* fashion: first by the design of individual modules[2] (aka intra-module design), then by the design of connections that links modules (aka connection design), and finally by the design of how modules are organized via the designed connections (aka inter-module design). Existing topologies can be viewed as instances with different design choices in each step.
2. **Concrete analytic tools for topology design:** For each step in the design procedure, the framework shows dedicated principles to instruct the corresponding design. Under the guidance of such design principles, the framework provides concrete analytic tools based on graph theory and combinatorial design theory. Particularly, the tools of intra-module and

2 In real-world large-scale networks, modules can be server racks, pods, device groups, containers, clusters, etc.

inter-module designs center around the construction of large-scale networks with dense interconnections, while the tools of connection and recursion design aim to achieve an ultra-low latency among modules. Notably, the framework provides a fine-grained distinction of the atomic modes and novel tools to extend connection design. With such tools, the framework not only opens a unified perspective to review existing topologies (e.g., *switch-centric* topologies and *server-centric* topologies in DCNs), but also enables network designers to develop network topologies in a systematic way.

3. **Quantitative analysis of performance metrics:** Based on the two parts and a quantitative analysis upon such design patterns, the framework offers a comprehensive understanding of the trade-offs among the primary performance metrics in topology design. The results justify the effectiveness of existing topology designs, illuminate their positions in the design space, and present an instructive guideline for further improvements.

1.3 Existing Works

In this section, we take a review of existing works on topology design as the preliminaries of our design framework. As mentioned earlier, our review basically follows two veins. One is related to the works of *concrete* designs for large-scale networks. The other vein involves the works that develop tools or frameworks to *search* topology designs subject to pre-specified requirements.

1.3.1 Concrete Topology Designs

Over the past decades, a wealth of concrete topologies have been proposed for large-scale networks, such as DCNs and HPCs. In such networks, part of devices take over computing functions (aka servers, hosts, computers, or terminals), while the other devices manage routing functions (aka switches, routers, or routing devices) (Chen et al. 2016). Depending on the randomness involved in the process of topology design, existing topologies can be divided into *deterministic* topologies and *random* (or *pseudo-random*) topologies. If a topology's design process does not involve any random operation (e.g., the selection of connections between devices), it is called a deterministic topology. Otherwise, it is a random topology.

1.3.1.1 Deterministic Topologies

Regarding deterministic topologies, depending on whether servers are responsible for networking and routing functions (Chen et al. 2016), deterministic topologies can be further divided into *server-centric* topologies and *switch-centric* topologies. If none of the servers are responsible for networking and routing functions in the topology, it is called a switch-centric topology. Otherwise, it is a server-centric topology.

Regarding *switch-centric* topology design, the representative topologies among existing works are demonstrated as follows.

Fat-Tree (Al-Fares et al. 2008) is an instance of the well-known Clos network topology (Scott et al. 2006). It has been widely adopted in various commercial data center networks. Overall, a Fat-Tree topology is constructed as a multi-layer network using a group of homogeneous commodity switches with the same port number. In particular, when the port number is equal to some even integer p, the topology is called a *p-ary Fat-Tree*. We take the classical three-layer p-ary Fat-Tree as an example. Such a topology contains three layers: edge layer, aggregation layer, and core layer. Switches in the edge and aggregation layers are divided into *p pods*. These pods are interconnected by $p^2/4$ core switches. Each pod contains $p/2$ aggregation switches and $p/2$ edge switches. Each edge switch is directly connected with $p/2$ servers and $p/2$ aggregation switches in the same pod. Meanwhile, each aggregation switch is directly connected with $p/2$ edge switches in the same pod and $p/2$ core switches. Specifically, the i-th aggregation switch in each pod is connected with the $\left[\frac{p}{2}(i-1)+j\right]$-th core switch, for $i, j = 1, 2, 3, \ldots, p/2$. In total, such a topology can support $p^3/4$ servers with $5p^2/4$ switches. Figure 1.1 shows a three-layer 4-ary Fat-Tree with twenty 4-port switches.

Aspen Tree (Walraed-Sullivan et al. 2013) is an improvement of multi-layer Fat-Tree for better fault-tolerance. We take the Aspen Tree with L layers as an example where $L \geq 3$. Compared to Fat-Tree, it removes links between switches in the l-th layer ($l = 2, 3, \ldots, L$) and part of switches in the $(l-1)$-th layer from some pods, leaving more redundant links for switches in the remaining pods. Such a

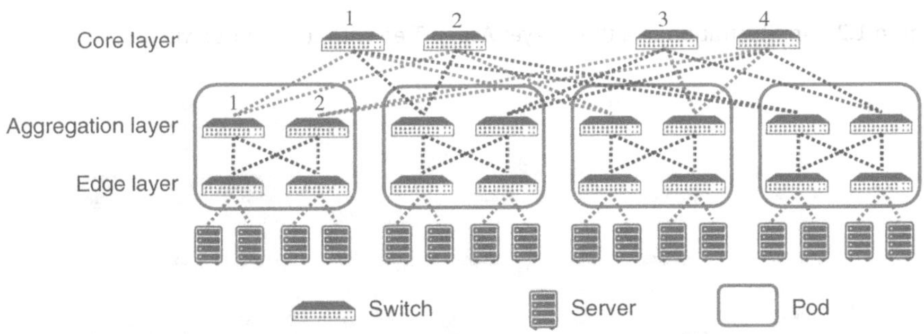

Figure 1.1 An illustration of a three-layer 4-ary Fat-Tree with 4-port switches. This topology contains 16 servers and 20 switches. In each pod, the first aggregation switch is connected to the first and the second core switches, while the second aggregation switch is connected to the third and the fourth core switches.

design improves fault-tolerance at the expense of a lower scalability. Figure 1.2 demonstrates a three-layer Aspen Tree with ten 4-port switches.

AB Fat-Tree (F10) (Liu et al. 2013) is another improvement of multi-layer Fat-Tree for better fault-tolerance. We take the AB Fat-Tree with L layers as an instance where $L \geq 3$. Unlike the homogeneous connection patterns between core switches and pods in Fat-Tree, AB Fat-Tree breaks the symmetry by introducing two distinct connection patterns, called *A-striping* and *B-striping*, respectively. For each $(l-1)$-th-layer switch, it follows the A-striping pattern to be connected with $p/2$ *consecutive* switches in the l-th layer. For each $(l-1)$-th-layer switch, it follows the B-striping pattern to be connected to $p/2$ switches in the l-th layer with a *stride* of $(p/2)^{(l-2)}$. Such a topology leads to a better fault-tolerance than Fat-Tree. Figure 1.3 gives an example of a three-layer AB Fat-Tree with twenty 4-port switches.

HyperX (Ahn et al. 2009) represents a family of topologies, including YARC (Scott et al. 2006), hypercube (Cheng and Chuang 1994), and Flattened Butterfly (Kim et al. 2007). Switches in a level-L HyperX (aka HyperX$_L$) are organized

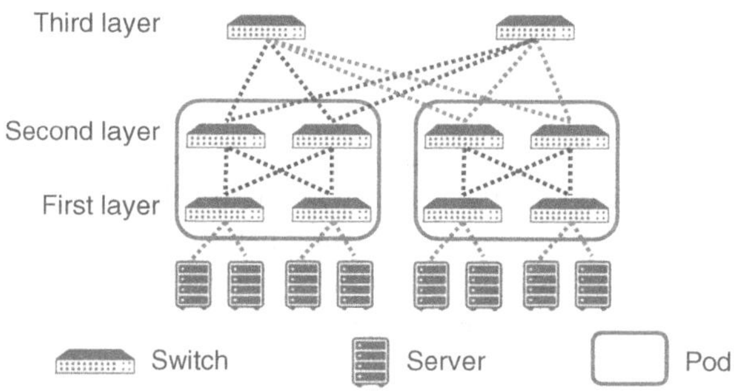

Figure 1.2 An illustration of a three-layer Aspen Tree with ten 4-port switches.

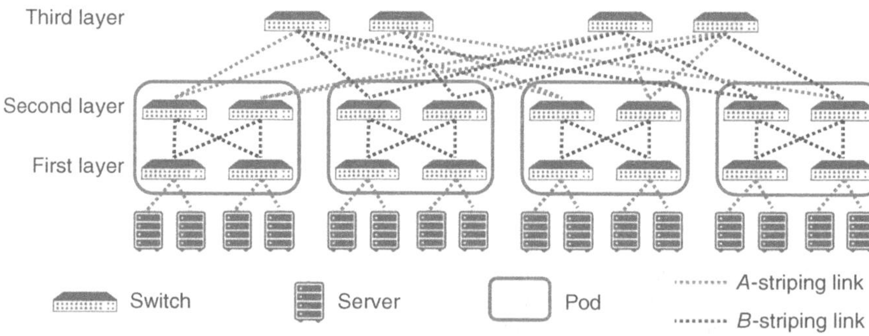

Figure 1.3 An illustration of a three-layer AB Fat-Tree with twenty 4-port switches.

as an L-dimensional integer lattice ($L \geq 1$). Each switch is identified by a coordinate vector $I \triangleq (I_1, I_2, ..., I_L)$ where $I_i = 1, 2, ..., s_i$ for $1 \leq i \leq L$. Such a lattice has a multi-dimensional nested structure. Each i-dimensional lattice contains s_i $(i-1)$-dimensional lattices. Then switches with the same coordinate sub-vector $(I_1, I_2, ..., I_{i-1})$ in all of such $(i-1)$-dimensional lattices are interconnected as a complete graph. Note that a zero-dimensional lattice is a switch. Following this way, HyperX with distinct values of L and s_i leads to different topologies. For example, hypercube can be viewed as an instance of HyperX with $s_i = 2$ for all $1 \leq i \leq L$. Another example is the topology of YARC high-radix switch. It can be viewed as an instance of HyperX with $L = 2$ and $s_1 = s_2 = 8$. Figure 1.4 shows an example of a HyperX with $L = 2$, $s_1 = 2$, and $s_2 = 4$. The topology is formed by eight 5-port switches. Each switch is connected with one and only one server.

As for *server-centric* topology, the typical designs are discussed as follows.

DCell (Guo et al. 2008) is a server-centric topology for DCNs, in which both switches and servers are involved with routing functions. DCell has a recursively-defined hierarchical structure. Each atomic component (called a *level-0 DCell* or *DCell*$_0$) contains p servers which are connected via a shared p-port switch. For each level-i DCell with $i \geq 1$ (aka DCell$_i$), it is constructed as a complete graph with m_i level-$(i-1)$ DCells. The value of m_i is equal to the number of servers in each level-$(i-1)$ DCell plus one. In this way, a level-i DCell requires each server to be

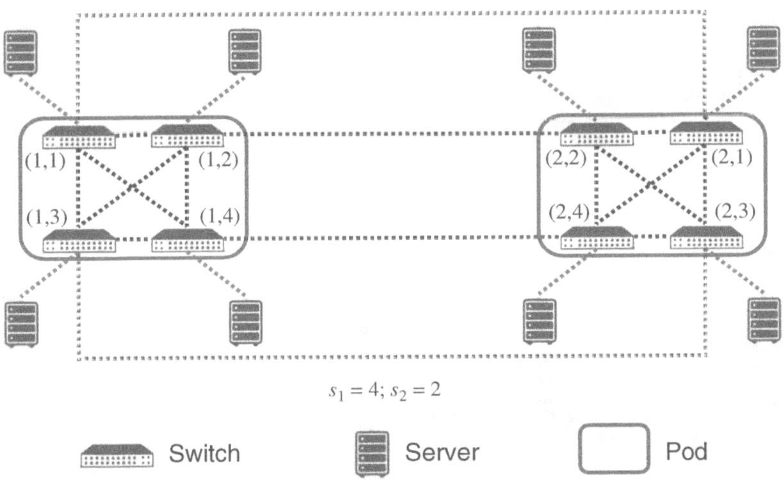

Figure 1.4 An illustration of a level-2 HyperX with eight 5-port switches.

equipped with $(i + 1)$ network interface cards (NICs).[3] Figure 1.5 demonstrates a level-1 DCell with five 4-port switches.

FiConn (Li et al. 2009) is a DCell-like topology. Compared to DCell topology, it requires each server to be equipped with only two NICs (an active NIC and a backup NIC) which makes it applicable for practical use. In such a topology, each atomic component (called a *level-*0 *FiConn* or $FiConn_0$) contains p servers that are connected via a shared p-port switch in the same way as DCell. For each level-i FiConn with $i \geq 1$ (aka FiConn_i), it is constructed as a complete graph with g_i level-$(i - 1)$ FiConns. The value of g_i is equal to half of the number of servers with available backup NICs in each level-$(i - 1)$ FiConn plus one. Compared with DCell, FiConn incurs lower wiring costs at the expense of a reduced network capacity. Figure 1.6 provides an instance of a level-1 FiConn with three 4-port switches.

BCube (Guo et al. 2009) is another recursively-defined hierarchical topology. Each atomic component (called a *level-*0 *BCube* or $BCube_0$) contains p servers which are connected via a shared p-port switch in the same way as DCell. For each level-i BCube with $i \geq 1$ (aka BCube_i), it is constructed with p level-$(i - 1)$ BCubes and p^i switches. Each level-$(i - 1)$ BCube contains p^i servers and each server is directly connected to one and only one switch in the level-i BCube. In this way, a level-i BCube requires each server to be equipped with $(i + 1)$ NICs. Figure 1.7 shows a level-1 BCube with eight 4-port switches.

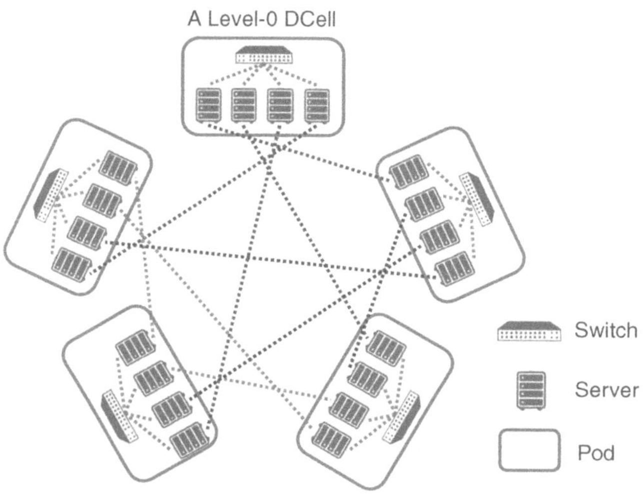

Figure 1.5 An illustration of a level-1 DCell with five 4-port switches.

3 An NIC is a control component of a server via which the server can establish communication links with other devices (e.g., servers and switches) in a network. For each server, the number of its NICs determines the maximum number of devices it can be connected to.

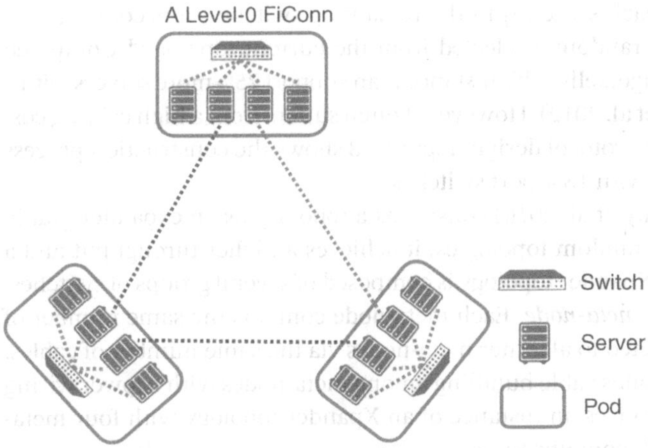

Figure 1.6 An illustration of a level-1 FiConn with three 4-port switches.

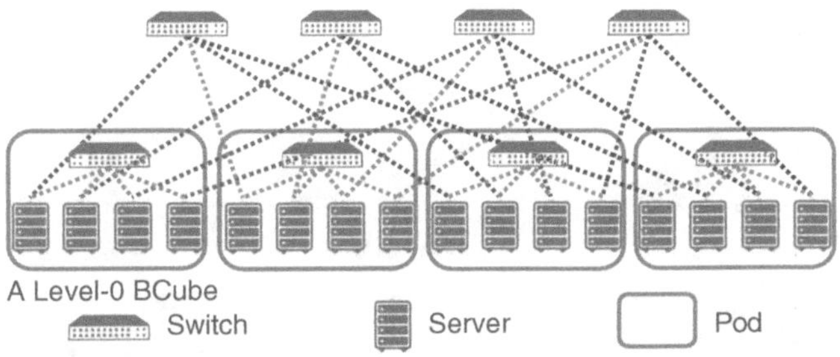

Figure 1.7 An illustration of a level-1 BCube with eight 4-port switches.

1.3.1.2 Random Topologies

Different from deterministic topologies, existing random topologies do not limit the type of routing devices; i.e., they can be either switches or servers with routing functions. Hence, in the following discussion, we do not make the distinction between switch-centric and server-centric designs for random topologies. The representative random topologies among existing works include Jellyfish (Singla et al. 2012), Scafida (Gyarmati and Trinh 2010), Small-World Data Center (Shin et al. 2011), Space Shuffle (Yu and Qian 2016), Xpander (Valadarsky et al. 2016), and Random Folded Clos (Guo and Yang 2015). We provide an exposition for two of them.

Jellyfish (Singla et al. 2012) is a random topology for DCNs by interconnecting switches as an approximately random regular graph. It can be constructed in the following way. Given a group of p-port switches, $(p + 1)$ switches are linked as a complete graph. For each switch s_i in the remaining switches, it is connected to two switches that are randomly selected from the current topology.[4] Compared with Fat-Tree, on average, Jellyfish's instances can support 25% more servers with a lower latency (Singla et al. 2012). However, it often suffers from a high cabling cost and a complex routing protocol design. Figure 1.8 shows the construction process of a Jellyfish topology with 18 4-port switches.

Xpander (Valadarsky et al. 2016) constructs a topology as an expander graph. Compared with other random topologies, it achieves a higher throughput and a higher scalability. An Xpander topology is composed of several groups of switches. Each group is called a *meta-node*. Each meta-node contains the same number of switches and is connected to all other meta-nodes via the same number of cables. Such a structure facilitates cable bundling among meta-nodes with a lower wiring complexity. Figure 1.9 gives an instance of an Xpander topology with four meta-nodes. Each meta-node contains three switches.

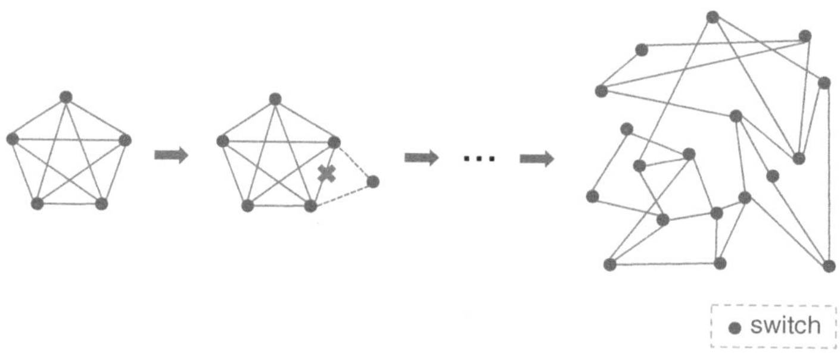

Figure 1.8 The construction process of a Jellyfish topology with 18 4-port switches.

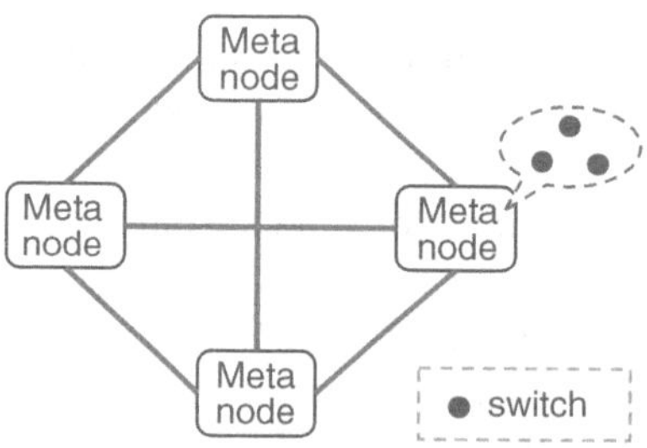

Figure 1.9 An illustration of an Xpander topology with four meta-nodes. Each meta-node contains three switches.

4 This is achieved by removing the link between the two selected switches and reconnecting them with switch s_i.

1.3.2 Search-based Topology Designs

The idea of this vein is to adopt search-based heuristic approaches to generate desired topologies with specific objectives and constraints.[5] To this end, a wealth of frameworks have been proposed in recent years (Schlinker et al. 2015; Curtis et al. 2010, 2012). These frameworks basically take design goals and constraints as inputs and generate a series of candidate topologies (See Figure 1.10). Some classical frameworks are introduced as follows.

LEGUP (Curtis et al. 2010) upgrades existing Clos-like topologies (e.g., Fat-Tree) with a higher bisection bandwidth. Compared to the classical Clos topology with homogeneous switches, LEGUP extends the design with heterogeneous switches and adopts a branch-and-bound algorithm to achieve near-optimal performances (e.g., bisection bandwidth and reliability) under host traffic constraints.

REWIRE (Curtis et al. 2012) is an optimization-based framework to design topologies for DCNs. In particular, REWIRE adopts a simulated annealing-based method to produce topologies that achieve a higher bisection bandwidth and a lower worst-case end-to-end latency than Fat-Tree. However, such a search process does not consider some critical deployment constraints in topology design (e.g., the limitation of physical distance between deployed devices) which may be a concern in practice.

Perseus (Mudigonda et al. 2011) is another optimization-based framework for the design of cost-effective topology in container-sized DCNs. To this end, by considering various primary metrics in topology design (e.g., cabling costs, cable installation costs, switch costs, and hop counts) and their related constraints,

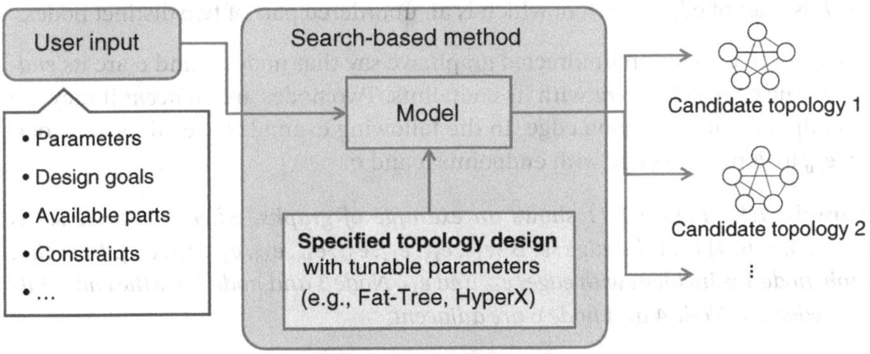

Figure 1.10 An illustration of search-based topology design.

5 Such approaches generally adopt deterministic optimization methods (Curtis et al. 2012) (e.g., linear programming, convex optimization) or machine learning techniques (Wang et al. 2018) (e.g., deep learning).

Perseus provides a workflow to quantify the costs and benefits of potential topologies (especially the extended generalized Fat-Trees [EGFTs] [Ohring et al. 1995] and HyperX topologies), leading designers toward more targeted topology choices.

Condor (Schlinker et al. 2015) is described as a pipeline that facilitates the rapid generation and evaluation of DCN topologies. Compared to concrete topologies, Condor allows designers to express requirements and goals via a declarative topology description language (TDL). According to the TDL descriptions, Condor can synthesize candidate designs with desired performances and support on-demand network expansion based on effective topology evaluations.

xWeaver (Wang et al. 2018) is a traffic-driven deep learning approach which enables high-performance global topology configurations in DCNs. At the core of xWeaver lie a set of properly-structured neural network models. Particularly, by adopting such models to learn the underlying mapping between traffic patterns and topology configurations, xWeaver can generate near-optimal topology configurations with continuous model improvements. Compared to Fat-Tree, xWeaver provides a more flexible and more adaptive way to design topologies that are better suited to ever-changing traffic dynamics.

1.4 Mathematical Background I: Graph Theory

1.4.1 Undirected Graph

An *undirected graph* G is a pair (V, E) such that

- V is a set of nodes;
- E is a set of edges, each of which is an unordered pair of two distinct nodes.

If (u, v) is an edge in an undirected graph, we say that nodes u and v are its *endpoints*. The edge is *incident* with its endpoints. Two nodes are *adjacent* if they are the endpoints of a common edge. In the following examples, we adopt the notation e_{uv} to denote an edge with endpoints u and v.

Example 1.1 *Figure 1.11 shows an example of graphs, where the node set is* $\{1, 2, 3, 4, 5, 6, 7\}$ *and the edge set is* $\{e_{12}, e_{13}, e_{34}, e_{24}, e_{36}, e_{46}, e_{45}, e_{67}, e_{57}\}$. *In such a graph, node* 1 *is incident with edges* e_{12} *and* e_{13}. *Node* 3 *and node* 4 *are the endpoints of the edge* e_{34}. *Node* 4 *and node* 6 *are adjacent.*

1.4.2 Order and Node Degree

For a graph, its *order* is the number of nodes in such a graph. For each node, its *degree* is the number of edges that are incident with it.

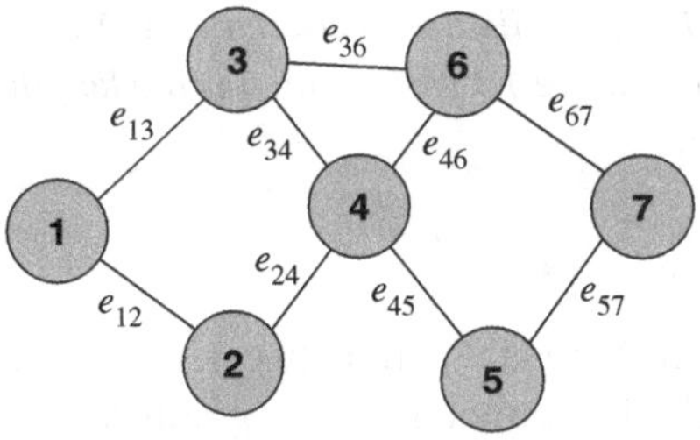

Figure 1.11 An undirected graph with node set V as $\{1, 2, 3, 4, 5, 6, 7\}$ and edge set E as $\{e_{12}, e_{13}, e_{34}, e_{24}, e_{36}, e_{46}, e_{45}, e_{67}, e_{57}\}$.

Example 1.2 *For the graph shown in Figure 1.11, its order is* 7 *and the degree of node* 4 *is* 4.

1.4.3 Path

Given two distinct nodes s and t in a graph $G = (V, E)$, a *path* between them is a sequence of nodes $\langle v_1, v_2, ..., v_k\rangle$ such that $v_1 = s$, $v_k = t$, and $(v_{i-1}, v_i) \in E$ for $i = 2, ..., k$. The *length* of a path is the number of nodes in the path minus one. The *shortest path* between two nodes is the path with the shortest length between them.

Example 1.3 *In the graph shown in Figure 1.11, we select three paths between node* 1 *and node* 7 *as examples, which are denoted by* P_1*,* P_2 *and* P_3*. The path* P_1 *is* $\langle 1, 3, 6, 7\rangle$*, the path* P_2 *is* $\langle 1, 2, 4, 5, 7\rangle$*, and the path* P_3 *is* $\langle 1, 3, 4, 6, 7\rangle$*. The lengths of paths* P_1*,* P_2*, and* P_3 *are* 3, 4, *and* 4*, respectively. The shortest path between node* 1 *and node* 7 *is the path* P_1 *with length* 3.

Given two paths $P_1 = \langle v_1, v_2, ..., v_m\rangle$ and $P_2 = \langle u_1, u_2, ..., u_n\rangle$, if $v_1 = u_1$ and $v_m = u_n$, we describe the disjointness of such two paths in the following two ways:

- the paths P_1 and P_2 are *edge-disjoint* if $(v_{i-1}, v_i) \neq (u_{j-1}, u_j)$ for any $1 < i \leq m$ and $1 < j \leq n$.
- the paths P_1 and P_2 are *node-disjoint* if $v_i \neq u_j$ for any $1 < i < m$ and $1 < j < n$.

Example 1.4 *For the paths* P_1*,* P_2*, and* P_3 *in Example 1.3, the paths* P_1 *and* P_2 *are both edge-disjoint and node-disjoint, while the paths* P_2 *and* P_3 *are edge-disjoint but not node-disjoint. The paths* P_1 *and* P_3 *are neither edge-disjoint nor node-disjoint.*

1.4.4 Distance and Diameter

The *distance* between two nodes is the length of the shortest path between them. For a graph, its *diameter* is the maximum distance between any two nodes in the graph.

Example 1.5 *In the graph shown in Figure 1.11, the distance between node* 1 *and node* 7 *is* 3. *The diameter of the graph shown in Figure 1.11 is* 5, *which is the length of the shortest path between node* 1 *and node* 5.

1.4.5 Connectivity

A graph with at least two nodes is *connected* if there is a path between each pair of nodes; otherwise, the graph is *disconnected*. The connectivity of a graph can be described in the following two ways.

- The *edge connectivity* of a graph is the minimum number of edges whose removal will disconnect the graph.
- The *node connectivity* of a graph is the minimum number of nodes whose removal will disconnect the graph.

Example 1.6 *For the graph shown in Figure 1.11, the removal of at least two edges (i.e.,* e_{67} *and* e_{57}*) will disconnect the graph. Hence, its edge connectivity is equal to* 2.

Example 1.7 *For the graph shown in Figure 1.11, the removal of at least two nodes (i.e., node* 4 *and node* 6*) will disconnect the graph. Hence, its node connectivity is equal to* 2.

In a graph $G = (V, E)$, for any $S \subset V$, we denote its complement as $\bar{S} = V - S$. The set of edges, whose one endpoint belongs to S and another endpoint belongs to $\bar{S}$, is called the *boundary* of S, denoted by ∂S.

Example 1.8 *For the graph shown in Figure 1.11, suppose* $S = \{1, 3, 6\}$, *then its complement* $\bar{S} = \{2, 4, 5, 7\}$ *and its boundary* $\partial S = \{e_{12}, e_{34}, e_{67}, e_{46}\}$.

1.4.6 Special Undirected Graphs

Undirected graphs have various classifications based on the properties. In the following parts, the definitions of some special graphs that are involved in this monograph are given.

1.4.6.1 Regular Graph

A *regular graph* is a graph where each node has the same degree. A regular graph with node degree d is called *d-regular graph*.

Example 1.9 *Figure 1.12 shows an example of* 2*-regular graphs.*

1.4.6.2 Complete Graph

Complete graph (CG) is a special case of regular graph where each pair of nodes is adjacent to each other. The complete graph with n nodes is denoted by K_n.

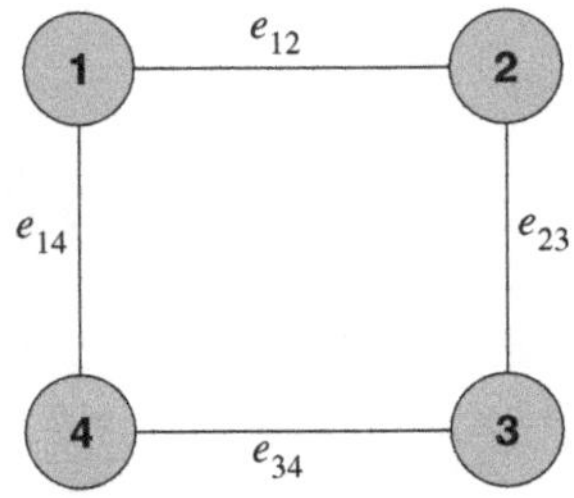

Figure 1.12 A 2-regular graph with node set V as $\{1, 2, 3, 4\}$ and edge set E as $\{e_{12}, e_{23}, e_{34}, e_{14}\}$.

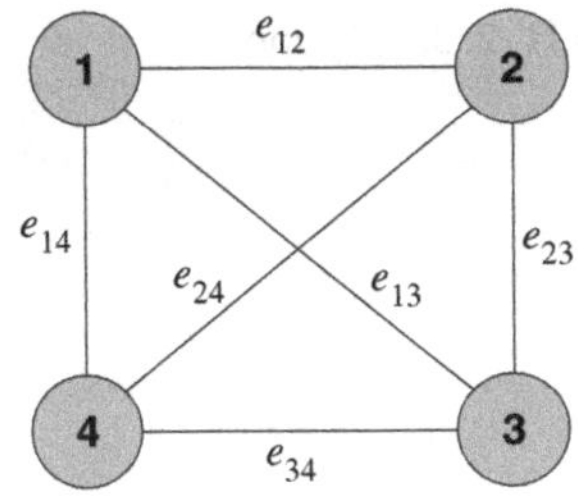

Figure 1.13 A complete graph K_4 (3-regular graph) with node set V as $\{1, 2, 3, 4\}$ and edge set E as $\{e_{12}, e_{23}, e_{34}, e_{14}, e_{13}, e_{24}\}$.

Remark 1.1 A CG K_n is a $(n-1)$-regular graph.

Remark 1.2 The diameter of a CG is *one*.

Example 1.10 *Figure 1.13 shows a CG K_4, which is also a 3-regular graph.*

1.4.6.3 Bipartite Graph

A graph is a *bipartite graph* if its nodes can be partitioned into two disjoint sets V_1 and V_2 (i.e., $V_1 \cap V_2 = \emptyset$), such that any pair of nodes in V_i is not adjacent to each other for $i = 1, 2$. We call each disjoint set as a *node partition*.

Example 1.11 *Figure 1.14 shows an example of bipartite graphs, where one node partition V_1 is $\{1, 2, 3, 4, 5\}$ and the other node partition V_2 is $\{6, 7, 8, 9\}$.*

1.4.6.4 Biregular Bipartite Graph

A *biregular bipartite graph* is a bipartite graph where each node in the same node partition has the same degree. Let d_1 and d_2 denote the degree of each node in the two node partitions, respectively. Then the graph is called (d_1, d_2)-*biregular bipartite graph*.

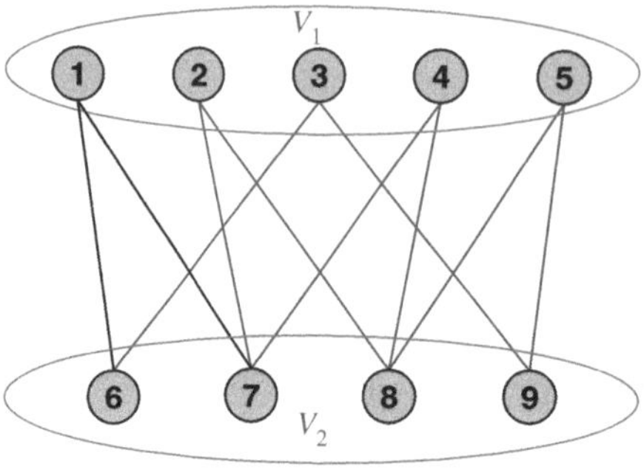

Figure 1.14 A bipartite graph with two node partitions $V_1 = \{1, 2, 3, 4, 5\}$ and $V_2 = \{6, 7, 8, 9\}$.

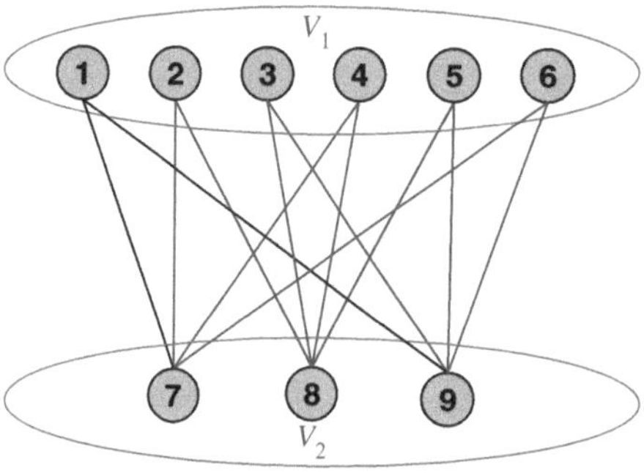

Figure 1.15 A (2, 4)-biregular bipartite graph with two partitions of nodes $V_1 = \{1, 2, 3, 4, 5, 6\}$ and $V_2 = \{7, 8, 9\}$. The degree of each node in V_1 is 2 and the degree of each node in V_2 is 4.

Example 1.12 *Figure 1.15 shows an example of biregular bipartite graphs, where one node partition V_1 is $\{1, 2, 3, 4, 5, 6\}$ and the other node partition V_2 is $\{7, 8, 9\}$. The degree of each node in V_1 is 2 and the degree of each node in V_2 is 4.*

1.4.6.5 Complete Bipartite Graph

Complete bipartite graph (CBG) is a special case of biregular bipartite graph where each node in one node partition is adjacent to all nodes in the other node partition. Suppose there are m and n nodes in each node partition of a complete bipartite graph, respectively. Then such a complete bipartite graph is denoted by $K_{m,n}$.

Remark 1.3 For a CBG $K_{m,n}$, the degrees of each node in two node partitions are n and m, respectively.

Remark 1.4 The diameter of a CBG is *two*.

Example 1.13 *Figure 1.16 shows an example of complete bipartite graphs $K_{4,3}$, where one node partition V_1 is $\{1, 2, 3, 4\}$ and the other node partition V_2 is $\{5, 6, 7\}$. The degree of each node in V_1 is 3 and the degree of each node in V_2 is 4.*

1.4.6.6 Incidence Graph

Incidence graphs are defined over incidence structures. An *incidence structure* I consists of a set P of points, a set L of lines (disjoint from P), and a relation set $R \subseteq P \times L$. For each pair of point $p \in P$ and line $l \in L$, they are said to be *incident* if $(p, l) \in R$. An *incidence graph* $G(I)$ is defined as a graph with incidence structure I whose node set is formed as $P \cup L$ and edge set contains all edges connecting two incident nodes.

Remark 1.5 An incidence graph is a bipartite graph, where all points and lines are represented by nodes. Particularly, nodes that correspond to points are contained in one partition while nodes that correspond to lines are contained in the other partition. Any two nodes with an incidence relation are connected by an edge.

Example 1.14 *Figure 1.17 shows an incidence graph which contains a point set $P = \{1, 2, 3\}$, a line set $L = \{a, b, c\}$, and a relation set $R = \{(1, a), (1, b), (2, b), (2, c), (3, c)\}$. The two node partitions in such a graph are represented by V_1 and V_2, respectively.*

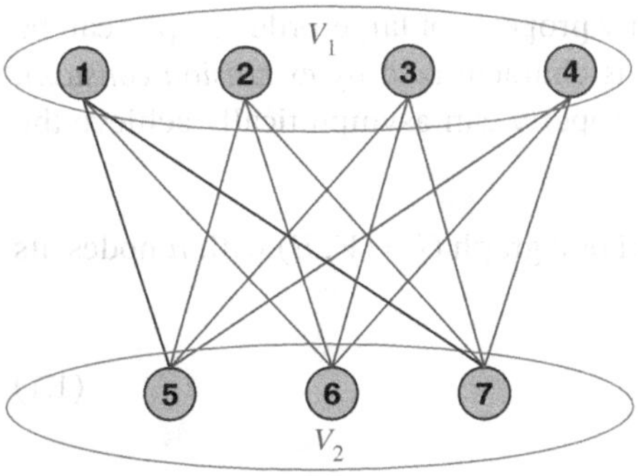

Figure 1.16 A complete bipartite graph $K_{4,3}$ ((3, 4)-biregular bipartite graph) with two node partitions $V_1 = \{1, 2, 3, 4\}$ and $V_2 = \{5, 6, 7\}$.

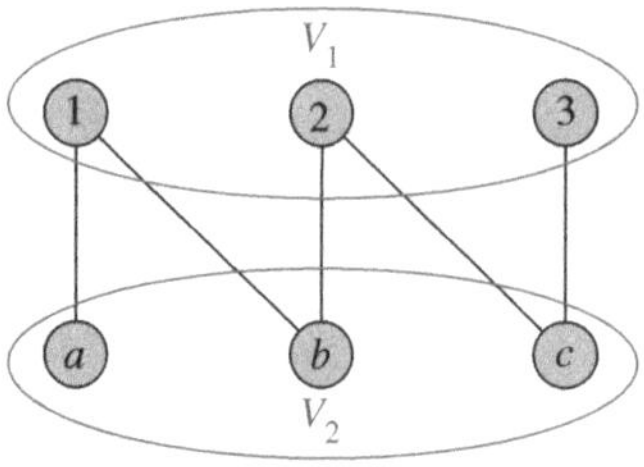

Figure 1.17 An incidence graph with two node partitions $V_1 = \{1, 2, 3\}$ and $V_2 = \{a, b, c\}$.

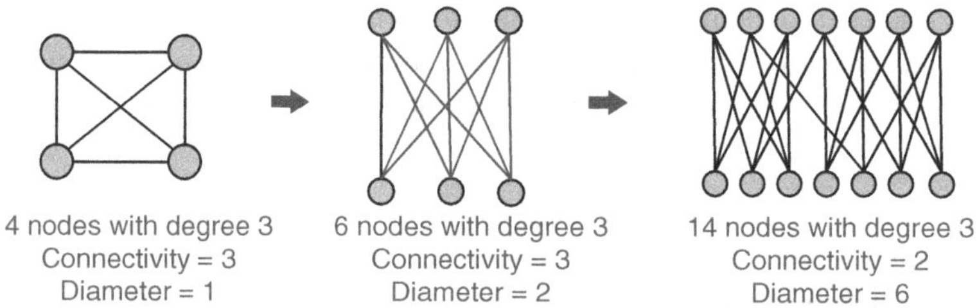

Figure 1.18 Three graphs with different orders given the node degree as three.

1.5 Mathematical Background II: Expander Graph

1.5.1 Expansion Constant

Intuitively, given the degree of each node in a graph, larger order of the graph leads to a tendency toward sparser connectivity and larger diameter. In Figure 1.18, three graphs with node degree as three show such a tendency. Correspondingly, we focus on special large-order graphs with dense connectivity and low diameter. In graph theory, the dense-connectivity property of large-order graphs can be described by expansion property, which is characterized by *expansion constant.* Moreover, graphs with good expansion property can asymptotically achieve the lowest possible diameter.

Definition 1.1 Expansion constant: For a graph $G = (V, E)$ with n nodes, its *expansion constant* $h(G)$ is defined as

$$h(G) = \min_{0<|S|\leq\frac{n}{2}} \frac{|\partial S|}{|S|}, \tag{1.1}$$

where $S \subset V$, $\bar{S} = V - S$, and ∂S denotes the boundary of S.

Example 1.15 *For a complete graph with n nodes, since each node is adjacent to all other nodes in the graph, the boundary of a node sub-set S satisfies* $|\partial S| = |S| * |\bar{S}| = |S| * (n - |S|)$. *Then the expansion constant of the complete graph is* $\frac{n}{2}$ *if n is an even number and* $\frac{n+1}{2}$ *if n is an odd number.*

Remark 1.6 A large value of $h(G)$ implies that there are many edges from any subset of nodes to its complement. Hence, good expansion property can be employed to characterize the dense connectivity of graph G.

Definition 1.2 **Expander graph:** A sequence of d-regular graphs $(G_m)_{m \geq 1}$ of size increasing with m is a family of expander graphs if there is a constant $\epsilon > 0$ such that $h(G_m) \geq \epsilon$ for all i.

Theorem 1.1 *Alon (1986): For a connected d-regular graph with n nodes, its diameter k is* $\Omega(log_{d-1}n)$, *while the diameter of expander graphs is* $O(logn)$.

Remark 1.7 The expander graphs can asymptotically achieve the lowest possible diameter.

1.5.2 Spectra of Graphs

It is known that the problem about determining the expansion constant of arbitrary graph is NP-complete problem (Høholdt and Janwa 2012). Instead, the spectra of graphs can be adopted to estimate the expansion constant because of easy calculability. The related definitions and properties are shown as follows.

Definition 1.3 **Adjacency matrix:** Suppose G is a graph with n nodes, which are indexed by $1, 2, ..., n$, respectively. The adjacency matrix of G is an n-by-n matrix $A(G)$ given by

$$A(G)_{i,j} = \begin{cases} 1, & \text{if node } i \text{ and node } j \text{ are adjacent;} \\ 0, & \text{otherwise.} \end{cases} \tag{1.2}$$

Example 1.16 *The adjacency matrix of a complete graph with three nodes is*

$$\begin{pmatrix} 0 & 1 & 1 \\ 1 & 0 & 1 \\ 1 & 1 & 0 \end{pmatrix}. \tag{1.3}$$

Definition 1.4 **Spectra of graph:** Let $A(G)$ be the adjacency matrix of a graph G with n nodes. Then $A(G)$ has n real eigenvalues, denoted by $\lambda_1 \geq \lambda_2 \geq ... \geq \lambda_n$, which are defined as the spectra of graph G.

Theorem 1.2 *Godsil and Royle (2001): For some special graphs with n nodes, their spectra ($\lambda_1 \geq \lambda_2 \geq ... \geq \lambda_n$) satisfy the following properties.*

- *If G is a d-regular graph, then $\lambda_1 = d$ and $\lambda_n \geq -d$.*
- *If G is a (d_1, d_2)-biregular bipartite graph, then $\lambda_1 = \sqrt{d_1 d_2}$ and $\lambda_n = -\sqrt{d_1 d_2}$.*

Example 1.17 *For any complete graph K_n with n nodes (a special $(n-1)$-regular graph), its spectra are $\lambda_1 = n - 1$ and $\lambda_2 = \lambda_3 = ... = \lambda_n = -1$.*

Example 1.18 *For any complete bipartite graph $K_{m,n}$ with m nodes and n nodes in two node partitions (a special (n, m)-biregular bipartite graph), its spectra are $\lambda_1 = \sqrt{mn}$, $\lambda_{m+n} = -\sqrt{mn}$, and $\lambda_2 = \lambda_3 = ...\lambda_{m+n-1} = 0$.*

Definition 1.5 **Nontrivial eigenvalues:** For a connected d-regular graph, any eigenvalue of its adjacency matrix $\lambda_i \neq \pm d$ is called a nontrivial eigenvalue. For a connected (d_1, d_2)-biregular bipartite graph, any eigenvalue of its adjacency matrix $\lambda_i \neq \pm\sqrt{d_1 d_2}$ is called as a nontrivial eigenvalue.

Example 1.19 *By Examples* 1.17, *the nontrivial eigenvalues of K_n include $\lambda_2 = \lambda_3 = ... = \lambda_n = -1$.*

Example 1.20 *By Examples* 1.18, *the nontrivial eigenvalues of $K_{m,n}$ include $\lambda_2 = \lambda_3 = ...\lambda_{m+n-1} = 0$.*

Theorem 1.3 *Alon (1986): Let G be a connected d-regular graph. Then its expansion constant $h(G)$ and the second largest eigenvalue (SLE) λ_2 of its adjacency matrix satisfy*

$$\frac{d - \lambda_2}{2} \leq h(G) \leq \sqrt{2d(d - \lambda_2)}. \tag{1.4}$$

Remark 1.8 For regular graphs, a smaller value of SLE implies a larger value of expansion constant.

1.5.3 Ramanujan Graph

Next, *Ramanujan graph* will be introduced, a special type of graph with good expansion property for regular graph and biregular bipartite graph, respectively. The related definitions and properties are recapped as follows.

Definition 1.6 Second largest eigenvalue modulus: For regular graphs and biregular bipartite graphs, their second largest eigenvalue modulus (SLEM) are defined as the largest eigenvalue modulus among all the nontrivial eigenvalues, denoted by $\bar{\lambda}_2$.

Remark 1.9 The second largest eigenvalue modulus (SLEM) $\bar{\lambda}_2$ of graphs satisfies the following properties.

- If G is a d-regular graph with n nodes, then $\bar{\lambda}_2 = max\{|\lambda_2|, |\lambda_n|\} \geq \lambda_2$.
- If G is a (d_1, d_2)-biregular bipartite graph with n nodes, $\bar{\lambda}_2 = max\{|\lambda_2|, |\lambda_{n-1}|\} = \lambda_2$.

Definition 1.7 Ramanujan graph: A d-regular graph G is called Ramanujan graph if its second largest eigenvalue modulus $\bar{\lambda}_2$ satisfies

$$\bar{\lambda}_2 \leq 2\sqrt{d-1}. \tag{1.5}$$

Example 1.21 *For any complete graph with n nodes (a special $(n-1)$-regular graph), its spectra are $\lambda_1 = n-1$ and $\lambda_2 = \lambda_3 = \ldots = \lambda_n = -1$, which implies $\bar{\lambda}_2 = 1$. When $n \geq 3$, we have $\bar{\lambda}_2 = 1 \leq 2\sqrt{(n-1)-1}$. Hence, complete graphs with at least three nodes are Ramanujan graphs.*

Definition 1.8 Bipartite Ramanujan graph: A (d_1, d_2)-biregular bipartite graph G is called bipartite Ramanujan graph if its second largest eigenvalue modulus $\bar{\lambda}_2$ satisfies

$$\bar{\lambda}_2 \leq \sqrt{d_1 - 1} + \sqrt{d_2 - 1}. \tag{1.6}$$

Example 1.22 *For any complete bipartite graph with m nodes and n nodes in two partitions of nodes (a special (n, m)-biregular bipartite graph), its SLEM is $\bar{\lambda}_2 = \lambda_2 = 0$. Since $\bar{\lambda}_2 = 0 \leq \sqrt{m-1} + \sqrt{n-1}$, complete bipartite graphs are bipartite Ramanujan graphs.*

Theorem 1.4 *Friedman (2003): Let $\langle G_m = (V_m, E_m)\rangle_{m\geq 1}$ be a family of connected d-regular graphs with $|V_m| \to +\infty$ as $m \to +\infty$. Then*

$$\liminf_{|V_m|\to+\infty} \bar{\lambda}_2 \geq 2\sqrt{d-1}, \tag{1.7}$$

which means the infimum of $\bar{\lambda}_2$ has a horizontal asymptote at $2\sqrt{d-1}$.

Theorem 1.5 *Feng (1996): Let $\langle G_m = (V_m, E_m)\rangle_{m\geq 1}$ be a family of connected (d_1, d_2)-biregular bipartite graphs with $|V_m| \to +\infty$ as $m \to +\infty$. Then*

$$\liminf_{|V_m|\to+\infty} \bar{\lambda}_2 \geq \sqrt{d_1 - 1} + \sqrt{d_2 - 1}, \tag{1.8}$$

which means the infimum of $\bar{\lambda}_2$ has a horizontal asymptote at $\sqrt{d_1 - 1} + \sqrt{d_2 - 1}$.

Remark 1.10 The expansion properties of Ramanujan graphs and bipartite Ramanujan graphs are optimal when the graphs are sufficiently large.

1.6 Mathematical Background III: Combinatorial Design Theory

1.6.1 Overview of Block Design

Combinatorial design theory studies how to select subsets from a finite set to satisfy some specified properties. The members of the finite set are called *elements* and the selected subsets are called *blocks*. If the blocks are simply unordered subsets of elements and there is no structural ordering or pattern between blocks, the set of elements and the collection of blocks constitute a *block design*. Some formal definitions of block designs are given as follows.

Definition 1.9 **Block design:** A *block design* is defined as a pair (X, B), where X is a set of elements and B is a collection (i.e., multi-set) of blocks, such that

- each block is a subset of X (i.e., contains some elements in X),
- there is no structural ordering or pattern between blocks.

Some examples of block design are given as follows. Blocks are described in the form abc instead of $\{a, b, c\}$ for simplicity.

Example 1.23 *An example of block design is given as follows:*

$$\begin{aligned} X &= \{1, 2, 3, 4\}; \\ B &= \{12, 34, 13, 24, 14, 23\}. \end{aligned} \tag{1.9}$$

Remark 1.11 Example 1.23 shows some special properties in block designs, which are listed here:

- *Uniform.* All blocks contain the same number of elements. For Example 1.23, each block contains *two* elements.
- *Regular*: All elements are contained in the same number of blocks. For Example 1.23, each element is contained in *three* blocks.
- *Pairwise balanced.* Each pair of distinct elements is contained in the same number of common blocks. For Example 1.23, each pair of distinct elements is contained in *one* common block.
- *Group divisible.* Elements can be partitioned into some groups, such that each pair of elements is contained in either the same group or the same number of

common blocks. For Example 1.23, elements are partitioned into four groups, i.e., [1], [2], [3], [4]. In this way, each pair of elements in distinct groups is contained in *one* common block.

- *Resolvable.* Blocks can be partitioned into some groups, such that each group of blocks contains each element exactly once. For Example 1.23, blocks are partitioned into three groups, i.e., [12, 34], [13, 24], [14, 23]. In this way, each group of blocks contains each element exactly once.

Remark 1.12 Any pairwise balanced block design is a trivial case of group divisible block design by partitioning elements into singleton groups, i.e., each group consists of exactly one element.

Example 1.24 *An example of block design is given as follows:*

$$X = \{1, 2, 3\};$$
$$B = \{123, 123, 123\}. \tag{1.10}$$

Example 1.24 is a trivial case of block design called *complete block design*, which is defined as follows.

Definition 1.10 **Complete and incomplete block design:** A block design is called complete block design (CBD) if each block contains all elements. Otherwise, the block design is called incomplete block design.

Remark 1.13 Any CBD has the uniform property, the regular property, the pairwise balanced property, the group divisible property, and the resolvable property.

Various block designs with different constraints on the properties are simply described in Table 1.1. Next, the most basic block design, i.e., balanced incomplete block design will be introduced first. Then we will show the extended block designs including pairwise balanced design, group divisible design, and transversal design.

Table 1.1 Various block designs.

Block design	Uniform	Regular	Pairwise balanced	Group divisible
Complete block design	•	•	•	•
Balanced incomplete block design	•	•	•	•
Pairwise balanced design	○	○	•	•
Group divisible design	○	○	○	•
Transversal design	•	•	•	•

1.6.2 Balanced Incomplete Block Design

Definition 1.11 Balanced incomplete block design: Let v, b, r, k, and λ be positive integers such that $v > k \geq 2$. A balanced incomplete block design-(v, b, r, k, λ) (BIBD-(v, b, r, k, λ)) is a block design (X, B) that satisfies the following properties:

- X is a set of v elements;
- B is a collection of b blocks; each of which contains k elements in X;
- each element is contained in exactly r blocks;
- every pair of distinct elements is contained in exactly λ common blocks.

Remark 1.14 A BIBD is called an *incomplete* block design because $k < v$. A complete block design (CBD) can be regarded as the counterpart of BIBD with $k = v$.

Example 1.25 *Example 1.23 is a BIBD-*(4, 6, 3, 2, 1) *with*

$$\begin{aligned} X &= \{1, 2, 3, 4\}; \\ B &= \{12, 34, 13, 24, 14, 23\}. \end{aligned} \tag{1.11}$$

Example 1.26 *An example of BIBD-*(7, 14, 6, 3, 2) *is given as follows:*

$$\begin{aligned} X &= \{1, 2, 3, 4, 5, 6, 7\}; \\ B &= \{123, 145, 167, 246, 257, 347, 356, \\ &\quad\ 123, 147, 156, 245, 267, 346, 357\}. \end{aligned} \tag{1.12}$$

Definition 1.12 Resolvable balanced incomplete block design: A BIBD satisfying resolvable property is called a resolvable balanced incomplete block design (RBIBD). To be specific, b blocks in an RBIBD-(v, b, r, k, λ) can be partitioned into r groups with the same number of blocks, such that each group of blocks contains each element exactly once.

Remark 1.15 For any RBIBD-(v, b, r, k, λ), $b \equiv 0 \ (mod\ r)$.

Example 1.27 *Example 1.23 is a resolvable BIBD-*(4, 6, 3, 2, 1) *with*

$$\begin{aligned} X &= \{1, 2, 3, 4\}; \\ B &= \{[12, 34], [13, 24], [14, 23]\}, \end{aligned} \tag{1.13}$$

in which the notation [ ] represents the group of blocks such that each group contains all elements exactly once.

Lemma 1.1 *Stinson (2007): The parameters of a BIBD-(v, b, r, k, λ) satisfy:*

- $r(k-1) = \lambda(v-1)$,
- $vr = bk$,
- $b > r > \lambda$,
- $b \geq v$, $r \geq k$.

Example 1.28 *For the BIBD-$(4, 6, 3, 2, 1)$ in Example 1.23, we have that $v = 4$, $b = 6$, $r = 3$, $k = 2$, and $\lambda = 1$. One can verify that $r(k-1) = \lambda(v-1) = 3$, $vr = bk = 12$, $b = 6 > r = 3 > \lambda = 1$, and $b = 6 \geq v = 4$.*

Remark 1.16 By Lemma 1.1, we have that $r = \dfrac{\lambda(v-1)}{k-1}$ and $b = \dfrac{\lambda v(v-1)}{k(k-1)}$. Hence, a BIBD-$(v, b, r, k, \lambda)$ can also be denoted by BIBD-(v, k, λ) for short.

Corollary 1.1 *The following equations are two necessary conditions for the existence of a BIBD-(v, b, r, k, λ):*

- $\lambda(v-1) \equiv 0 \ (mod\ k-1)$;
- $\lambda v(v-1) \equiv 0 \ (mod\ k(k-1))$.

Remark 1.17 In the following cases, the necessary conditions are also sufficient for the existence of a BIBD-$(v, k, 1)$:

- $k = 2, v \geq 2$;
- $k = 3, v \equiv 1, 3 \ (mod\ 6)$;
- $k = 4, v \equiv 1, 3 \ (mod\ 12)$;
- $k = 5, v \equiv 1, 5 \ (mod\ 20)$.

In the following theorems and definitions, some special types of BIBD with existence guarantee will be introduced.

Theorem 1.6 *Stinson (2007): Let B be a collection consisting of all k-subsets of X. Then (X, B) is a BIBD-$(v, k, \binom{v-2}{k-2})$.*

Example 1.29 *An example of the BIBD-$(v, k, \binom{v-2}{k-2})$ with $v = 3$ and $k = 2$ is a BIBD-$(3, 3, 2, 2, 1)$ with*

$$\begin{aligned} X &= \{1, 2, 3\}; \\ B &= \{12, 13, 23\}. \end{aligned} \tag{1.14}$$

Theorem 1.7 *Stinson (2007): When $k = 2$, the BIBD-$(v, k, \binom{v-2}{k-2})$ is resolvable if and only if v is an even number and $v \geq 4$.*

Example 1.30 *An example of the BIBD-*$(v, k, \binom{v-2}{k-2})$ *with* $v = 4$ *and* $k = 2$ *is the resolvable BIBD-*$(4, 6, 3, 2, 1)$ *appeared in Examples 1.23 and 1.27.*

Example 1.31 *An example of the BIBD-*$(v, k, \binom{v-2}{k-2})$ *with* $v = 6$ *and* $k = 2$ *is a resolvable BIBD-*$(6, 15, 5, 2, 1)$ *with*

$$\begin{aligned} X &= \{1, 2, 3, 4, 5, 6\}; \\ B &= \{[12, 34, 56], [13, 25, 46], [14, 26, 35], [15, 24, 36], [16, 23, 45]\}, \end{aligned} \tag{1.15}$$

in which the notation [ ] represents the group of blocks such that each group contains all elements exactly once.

Definition 1.13 **Projective plane:** Given an integer $n \geq 2$, a BIBD-$(n^2 + n + 1, n^2 + n + 1, n + 1, n + 1, 1)$ is called a projective plane of order n.

Theorem 1.8 *Kirkman (1850): For every prime power* $n \geq 2$*, there exists a projective plane of order* n*.*

Example 1.32 *A projective plane of order* 2 *is a BIBD-*$(7, 7, 3, 3, 1)$ *with*

$$\begin{aligned} X &= \{1, 2, 3, 4, 5, 6, 7\}; \\ B &= \{123, 145, 167, 246, 257, 347, 356\}. \end{aligned} \tag{1.16}$$

Example 1.33 *A projective plane of order* 4 *is a BIBD-*$(21, 21, 5, 5, 1)$ *with*

$$\begin{aligned} X = \{&A, B, C, D, E, F, G, H, I, J, K, L, M, N, O, P, Q, R, S, T, U\}; \\ B = \{&ABCDE, AFGHI, AJKLM, ANOPQ, ARSTU, BFJNS, BGLOR, \\ &BHMOU, BIKPT, CFKQU, CGMNT, CHLPS, CIJOR, DFMPR, \\ &DGKOS, DHJQT, DILNU, EFLOT, EGJPU, EHKNR, EIMQS\}. \end{aligned} \tag{1.17}$$

Definition 1.14 **Affine plane:** Let $n \geq 2$. A BIBD-$(n^2, n^2 + n, n + 1, n, 1)$ is called an affine plane of order n.

Theorem 1.9 *Kirkman (1850): For every prime power* $n \geq 2$*, there exists an affine plane of order* n*.*

Theorem 1.10 *Bose (1942): Affine plane is a resolvable BIBD.*

Example 1.34 *Example 1.23 is an affine plane of order 2 (i.e., BIBD-(4, 6, 3, 2, 1)) with*

$$\begin{aligned} X &= \{1, 2, 3, 4\}; \\ B &= \{[12, 34], [13, 24], [14, 23]\}, \end{aligned} \tag{1.18}$$

where the notation [ ] represents the group of blocks such that each group contains all elements exactly once.

Example 1.35 *An affine plane of order 4 is a BIBD-(16, 20, 5, 4, 1) with*

$$\begin{aligned} X &= \{A, B, C, D, E, F, G, H, I, G, K, L, M, N, O, P\}; \\ B &= \{[ABCD, EFGH, IJKL, MNOP], \\ &\quad [AEIN, BGLM, CHJP, DFKO], \\ &\quad [AFLP, BHIO, CGKN, DEJM], \\ &\quad [AHKM, BFJN, CELO, DGIP], \\ &\quad [AGJO, BEKP, CFIM, DHLN]\}, \end{aligned} \tag{1.19}$$

where the notation [ ] represents the group of blocks such that each group contains all elements exactly once.

Definition 1.15 **Hadamard matrix:** Let I_n denote the identity matrix of size n. A Hadamard matrix of order n is an n-by-n matrix H in which every entry is 1 or −1 such that $HH^T = nI_n$.

Example 1.36 *A Hadamard matrix of order 2 is given by*

$$\begin{pmatrix} 1 & 1 \\ 1 & -1 \end{pmatrix}. \tag{1.20}$$

Example 1.37 *A Hadamard matrix of order 4 is given by*

$$\begin{pmatrix} 1 & 1 & 1 & 1 \\ 1 & -1 & 1 & -1 \\ 1 & 1 & -1 & -1 \\ 1 & -1 & -1 & 1 \end{pmatrix}. \tag{1.21}$$

Theorem 1.11 *Hadamard (1893): If there exists a Hadamard matrix of order $n > 2$, then $n \equiv 0$ (mod 4).*

Remark 1.18 The smallest order $n \equiv 0$ $(mod\ 4)$ for which a Hadamard matrix is not currently known to exist is $n = 668$.

Theorem 1.12 *Stinson (2007): Suppose $m > 1$. Then there exists a Hadamard matrix of order $4m$ if and only if there exists a BIBD-$(4m-1, 2m-1, m-1)$.*

Example 1.38 *Example 1.32 is a BIBD induced by a Hadamard matrix of order 8 (i.e., BIBD-$(7, 7, 3, 3, 1)$) with*

$$\begin{aligned} X &= \{1, 2, 3, 4, 5, 6, 7\}; \\ B &= \{123, 145, 167, 246, 257, 347, 356\}. \end{aligned} \tag{1.22}$$

Theorem 1.13 *Stinson (2007): Suppose there is a Hadamard matrix of order $4m$. If $m \geq 3$, then there exists a BIBD-$(2m-1, 4m-2, 2m-2, m-1, m-2)$; if $m \geq 2$, then there exists a BIBD-$(2m, 4m-2, 2m-1, m, m-1)$ and such a BIBD is resolvable.*

Example 1.39 *By Theorem 1.13, when $m = 3$, these exists a BIBD-$(5, 10, 4, 2, 1)$ with*

$$\begin{aligned} X &= \{1, 2, 3, 4, 5\}; \\ B &= \{12, 13, 14, 15, 23, 24, 25, 34, 35, 45\}. \end{aligned} \tag{1.23}$$

Example 1.40 *By Theorem 1.13, when $m = 2$, these exists a resolvable BIBD-$(4, 6, 3, 2, 1)$ with*

$$\begin{aligned} X &= \{1, 2, 3, 4\}; \\ B &= \{[12, 34], [13, 24], [14, 23]\}, \end{aligned} \tag{1.24}$$

where the notation [ ] represents the group of blocks such that each group contains all elements exactly once.

By the following theorems and definitions, we present two methods to construct new BIBDs from old ones.

Theorem 1.14 *Stinson (2007): Suppose there exist a BIBD-(v, k, λ_1) and a BIBD-(v, k, λ_2). Then there exists a BIBD-$(v, k, \lambda_1 + \lambda_2)$.*

Example 1.41 *Given a BIBD-$(3, 2, 1)$ parameterized as (X_1, B_1) with*

$$\begin{aligned} X_1 &= \{1, 2, 3\}; \\ B_1 &= \{12, 23, 13\}, \end{aligned} \tag{1.25}$$

and a BIBD-$(3, 2, 2)$ parameterized as (X_2, B_2) with

$$\begin{aligned} X_2 &= \{1, 2, 3\}; \\ B_2 &= \{12, 23, 13, 12, 23, 13\}, \end{aligned} \tag{1.26}$$

the following (X_3, B_3) *is a BIBD-*$(3, 2, 3)$ *with*

$$\begin{aligned} X_3 &= \{1, 2, 3\}; \\ B_3 &= \{12, 23, 13, 12, 23, 13, 12, 23, 13\}. \end{aligned} \tag{1.27}$$

Corollary 1.2 *Suppose there exists a BIBD-*(v, k, λ)*. Then there exists a BIBD-*$(v, k, s\lambda)$ *for any integer* $s \geq 1$.

Definition 1.16 Complementary design: Given a BIBD-(v, b, r, k, λ) with element set X and blocks $\{B_1, B_2, ..., B_b\}$, its complementary design is a block design with element set X and blocks $\{X \setminus B_1, X \setminus B_2, ..., , X \setminus B_b\}$.

Example 1.42 *The original BIBD-*$(5, 10, 4, 2, 1)$ *in Example 1.39 is parameterized as* (X_1, B_1) *with*

$$\begin{aligned} X_1 &= \{1, 2, 3, 4, 5\}; \\ B_1 &= \{12, 13, 14, 15, 23, 24, 25, 34, 35, 45\}, \end{aligned} \tag{1.28}$$

and its complementary design is parameterized as (X_2, B_2) *with*

$$\begin{aligned} X_2 &= \{1, 2, 3, 4, 5\}; \\ B_2 &= \{345, 245, 235, 234, 145, 135, 134, 125, 124, 123\}. \end{aligned} \tag{1.29}$$

Theorem 1.15 *Stinson (2007): The complementary design of a BIBD-*(v, b, r, k, λ) *is a BIBD-*$(v, b, b-r, v-k, b-2r+\lambda)$*, provided that* $b - 2r + \lambda > 0$.

Example 1.43 *The design* (X_2, B_2) *in Example 1.42 is a BIBD-*$(5, 10, 6, 3, 3)$*, which is the complementary design of a BIBD-*$(5, 10, 4, 2, 1)$.

1.6.3 Pairwise Balanced Design and Group Divisible Design

1.6.3.1 Pairwise Balanced Design

Definition 1.17 Pairwise balanced design: Suppose $v \geq 2$, $\lambda \geq 1$, and $K \subseteq \{n \in Z : n \geq 2\}$. A pairwise balanced design-(v, K, λ) (PBD-(v, K, λ)) is a block design (X, B) such that the following properties are satisfied:

- X is a set of v elements;
- B is a collection of blocks, each of which is a sub-set of elements in X;
- The number of elements contained in each block belongs to the set K;
- Every pair of distinct elements is contained in exactly λ blocks.

Remark 1.19 A PBD-$(v, K, 1)$ is often simply denoted as a PBD-(v, K).

Example 1.44 *An example of PBD-*$(5, \{2, 3\}, 1)$ *is given as follows:*

$$X = \{1, 2, 3, 4, 5\};$$
$$B = \{125, 134, 23, 24, 35, 45\}. \tag{1.30}$$

Definition 1.18 Regular pairwise balanced design: A PBD-(v, K, λ) is a *regular* pairwise balanced design if each element is contained in exactly r blocks, where r is a positive integer.

Example 1.45 *The following design* (X, B) *is a regular PBD-*$(6, \{2, 3\}, 1)$ *with*

$$X = \{1, 2, 3, 4, 5, 6\};$$
$$B = \{123, 456, 14, 15, 16, 24, 25, 26, 34, 35, 36\}, \tag{1.31}$$

where each element is contained in exactly four blocks.

Definition 1.19 Uniform pairwise balanced design: A PBD-(v, K, λ) is a *uniform* pairwise balanced design if each block contains exactly k elements, i.e., $K = \{k\}$.

Remark 1.20 A complete block design (CBD) is a special case of uniform PBD-$(v, \{k\}, \lambda)$ with $k = v$.

Theorem 1.16 *Stinson (2007): A uniform PBD-*$(v, \{k\}, \lambda)$ *with* $k < v$ *is a BIBD-*(v, k, λ).

Remark 1.21 A uniform PBD is also a regular PBD. However, a regular PBD is not necessarily a uniform PBD.

Theorem 1.17 *Stinson (2007): For a set of positive integers K, let* $\alpha(K) = \gcd\{k - 1 : k \in K\}$ *and* $\beta(K) = \gcd\{k(k-1) : k \in K\}$. *Then the necessary conditions for the existence of a PBD-*(v, K, λ) *are:*

- $\lambda(v - 1) \equiv 0 \ (mod\ \alpha(K))$;
- $\lambda v(v - 1) \equiv 0 \ (mod\ \beta(K))$.

Example 1.46 *Given* $K = \{3, 4, 6\}$, *we can compute that*

$$\alpha(K) = \gcd\{2, 3, 5\} = 1, \tag{1.32}$$

and

$$\beta(K) = \gcd\{6, 12, 30\} = 6, \tag{1.33}$$

then the necessary conditions for the existence of a PBD-$(v, K, 1)$ *are* $v(v-1) \equiv 0 \ (mod\ 6)$, *which can be simplified to* $v \equiv 0$ or $1 \ (mod\ 3)$.

Theorem 1.18 *De Bruijn and Erdös (1948): For a PBD-$(v, K, 1)$ with b blocks, where $b > 1$, we have $b \geq v$. If $b = v$, then either the PBD-$(v, K, 1)$ has one block of size $v - 1$ and the rest of size 2, or else $K = \{k\}$ and $b = v = k^2 - k + 1$.*

Example 1.47 *The PBD-$(v, K, 1)$ with $v = b = 7$ only exists as the following two cases:*

$$\begin{aligned} X_1 &= \{1, 2, 3, 4, 5, 6, 7\}; \\ B_1 &= \{12, 13, 14, 15, 16, 17, 234567\}. \end{aligned} \tag{1.34}$$

$$\begin{aligned} X_2 &= \{1, 2, 3, 4, 5, 6, 7\}; \\ B_2 &= \{123, 145, 167, 246, 257, 347, 356\}. \end{aligned} \tag{1.35}$$

Theorem 1.19 *Stinson (2007): Given a resolvable BIBD-$(v, k, 1)$, there exists a PBD-$(v + r, \{k + 1, r\}, 1)$, where $r = (v - 1)/(k - 1)$.*

Example 1.48 *Given the resolvable BIBD-$(4, 2, 1)$ in Example 1.27, there exists a PBD-$(7, 3, 1)$ with*

$$\begin{aligned} X &= \{1, 2, 3, 4, 5, 6, 7\}; \\ B &= \{123, 145, 167, 246, 257, 347, 356\}. \end{aligned} \tag{1.36}$$

Corollary 1.3 *For any even integer $v \geq 4$, given the resolvable BIBD-$(v, 2, 1)$, there exists a PBD-$(2v - 1, \{3, v - 1\}, 1)$.*

1.6.3.2 Group Divisible Design

Definition 1.20 **Group divisible design:** Suppose $t \geq 2$, $\mu \geq 1$, $K \subseteq \{n \in Z : n \geq 2\}$, and $G \subseteq \{n \in Z : n \geq 1\}$. A group divisible design (GDD-(t, K, G, μ)) is a triple $(X, \mathcal{G}, B)$ that satisfies the following properties:

- X is a set of t elements;
- $\mathcal{G}$ is a partition of X into at least two groups;
- B is a collection of blocks, each of which is a sub-set of elements in X;
- the number of elements contained in each block belongs to the set K;
- the number of elements in each group belongs to the set G;
- every pair of distinct elements is contained in exactly μ common blocks or one group, but not both.

Remark 1.22 For a GDD, it is allowed that each group consists of only one element.

Given a GDD-(t, K, G, λ) with a_i groups of g_i elements, $i = 1, 2, ..., s$, which implies $\sum_{i=1}^{s} a_i g_i = t$, we use the notation $g_1^{a_1} g_2^{a_2} ... g_s^{a_s}$ for the group type of such a GDD.

Example 1.49 *A GDD-*$(10, \{3, 4\}, \{1, 3\}, 1)$ *with group type* $1^1 3^3$.

$$\begin{aligned} X &= \{0, 1, 2, 3, 4, 5, 6, 7, 8, 9\};\\ \mathcal{G} &= \{\{0, 1, 2\}, \{3, 4, 5\}, \{6, 7, 8\}, \{9\}\};\\ B &= \{0369, 1479, 2589, 048, 156, 237, 057, 138\}. \end{aligned} \tag{1.37}$$

Theorem 1.20 *Colbourn and Dinitz (2006): Suppose that* $v > k \geq 2$ *and* $u = \dfrac{v-1}{k-1}$. *Then there exists a BIBD-*$(v, k, 1)$ *if and only if there exists a GDD-*$(v - 1, \{k\}, \{k-1\}, 1)$ *with group type* $(k-1)^u$.

Example 1.50 *Suppose* $v = 7$ *and* $k = 3$, *then we can calculate* $u = 3$. *Since there exists a BIBD-*$(7, 3, 1)$ *in Example 1.32, there exists a GDD-*$(6, \{3\}, \{2\}, 1)$ *with group type* 2^3 *as follows.*

$$\begin{aligned} X &= \{1, 2, 3, 4, 5, 6\};\\ \mathcal{G} &= \{\{1, 2\}, \{3, 4\}, \{5, 6\}\};\\ B &= \{135, 146, 236, 245\}. \end{aligned} \tag{1.38}$$

Theorem 1.21 *Zhu (1993): Let* k, m *be integers with* $k \geq 2$ *and* $m \geq 1$. *The necessary conditions for the existence of a GDD-*$(t, \{k\}, \{m\}, \mu)$ *with group type* m^u *are:*

- $u \geq k$;
- $\mu(u-1)m \equiv 0 \pmod{k-1}$;
- $\mu u(u-1) \equiv 0 \pmod{k(k-1)}$.

Example 1.51 *Suppose* $k = 3$ *and* $m = 2$, *then the necessary conditions for the existence of a GDD-*$(t, \{k\}, \{m\}, 1)$ *with group type* m^u *are* $u \geq 3$, $2(u-1) \equiv 0 \pmod 2$, *and* $u(u-1) \equiv 0 \pmod 6$, *which can be simplified to* $u \geq 3$ *and* $u \equiv 0$ or $1 \pmod 3$.

1.6.3.3 Transversal Design

Definition 1.21 **Transversal Design:** Suppose $n \geq 2$, $\mu \geq 1$, and $v \geq 2$. A transversal design (TD-(n, μ, v)) is a triple $(X, \mathcal{G}, B)$ that satisfies the following properties:

- X is a set of nv elements;
- $\mathcal{G}$ is a partition of X into n groups;

- Each group contains exactly v elements;
- B is a collection of μv^2 blocks, each of which is a sub-set of elements in X;
- Each block contains exactly n elements;
- Every pair of distinct elements is contained in exactly μ common blocks or one group, but not both.

Remark 1.23 A TD-(n, μ, v) is a GDD-$(nv, \{n\}, \{v\}, \mu)$ with group type v^n.

Example 1.52 *An example of TD-(3, 1, 2) is given as follows:*

$$X = \{1, 2, 3, 4, 5, 6\};$$
$$\mathcal{G} = \{\{1, 2\}, \{3, 4\}, \{5, 6\}\}; \tag{1.39}$$
$$B = \{135, 146, 236, 245\}.$$

Remark 1.24 Given a TD-(n_1, μ, v), then there exists a TD-(n_2, μ, v) for any integer n_2 with $2 \leq n_2 \leq n_1$ by deleting $(n_2 - n_1)$ groups of elements in the original TD-(n_1, μ, v).

Lemma 1.2 *MacNeish (1922), Stinson (2007): Let an integer $v > 1$ with its prime factorization as $p_1^{e_1} p_2^{e_2} \dots p_l^{e_l}$, where the p_is are distinct primes and $e_i \geq 1$ for all i. Let $n = \min\{p_i^{e_i} + 1 : 1 \leq i \leq l\}$. Then there exists a TD-(n, μ, v).*

Example 1.53 *If v is a prime power, TD-(n, μ, v) exists if and only if $2 \leq n \leq v + 1$.*

Definition 1.22 **Resolvable transversal design:** A TD satisfying resolvable property is called a resolvable transversal design. To be specific, blocks in a resolvable transversal design can be partitioned into some groups, such that each group of blocks contains each element exactly once.

Theorem 1.22 *Stinson (2007): Given a prime power v and an integer n such that $2 \leq n \leq v$, then TD-(n, μ, v) is resolvable.*

Example 1.54 *Given $v = n = 2$, a resolvable TD-(2, 1, 2) is provided as follows:*

$$X = \{1, 2, 3, 4\};$$
$$\mathcal{G} = \{\{1, 2\}, \{3, 4\}\}; \tag{1.40}$$
$$B = \{[13, 24], [14, 23]\},$$

where the notation [ ] represents the group of blocks such that each group contains all elements exactly once.

1.6.4 The Relationships Among Extended Block Designs

Theorem 1.23 *Stinson (2007): If $(X, \mathcal{G}, A)$ is a group divisible design, then (X, B) is a pairwise balanced design with $\lambda = 1$, where*

$$B = A \cup \{g \in \mathcal{G} : |g| \geq 2\}. \tag{1.41}$$

Example 1.55 *Given the GDD-$(10, \{3, 4\}, \{1, 3\}, 1)$ in Example 1.49, the induced PBD-$(10, \{3, 4\}, 1)$ is provided as follows.*

$$\begin{aligned} X &= \{0, 1, 2, 3, 4, 5, 6, 7, 8, 9\}; \\ B &= \{0369, 1479, 2589, 048, 156, 237, 057, 138, 012, 345, 678\}. \end{aligned} \tag{1.42}$$

Theorem 1.24 *Stinson (2007): Suppose that $(X, \mathcal{G}, A)$ is a group divisible design and $\infty \notin X$, define $Y = X \cup \{\infty\}$ and*

$$B = A \cup \{g \cup \{\infty\} : g \in \mathcal{G}\}. \tag{1.43}$$

Then (Y, B) is a PBD with $\lambda = 1$.

Example 1.56 *Given the GDD-$(10, \{3, 4\}, \{1, 3\}, 1)$ in Example 1.49, the induced PBD-$(11, \{2, 3, 4\}, 1)$ is provided as follows.*

$$\begin{aligned} Y &= \{0, 1, 2, 3, 4, 5, 6, 7, 8, 9, \infty\}; \\ B &= \{0369, 1479, 2589, 048, 156, 237, 057, 138, 012\infty, 345\infty, 678\infty, 9\infty\}. \end{aligned} \tag{1.44}$$

Theorem 1.25 *Stinson (2007): If (X, A) is a PBD with $\lambda = 1$, then $(X, \mathcal{G}, A)$ is a GDD, where*

$$\mathcal{G} = \{\{x\} : x \in X\}. \tag{1.45}$$

Remark 1.25 Any pairwise balanced design is a group divisible design by partitioning elements into singleton groups, i.e., each group consists of exactly one element.

Theorem 1.26 *Stinson (2007): Given an integer $n \geq 2$ and a TD-$(n + 1, 1, v)$, then the following pairwise balanced designs exist:*

- *A PBD-$(nv + 1, \{n, n + 1, v\}, 1)$;*
- *A PBD-$((n + 1)v, \{n + 1, v\}, 1)$;*
- *A PBD-$(nv + u, \{n, n + 1, v, u\}, 1)$ for all u such that $2 \leq u \leq v - 1$.*

Example 1.57 *Suppose $n = 2$ and $v = 3$, then the following $(X, \mathcal{G}, B)$ is an example of TD-(3, 1, 3) with*

$$X = \{1, 2, 3, 4, 5, 6, 7, 8, 9\};$$
$$\mathcal{G} = \{123, 456, 789\}; \tag{1.46}$$
$$B = \{147, 159, 168, 249, 258, 267, 348, 357, 369\},$$

According to Theorem 1.26, there exist the following pairwise balanced designs:

- *A PBD-(7, {2, 3}, 1) with*

$$X_1 = \{1, 2, 3, 4, 5, 6, 7\}; \tag{1.47}$$
$$B_1 = \{123, 456, 147, 15, 16, 24, 25, 267, 34, 357, 36\}$$

- *A PBD-(9, {3}, 1) with*

$$X_2 = \{1, 2, 3, 4, 5, 6, 7, 8, 9\}; \tag{1.48}$$
$$B_2 = \{123, 456, 789, 147, 159, 168, 249, 258, 267, 348, 357, 369\}$$

- *A PBD-(8, {2, 3}, 1) (i.e., $u = 2$) with*

$$X_3 = \{1, 2, 3, 4, 5, 6, 7, 8\}; \tag{1.49}$$
$$B_3 = \{123, 456, 789, 147, 15, 168, 24, 258, 267, 348, 357, 36\}$$

1.7 Subgraph Options

In the design framework, the family of Ramanujan Graph is selected as the main option of subgraphs. Complete graph(CG) and complete bipartite graph(CBG) are members of this family. Next we will introduce two novel members of such family: block incidence graph(BIG) and transversal design graph(TDG).

1.7.1 Block Incidence Graph

Definition 1.23 A *block incidence graph*, denoted by BIG-(θ, r, k, λ), is defined as a biregular bipartite graph that contains θk nodes with degree r (called *element nodes*) and θr nodes with degree k (called *block nodes*), such that

- all element nodes are contained in one partition of the graph, while the other partition contains all of the blocks nodes;
- each pair of element nodes are adjacent to a number of λ common block nodes;
- the parameters satisfy $\theta, r, \lambda \geq 1$ and $k \geq 2$.

Examples of block incidence graph are shown in the following figure:

The name "block incidence graph" comes from the fact that such a graph can be regarded as the incidence graph of a complete block design (CBD) or a balanced incomplete block design (BIBD) (*cf.* Section 1.6.2). Specifically, given a CBD or a BIBD, its corresponding BIG can be constructed by mapping the elements and blocks in such a design as element nodes and block nodes, respectively. For each element in the CBD (or BIBD), if it is contained in a block, then its corresponding element node and block node are adjacent to each other. For example, a complete bipartite graph (CBG) can be generated by a CBD as a special case of BIG with $\theta = 1$ (e.g., the BIG-(1, 3, 3, 3) in Figure 1.19). It is not hard to obtain the following lemma.

Lemma 1.3 *If $\theta = 1$, any BIG-(θ, r, k, λ) is an incidence graph of a CBD with k elements and r blocks (i.e., a CBG $K_{r,k}$). Otherwise, if $\theta > 1$, then it is an incidence graph of a BIBD-(v, b, r, k, λ) with $v = \theta k$ and $b = \theta r$. Conversely, for any CBD with at least two elements, its incidence graph is a BIG-(θ, r, k, λ) with $\theta = 1$. For any BIBD-(v, b, r, k, λ), its incidence graph is a BIG-(θ, r, k, λ) with $\theta > 1$, $v = \theta k$, and $b = \theta r$.*

Properties of BIGs: To employ BIG as the internal graph of a module, we investigate its expansion property and diameter in the following two lemmas.

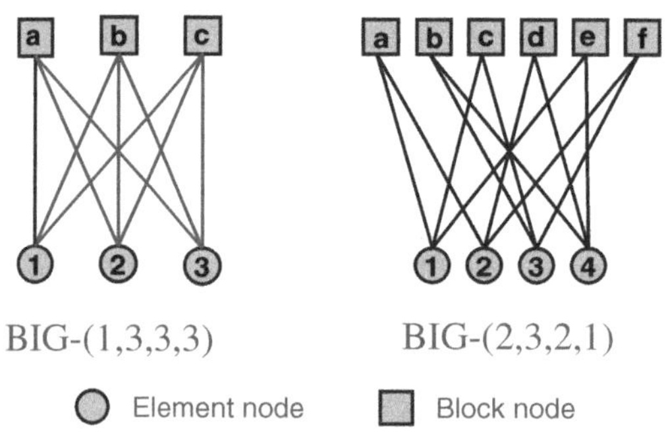

Figure 1.19 Two instances of block incidence graph (denoted by BIG-(θ, r, k, λ)). In the BIG-(1, 3, 3, 3), the number of element nodes θk and the number of block nodes θr are both three. Each pair of element nodes are adjacent to three common block nodes ($\lambda = 3$), including block nodes a, b, and c. As for the BIG-(2, 3, 2, 1), it contains four element nodes ($\theta k = 4$) and six block nodes ($\theta r = 6$). Each pair of element nodes are adjacent to only one common block node ($\lambda = 1$). For example, element nodes 1 and 2 are both adjacent to block node a.

Lemma 1.4 *For a BIG-(θ, r, k, λ), the second largest eigenvalue (SLE) of its adjacency matrix satisfies that*

$$\lambda_2^{eig}(BIG) = \sqrt{\frac{rk(\theta - 1)}{\theta k - 1}},$$

which achieves the tightest lower bound of SLEs for all biregular bipartite graphs given the same number of nodes and the same node degrees in such a BIG.

The proof of Lemma 1.4 is provided in Section A.1.1. The lemma leads to the following corollary whose proof is given in Section A.1.2.

Corollary 1.4 *A BIG-(θ, r, k, λ) is a bipartite Ramanujan graph.*

The diameter of BIG is demonstrated by the following lemma, which is proved in Section A.1.3.

Lemma 1.5 *The diameter of a BIG-(θ, r, k, λ) is at most four. Specifically, the distance between two nodes in a BIG is discussed in the following three cases.*

- *The distance between each pair of element nodes is two.*
- *The distance between an element node and a block node is one or three.*
- *The distance between each pair of block nodes is two or four.*

Based on the properties of CBD and BIBD, it can be shown that the parameters of BIG satisfy the following lemma, which is proved in Section A.1.4.

Lemma 1.6 *For a BIG-(θ, r, k, λ), we have*

$$\lambda = \frac{r(k - 1)}{\theta k - 1}. \tag{1.50}$$

Note that BIG does not always exist under any parameter settings (i.e., θ, r, k, λ). When $\theta = 1$, we have $\lambda = r$ according to Lemma 1.6. Such a BIG-($1, r, k, r$) is a CBG and universally exists for any integers r, k with $r \geq 1$ and $k \geq 2$. However, when $\theta > 1$, the BIG is induced by some BIBD which does not necessarily exist. By the sufficient conditions for the existence of BIBD, there are some sufficient conditions for the existence of BIG. Such conditions are demonstrated by the following lemma, which is proved in Section A.1.5. Figure 1.20 shows some examples for illustration.

Lemma 1.7 *A number of BIGs can be induced by some well-known BIBDs Colbourn and Dinitz (2006). The sufficient conditions with respect to the existence of the BIGs with $\theta > 1$ and $r = q$ are given as follows.*

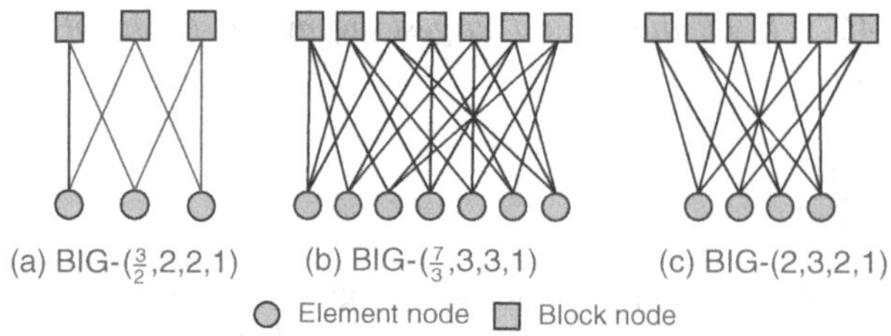

Figure 1.20 Some examples to illustrate the BIGs specified in Lemma 1.7. (a) When $q = 2$, a BIG-$(\frac{3}{2}, 2, 2, 1)$ can be viewed as a special case of a BIG-$(1 + \frac{1}{q}, q, q, q - 1)$, a BIG-$(2 - \frac{1}{q}, q, q, \frac{q}{2})$, a BIG-$(q - 1 + \frac{1}{q}, q, q, 1)$, and a BIG-$(\frac{q+1}{2}, q, 2, 1)$. (b) When $q = 3$, a BIG-$(\frac{7}{3}, 3, 3, 1)$ can be viewed as a special case of a BIG-$(2 + \frac{1}{q}, q, q, \frac{q-1}{2})$ and a BIG-$(q - 1 + \frac{1}{q}, q, q, 1)$. (c) When $q = 3$, a BIG-$(2, 3, 2, 1)$ as a special case of a BIG-$(\frac{q+1}{2}, q, 2, 1)$ and a BIG-$(q - 1, q, q - 1, 1)$.

- *Given an integer q such that $q \geq 2$, there exists a BIG-$(1 + \frac{1}{q}, q, q, q - 1)$.*
- *Given an even number q such that $2 \leq q \leq 212$, a BIG-$(2 - \frac{1}{q}, q, q, \frac{q}{2})$ can be induced by a Hadamard BIBD.*
- *Given an odd number q such that $3 \leq q \leq 211$, a BIG-$(2 + \frac{1}{q}, q, q, \frac{q-1}{2})$ can be induced by a Hadamard BIBD.*
- *Given an integer q such that $q \geq 2$, there exists a BIG-$(\frac{q+1}{2}, q, 2, 1)$.*
- *Given a prime power $q - 1$, a BIG-$(q - 1, q, q - 1, 1)$ can be induced by an affine BIBD.*
- *Given an integer q such that $q = 2$ or $q - 1$ is a prime power, a BIG-$(q - 1 + \frac{1}{q}, q, q, 1)$ can be induced by a projective BIBD.*

1.7.2 Transversal Design Graph

Definition 1.24 A *transversal design graph*, denoted by TDG-(n, μ, υ), is defined as a biregular bipartite graph, where nodes in one node partition are called *element nodes* and nodes in the other node partition are called *block nodes*, such that

- there are n groups of υ element nodes with degree $\mu\upsilon$ and $\mu\upsilon^2$ block nodes with degree n;
- two element nodes in distinct groups are adjacent to μ common block nodes;

- two element nodes in the same group have no common adjacent block nodes;
- the parameters satisfy $n, v \geq 2$ and $\mu \geq 1$.

Examples of block incidence graph are shown in the following Figure 1.21.

The name "transversal design graph" comes from the fact that such a graph can be regarded as the incidence graph of a transversal design (TD). Specifically, given a TD, its corresponding TDG can be constructed by mapping the elements and blocks in such a TD to element nodes and block nodes, respectively. For each element in the TD, if it is contained in a block, its corresponding element node and block node are adjacent to each other. Such a mapping is concluded by the following lemma, whose proof is provided in Section A.1.6.

Lemma 1.8 *Given a TDG-(n, μ, v), it can be viewed as the incidence graph of a TD-(n, μ, v). Conversely, for any TD-(n, μ, v), its incidence graph is a TDG-(n, μ, v).*

Properties of TDGs:

Lemma 1.9 *For a TDG-(n, μ, v), the SLE of its adjacency matrix satisfies*

$$\lambda_2^{eig}(\text{TDG}) = \sqrt{\mu v}. \tag{1.51}$$

The proof of Lemma 1.9 is provided in Section A.1.7. The lemma leads to the following corollary whose proof is given in Section A.1.8.

Corollary 1.5 *A TDG-(n, μ, v) is a bipartite Ramanujan graph.*

It is not hard to obtain the following lemma:

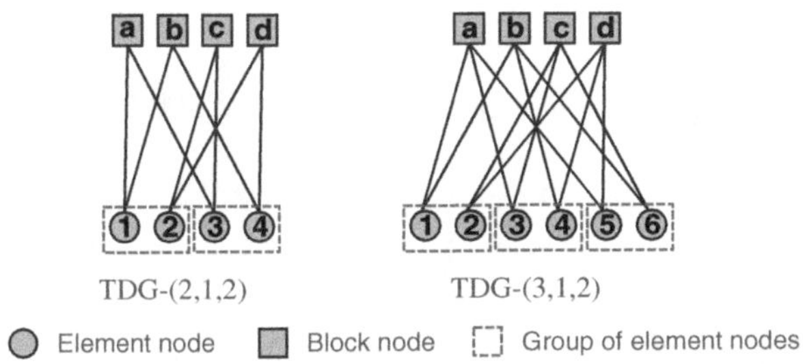

Figure 1.21 Two instances of transversal design graph (denoted by TDG-(n, μ, v)). In the TDG-(2, 1, 2), the numbers of element nodes nv and block nodes μv^2 are both four. Any two element nodes in distinct groups are adjacent to only one common block node ($\mu = 1$). For example, element nodes 1 and 4 are both adjacent to block node b. As for the TDG-(3, 1, 2), it contains six element nodes ($nv = 6$) and four block nodes ($\mu v^2 = 4$). Any two element nodes in distinct groups are adjacent to only one common block node ($\mu = 1$). For example, element nodes 1 and 3 are both adjacent to block node a.

Lemma 1.10 *The diameter of a TDG-(n, μ, υ) is at most four. Specifically, the distance between two nodes in a TDG is considered in the following three cases.*

- *For each element node, its distance with any element node in distinct groups is two, while its distance with another element node in the same group is four.*
- *The distance between an element node and a block node is one or three.*
- *For each block node, there exist at most $\mu n(\upsilon - 1) + \mu - 1$ block nodes at a distance of two from such a block node. For the other block nodes, their distances from the block node are all equal to four.*

Moreover, the existence conditions of TDG can be implied from the existence conditions of TD, as given by the following lemma. This lemma is proved in Section A.1.9.

Lemma 1.11 *Regarding the existence of TDG, two sufficient conditions are given as follows, respectively.*

- *If υ is a prime power, there exists a TDG-(n, μ, υ) for $2 \leq n \leq \upsilon + 1$.*
- *If $n = 2$ or $n = 3$, there exists a TDG-(n, μ, υ) for any integer $\upsilon \geq 2$.*

2

A Unified Design Framework

For a large-scale computer network, its network topology defines the way how all of the devices in the network are arranged via communication links.[1] Such an arrangement is jointly determined by two factors. One is the interconnection relationships among the devices. Such relationships are usually abstracted as a graph in which the nodes and the edges represent the devices and the communication links between them, respectively. The other factor is concerned with the *mapping* between each node in the graph and the type of the device it corresponds to. In fact, even given the same graphical structure, a modification in the assigned device type of some node can lead to different network topologies and performances. Figure 2.1 shows an example of an illustration. Considering the impact of such a mapping, it is necessary to distinguish the node types in the graph. To this end, in Section 2.1, we present a unified categorization of nodc types. The specified types serve as the primitive abstractions in the subsequent discussion of topology design.

On account of the interplay between such two factors, the topology design of large-scale computer networks is often confronted with a tremendous number of possible arrangements in the design space. Inspired by Shannon's style thinking of modularization and decoupling, we adopt a *modular design* perspective to decouple topology design at three levels of granularities and propose a novel unified procedure to describe the design of network topologies. In Section 2.2, we first introduce the ingredients of the unified procedure, including the components of topology design and the abstractions of their interconnections. Based on such ingredients, we demonstrate the details of the design procedure in Section 2.3. Next, in Section 2.4, for each step in the procedure, we propose general guiding principles and develop concrete tools as options to facilitate the design. Finally, in Section 2.5, we revisit a few classical topologies and demonstrate how they can be

1 In this monograph, network devices refer to electrical equipments providing either or both of routing and computing functions, such as switches (with pure routing function), hosts (with pure computing function), and edge servers (with both routing and computing functions). The communication links can be wired or wireless connections in practice.

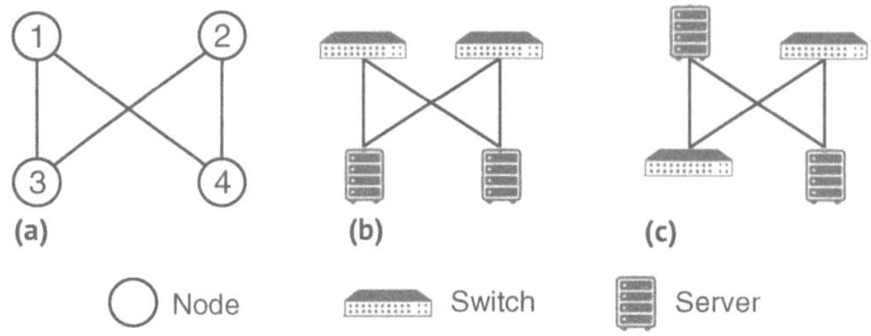

Figure 2.1 (a) A complete bipartite graph (CBG) $K_{2,2}$ with two nodes in each node partition.[2] (b) Topology I, a topology induced by the graph in subfigure (a). Nodes 1 and 2 correspond to switches with two ports, while nodes 3 and 4 correspond to servers with two NICs. (c) Topology II, a topology induced by the graph in subfigure (a). Nodes 2 and 3 are mapped to switches with two ports, while nodes 1 and 4 are mapped to servers with two NICs. Topologies I and II are generated by mapping nodes in the same graph to different types of devices, leading to distinct performances. For example, the hop-count of the shortest path between two servers (i.e., the minimum number of devices through which data pass through between such two servers) is one in Topology I, whereas such a hop-count is zero in Topology II.

constructed by following the procedure with different combinations of concrete tools. All proofs are relegated to Section A.2.

2.1 Categorization of Nodes

In large-scale computer networks, network devices generally provide two kinds of function: *computing* function and *routing* function. Computing function refers to the ability of a device to provide task-oriented data processing in an algorithmic and automated fashion (Shackelford et al. 2006). Routing function refers to the ability of a device to find a path and forward data for the communication between devices in a network (Peterson and Davie 2007). Accordingly, in a network topology, the corresponding nodes of such network devices can be classified by their functions, which are summarized in Table 2.1.

First, we consider the nodes with computing function. For such nodes, in some network topology, they are also employed to execute routing function (e.g., servers in DCell [Guo et al. 2008] and BCube [Guo et al. 2009] topologies). To clarify such a distinction, we refer to nodes corresponding to devices with *both* computing and routing functions as *hybrid nodes* (HNs), and nodes corresponding to devices with only computing function as *computing nodes* (CNs). For example, servers in

2 In a CBG, nodes are partitioned into two disjoint sets, such that any two nodes in the same set are not adjacent to each other. Each set is called a node partition. More details can be found in Section 1.4.6.3.

Table 2.1 The categorization of nodes in computer networks.

Node name	Function in network		
Computing node (CN)	Computing only		
Hybrid node (HN)	Computing and routing		Node with Routing Function
Gateway node (GN)	Routing	Direct access to CN	
Routing node (RN)	Only	No direct access to CN	

a server-centric topology are HNs since they also function as routing devices in the network, whilst all servers in a switch-centric topology are CNs because they are not employed as routing devices. In most existing topologies (Chen et al. 2016), each server with pure computing function is equipped with one and only one NIC. Correspondingly, in the framework, unless otherwise stated, each CN is assumed to have a degree of one by being attached to one and only one node with routing function (i.e., an HN or a node with pure routing function).

Next, we focus on the nodes with only routing function. Depending on whether they have direct access to CNs in a network topology, we divide them into two categories. If they have direct access to some CNs, we call them *gateway nodes* (GNs); otherwise, we call them *routing nodes* (RNs). For example, as presented in Figure 1.1, in a Fat-Tree topology, edge switches are GNs since they are directly connected to CNs (servers without routing function), while the aggregation switches and core switches are RNs.

Based on the categorization, a network topology is modeled as a *simple graph*[3] with nodes of different types (i.e., CNs, HNs, GNs, and RNs) and edges as the communication links between nodes. Defined in this way, the construction of a network topology consists of interconnecting nodes with routing function (i.e., HNs, GNs, and RNs), and interconnections involving CNs.

In this process, an essential concern lies in how to design the interconnection of HNs, GNs, and RNs to facilitate communications among nodes with computing function (i.e., CNs and HNs). Based on such a consideration, in the next subsection, we present the basic ingredients of topology design in the framework.

2.2 Ingredients of Topology Design

Faced with the great challenges brought by the deployment and management of large-scale computer networks, a widely adopted topology design methodology is *modular design*. Under modular design, a network's topology is constructed as a

3 A simple graph is defined as an unweighted, undirected graph with neither self-loops nor multiple edges between any pair of nodes (*cf.* Section 1.4.1).

group of loosely coupled components. The design, deployment, and management of each component are conducted separately with a lower complexity. Meanwhile, distinct components are interconnected with each other via well-defined interfaces (e.g., groups of routing devices) to form the network. A well-known example is the Fat-Tree topology (Al-Fares et al. 2008). As shown in Figure 2.2, such a topology is constructed by interconnecting independently deployed pods with a group of core switches.

Another example is DCell topology (Guo et al. 2008). As shown in Figure 2.3, for a level-l DCell ($l \geq 1$), it is composed of multiple interconnected level-($l-1$) DCells. The interconnection is formed by a set of links between servers in distinct level-($l-1$) DCells. In the next subsections, we illustrate how such ingredients (i.e., components, interfaces, and connections among components) are defined in the framework.

2.2.1 Components and Their Interfaces

In this framework, two categories of components are considered, namely *modules* and *RN clusters*. Their definitions and interfaces are specified as follows.

Definition 2.1 A module is defined as a connected simple graph that must contain either or both of HNs and GNs, without any CNs.

Some examples of modules are presented in Figure 2.4. As shown in Figure 2.4a, a level-0 DCell can be viewed as a module which contains servers as HNs and switches as RNs. The second example is Fat-Tree topology (see Figure 2.4b) in which the pods can be viewed as modules. Each module contains edge switches as GNs and aggregation switches as RNs. Each edge switch is connected to several servers, with each server as a CN. Another example is HyperX topology (see Figure 2.4c). In such a topology, each pod can also be viewed as a

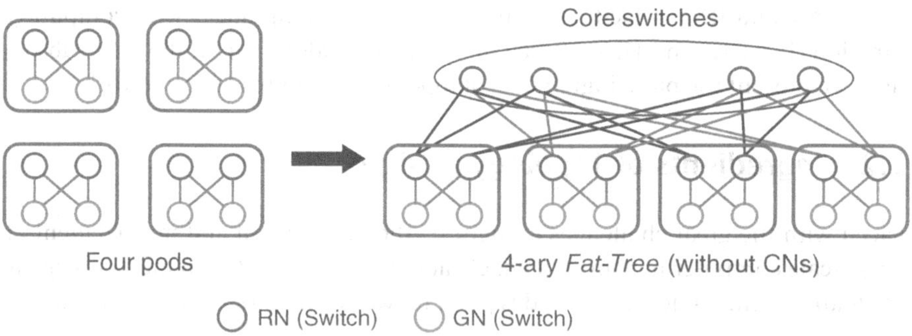

Figure 2.2 The interconnection of four pods in a 4-ary Fat-Tree topology.

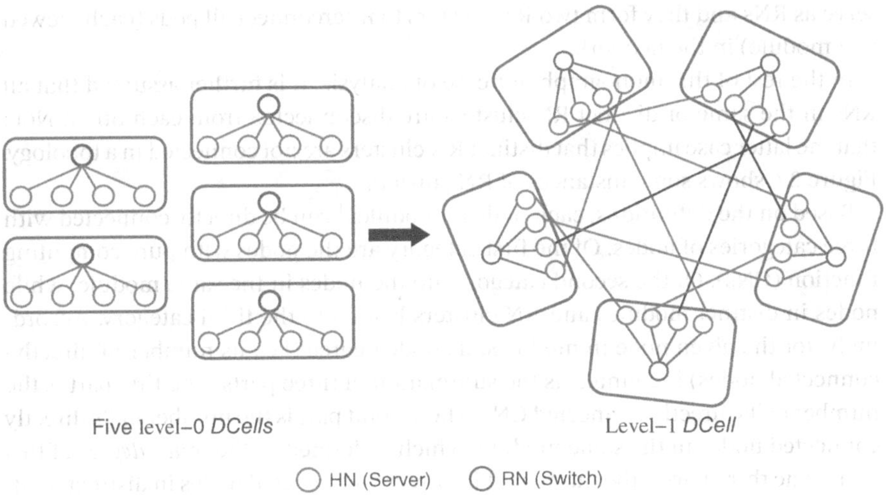

Figure 2.3 The interconnection of five level-0 DCells as a level-1 DCell with 4-port switches.

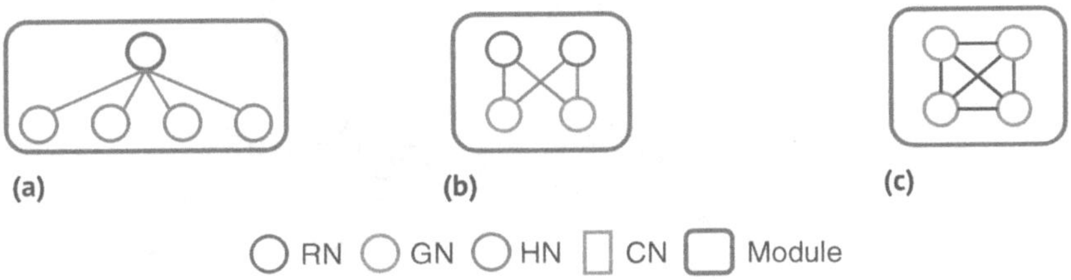

Figure 2.4 Modules in different topologies. (a) A module in a level-1 DCell topology. (b) A module in a 4-ary Fat-Tree topology. (c) A module in a level-1 HyperX topology.

module. Each switch is regarded as a GN since it is directly connected to several servers (CNs).

By this definition, modules can be viewed as distinct groups of nodes with routing function (i.e., HNs, GNs, and RNs) that are deployed and managed separately. Note that modules may also contain RNs, but their involvements are not necessary.

Definition 2.2 An RN cluster is defined as a group of RNs that are *not* contained in any modules.

RN clusters are employed in network topologies for communications among modules. For example, in a 4-ary Fat-Tree topology (see Figure 2.5), core switches serve as RNs and they form two RN clusters to interconnect all pods (each viewed as a module) in the network.

In the rest of this monograph, for ease of analysis, it is further assumed that all RNs in the same or distinct RN clusters are disconnected from each other. Note that the latter case implies that distinct RN clusters are not connected in a topology. Figure 2.6 shows some instances of RN clusters.

Based on the definitions, each node in modules[4] can be directly connected with three categories of nodes. Of the first category are the nodes with pure computing function (CNs). Of the second category are the nodes in the same module, while nodes in distinct modules and RN clusters belong to the third category. Accordingly, for the given node in modules, its node degree (i.e., its number of directly-connected nodes) is counted as the summation of three parts. The first part is the number of its directly-connected CNs. The second part is the number of its directly connected nodes in the same module, which is defined as the *inner-degree* of this node. The third part is the number of its directly-connected nodes in distinct modules or RN clusters, which is defined as the *outer-degree* of this node. Under the

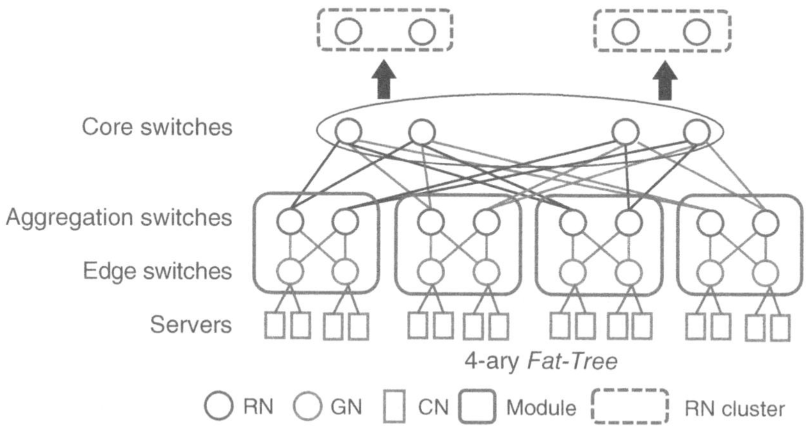

Figure 2.5 An illustration of RN clusters in a 4-ary Fat-Tree topology.

Figure 2.6 An illustration of RN clusters with different numbers of RNs.

4 Note that each node contained in modules can be an RN, a GN, or an HN.

setting, only nodes with positive outer-degrees are selected to establish communications between this module and other components. We refer to the selected nodes as *dock nodes* of each module, which are formally defined as follows.

Definition 2.3 For each module, its dock nodes are defined as the nodes in such a module with positive outer-degrees.

Next, we take a 4-ary Fat-Tree topology and a level-1 DCell with 4-port switches as examples for illustration.

In the 4-ary Fat-Tree topology (see Figure 2.7), each aggregation switch (RN) is directly connected with two edge switches (GNs) in the same module and two core switches (RNs) in RN clusters. Hence, the inner-degree and outer-degree of each aggregation switch (RN) are both two. As for the edge switches (GNs), they are directly connected with two aggregation switches in the same module and two CNs. Hence, for each edge switch (GN), its inner-degree is two and outer-degree is zero. Since the outer-degree of each aggregation switch (RN) is positive and the outer-degree of each edge switch (GN) is zero, only RNs are selected as dock nodes. These settings are summarized in Table 2.2.

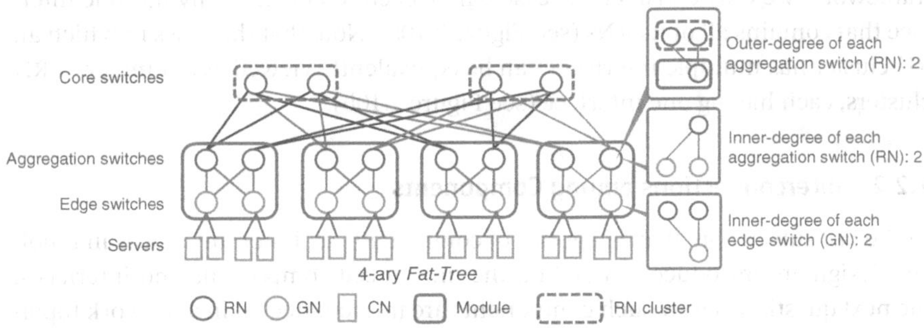

Figure 2.7 An illustration of node degrees in a 4-ary Fat-Tree topology. In this case, the inner-degree and outer-degree of each aggregation switch are both two. Meanwhile, the inner-degree and outer-degree of each edge switch are two and zero, respectively.

Table 2.2 The settings of degrees of nodes in the modules of a 4-ary Fat-Tree topology.

	Degree	**# of directly-connected CNs**	**Inner-degree**	**Outer-degree**	**Selected as dock nodes**
Aggregation switch (RN)	4	0	2	2	
Edge switch (GN)	4	2	2	0	×

In the level-1 DCell topology with 4-port switches (see Figure 2.8), each switch (RN) is directly connected with four servers (HNs) in the same module. Hence, for each switch (RN), its inner-degree is four and its outer-degree is zero. As for the servers (HNs), they are directly connected with one switch in the same module and one HN in distinct modules. Hence, the inner-degree and outer-degree of each server (HN) are both one. Since the outer-degree of each switch (RN) is zero and the outer-degree of each server (HN) is positive, only HNs are selected as dock nodes. These settings are summarized in Table 2.3.

Recall that under the modular design methodology, distinct components are interconnected with each other via well-defined interfaces. Next, we introduce the interfaces of modules and RN clusters as follows, respectively.

Interfaces of modules: In each module, a number of dock nodes are selected to connect with other components. Such dock nodes are further divided into distinct groups.[5] Each group is called an *intermediate super-node* (ISN), which is regarded as an *interface* of the module. Figure 2.9 provides an example for illustration.

Interfaces of RN clusters: By Definition 2.2, an RN cluster is formed by a group of RNs outside modules. Such RNs are always connected with dock nodes in modules and thus can be used to form the interfaces of the RN cluster. In this framework, we only consider the case where each RN cluster only has one interface that contains all of its RNs (see Figure 2.10a). Note that the cases in which an RN cluster has multiple interfaces can be equivalently viewed as a number of RN clusters, each having one interface (see Figure 2.10b).

2.2.2 Interconnections among Components

In the last subsection, two categories of components and their interfaces in topology design are introduced. Based on the introduced components and interfaces, the next question is how such components are interconnected in a network topology. To this end, we first identify two atomic modes to interconnect distinct components in Section 2.2.2.1. Then some definitions are introduced to describe the

Table 2.3 The settings of degrees of nodes in the modules of a level-1 DCell topology.

	Degree	# of directly-connected CNs	Inner-degree	Outer-degree	Selected as dock nodes
Switch (RN)	4	0	4	0	×
Server (HN)	2	0	1	1	

5 The grouping of dock nodes is left as a degree of freedom in topology design. This will be elaborated in Section 2.4.3.2.

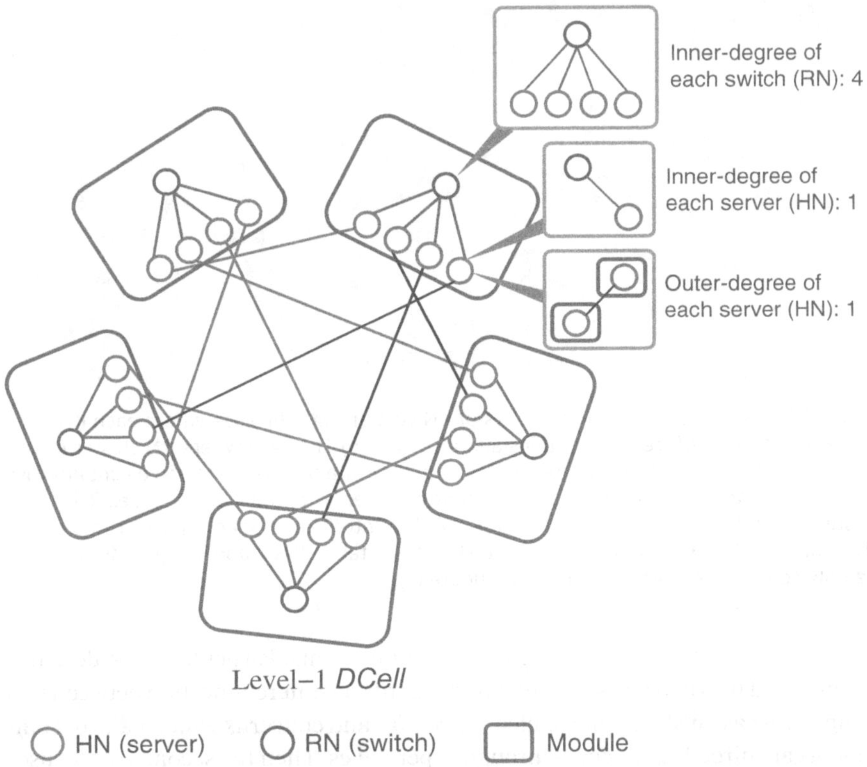

Figure 2.8 An illustration of node degrees in a level-1 DCell topology with 4-port switches. In this case, the inner-degree and outer-degree of each server are both one. Meanwhile, the inner-degree and outer-degree of each switch are four and zero, respectively.

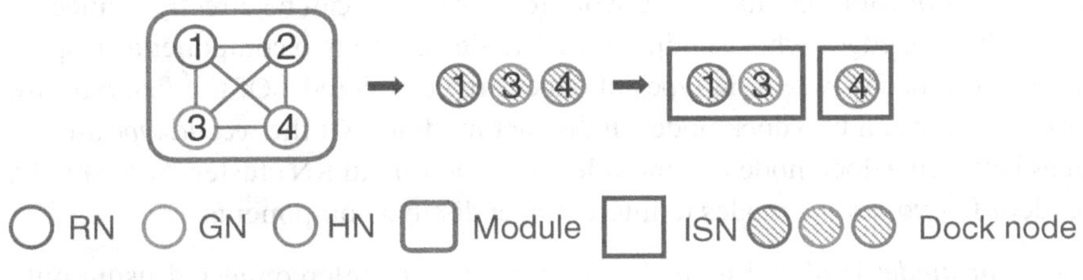

Figure 2.9 An illustration of dock node selection. In the given module, nodes 1, 3, and 4 are selected as dock nodes are partitioned into two ISNs. One ISN contains nodes 1 and 3, while the other ISN contains node 4.

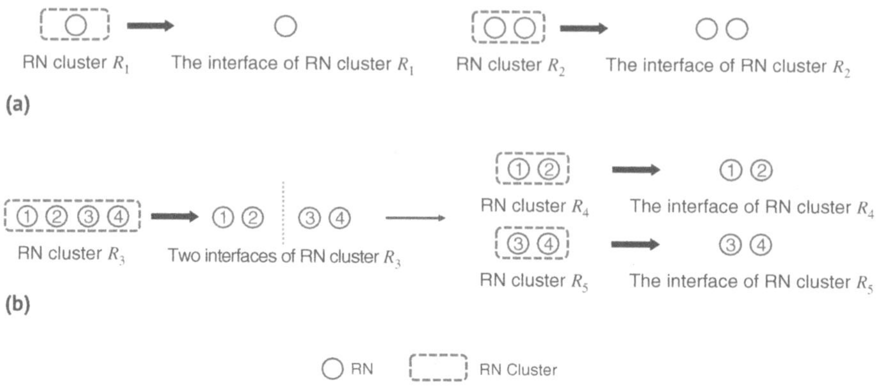

Figure 2.10 An illustration of interfaces of RN clusters. (a) The case where each RN cluster has one interface that contains all of its RNs. (b) The case where RN cluster R_3 has two interfaces. One interface contains RNs 1 and 2, while the other interface contains RNs 3 and 4. Such an RN cluster can be equivalently viewed as two RN clusters (i.e., RN clusters R_4 and R_5), each having one interface. The interface of RN cluster R_4 contains RNs 1 and 2, while the interface of RN cluster R_5 contains RNs 3 and 4. In this framework, we only consider the case shown in sub-figure (a).

design of interconnections among distinct components. Particularly, the design is decomposed into two parts. The first part regards the interconnection between two components as an abstraction called *super-edge* and concerns which pairs of components are directly connected through super-edges. Then the second part focuses on the construction of super-edges by considering how the interfaces of distinct components are interconnected. The details of these two parts are presented in Sections 2.2.2.2 and 2.2.2.3, respectively.

2.2.2.1 Interconnection Modes

For any pair of components in a network topology, they can be directly connected by establishing edges whose endpoints are nodes in the two components, respectively. Particularly, two basic types of edges are considered.[6] Of the *first type* are the edges between two dock nodes in distinct modules. Of the *second type* are the edges between a dock node in a module and an RN in an RN cluster. Accordingly, we identify two *atomic* modes to interconnect distinct components.

- *Plain mode*: Under this mode, components are interconnected using only edges of the first type.
- *Bipartite mode:* Under this mode, components are interconnected using only edges of the second type.

---

6 Note that there is assumed to be no edge between two RNs in distinct RN clusters.

To illustrate such two atomic modes, we take a level-1 DCell topology and a 4-ary Fat-Tree topology as examples. In the DCell topology (see Figure 2.11a), HNs in modules play the role of dock nodes. Each dock node is directly connected to dock nodes in other modules (note that there is no RN cluster in the topology). In other words, in DCell, all components (i.e., modules) are interconnected under the plain mode. In the Fat-Tree topology (see Figure 2.11b), RNs in modules play the role of dock nodes. Such dock nodes are only directly connected to RNs in RN clusters. In other words, in Fat-Tree, distinct components (i.e., modules and RN clusters) are interconnected under the bipartite mode.

Remark 2.1 The name "bipartite mode" comes from the following reasoning. Under this mode, only edges of the second type are used. Hence, dock nodes in distinct modules are not directly connected. Instead, they are directly connected to RNs in the RN clusters. Given this setting, dock nodes in modules and RNs in RN clusters are interconnected as a bipartite graph in which one partition contains all dock nodes while the other partition consists of all RNs in RN clusters.[7] Therefore, such an interconnection mode is called "bipartite mode".

2.2.2.2 Super-edges and Skeleton Graph

In this framework, for each pair of components, we say there is a *super-edge* between them if there exists one or more edges whose endpoints are contained

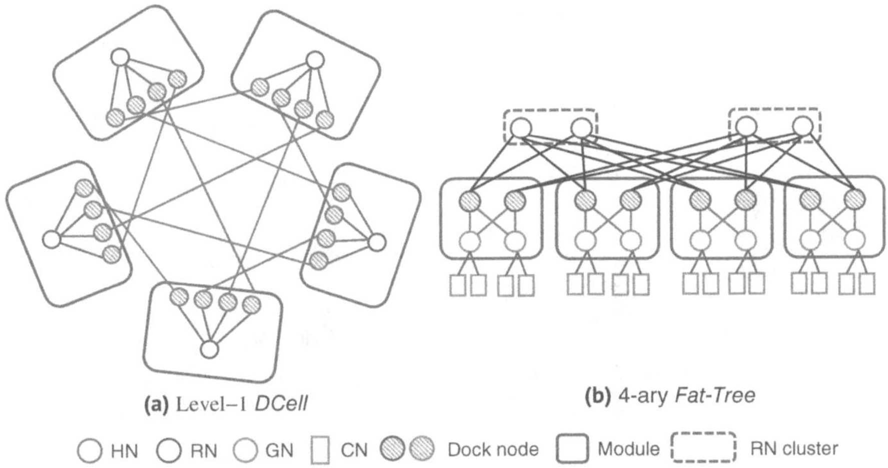

Figure 2.11 An illustration of interconnection modes of (a) level-1 DCell and (b) 4-ary Fat-Tree.

---

7 Note that the discussion of interconnections among distinct components does not involve edges between nodes in the same component.

in such two components, respectively. Recall that there are two atomic modes to interconnect distinct components. Correspondingly, the super-edges under such two modes are considered as follows, respectively.

- *Super-edge under the plain mode.* A super-edge under this mode interconnects two modules. Such a super-edge is defined as the set of all edges whose endpoints are dock nodes in the two modules, respectively. In Figure 2.12, we take a level-1 DCell topology with 4-port switches as an example for illustration.
- *Super-edge under the bipartite mode.* A super-edge under this mode interconnects a module and an RN cluster. Such a super-edge is defined as the set of all edges between a dock node in the module and an RN in the RN cluster. In Figure 2.13, we take a 4-ary Fat-Tree topology as an example for illustration.

By employing the abstraction of super-edges, a network topology can be described as a *skeleton graph*. In the graph, each node denotes a component (i.e., either a module or an RN cluster) while each edge denotes a super-edge between two components. For example, the skeleton graph of a level-1 DCell topology with 4-port switches is a complete graph (CG) with five modules as nodes (see Figure 2.14a). Each pair of modules are directly connected by a super-edge under the plain mode. Regarding a 4-ary Fat-Tree topology, its skeleton graph is a

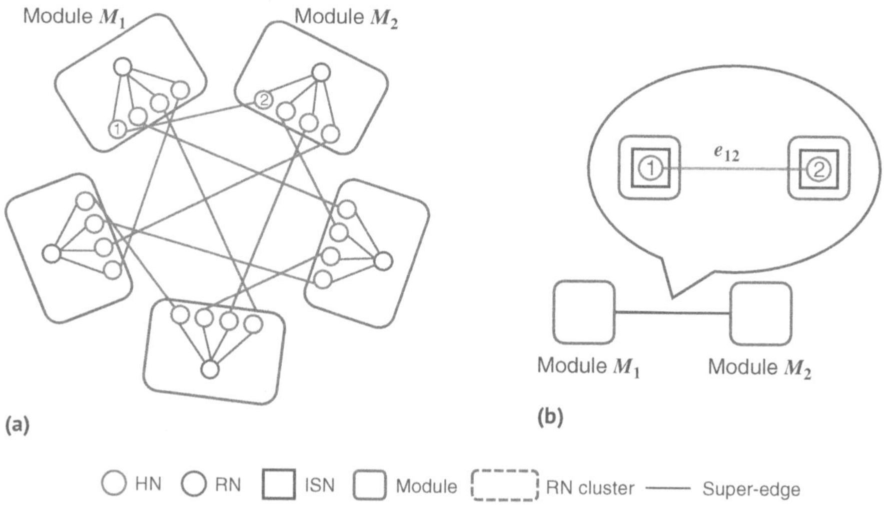

Figure 2.12 An illustration of super-edges under the plain mode for a level-1 DCell topology with 4-port switches. (a) An overview of the level-1 DCell topology. (b) A super-edge between modules M_1 and M_2, which contains an edge e_{12}.

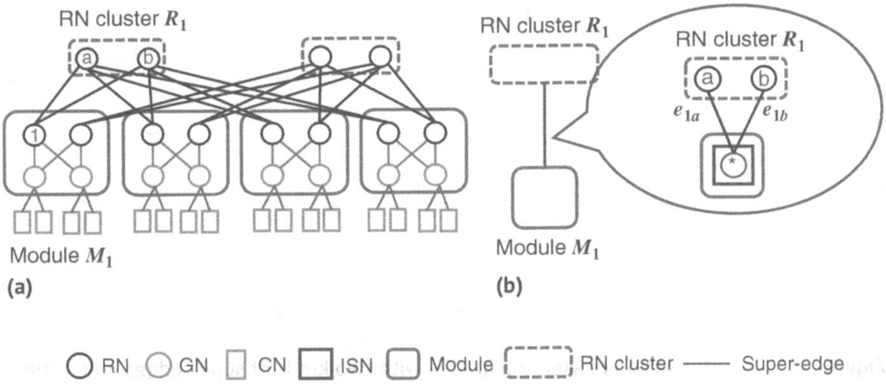

Figure 2.13 An illustration of super-edges under the bipartite mode for a 4-ary Fat-Tree topology. (a) An overview of the 44-ary Fat-Tree topology. (b) A super-edge between module M_1 and RN cluster R_1, which contains two edges e_{1a} and e_{1b}.

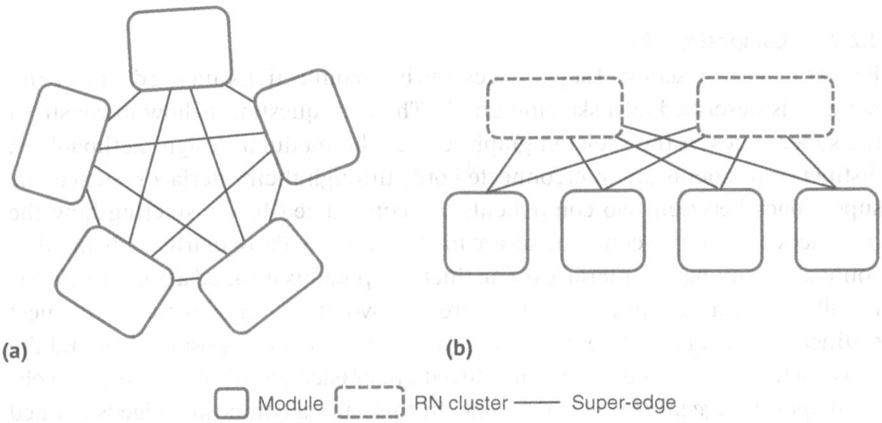

Figure 2.14 An illustration of skeleton graph. (a) The skeleton graph of a level-1 DCell topology with 4-port switches. (b) The skeleton graph of a 4-ary Fat-Tree topology.

complete bipartite graph (CBG) (see Figure 2.14b), where one node partition consists of four modules and the other node partition consists of two RN clusters. Each module is directly connected to each RN cluster by a super-edge under the bipartite mode.

In addition, a topology can simultaneously involve super-edges under the two atomic modes. Figure 2.15 shows an example for illustration.

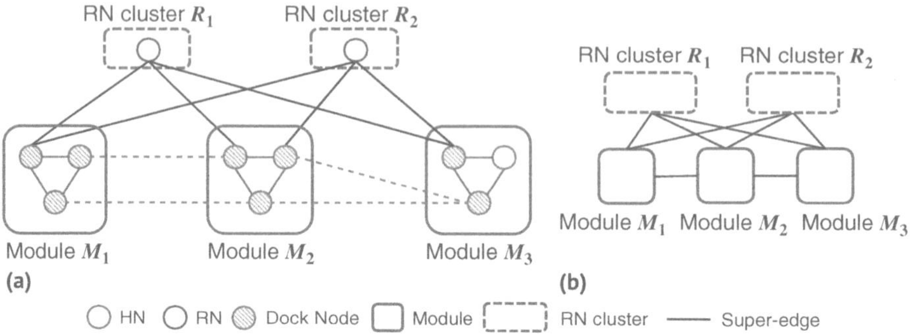

Figure 2.15 An illustration of skeleton graphs with two kinds of super-edges under the two atomic modes, respectively. (a) An overview of the topology. Note that the two kinds of edges between nodes that are employed to interconnect components under the plain mode and the bipartite mode are indicated by dashed and solid lines, respectively. (b) The corresponding skeleton graph. Each super-edge between two components is indicated by solid lines.

2.2.2.3 Composite-edge

Based on the abstraction of super-edges, the interconnections among distinct components is described as a skeleton graph. The next question is how to construct the super-edges in the skeleton graph. Under the modular design methodology, distinct components are interconnected only through their interfaces. Hence, the super-edges between two components are constructed by considering how the interfaces of such two components are interconnected. Particularly, such an interconnection among the interfaces of distinct components is called a *composite-edge*. Recall that in a network topology, there are two atomic modes to interconnect distinct components. Correspondingly, the definition of composite-edges and the construction of super-edges are introduced under each atomic mode, respectively.

Composite-edge under the plain mode. In this case, a composite-edge is defined as the interconnection between two ISNs in distinct modules. It contains the set of all edges whose endpoints are in the two ISNs, respectively. In Figure 2.16, we provide two examples to illustrate the composite-edges under the plain mode.

The composite-edges under this mode are adopted to construct super-edges under the plain mode. Specifically, for each super-edge between two modules, it is constructed by directly connecting some pairs of ISNs across the two modules using composite-edges under the plain mode.

For example, in a level-1 DCell topology with 4-port switches, each dock node (i.e., HN) forms an ISN and the composite-edge contains only one edge between two dock nodes in the two ISNs (see Figure 2.17). For each pair of directly connected modules, the super-edge between them is constructed by selecting one ISN in each module and interconnecting the selected two ISNs using the composite-edge.

Remark 2.2 Given two modules that are directly connected in a skeleton graph, the selected pair of ISNs in the two modules for each composite-edge should be subject to the following two constraints.

- The selected pair of ISNs in the two modules should have the same number of dock nodes as required by the composite-edge.
- For each dock node in the selected pair of ISNs, its remaining outer-degree should be no smaller than the number of its directly-connected dock nodes as required by the composite-edge.

In the skeleton graph of a level-1 DCell topology with 4-port switches, there are ten super-edges under the plain mode. Each of such super-edges is constructed using the composite-edge shown in Figure 2.17.

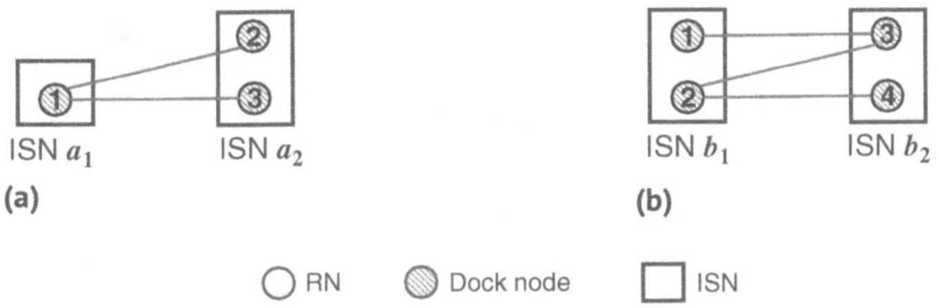

Figure 2.16 An illustration of composite-edges under the plain mode. (a) A composite-edge between two ISNs a_1 and a_2. (b) A composite-edge between two ISNs b_1 and b_2.

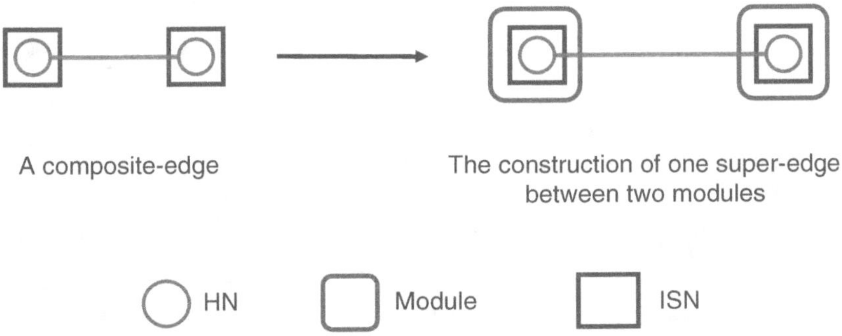

Figure 2.17 An illustration of super-edge construction in a level-1 DCell topology with 4-port switches.

Composite-edge under the bipartite mode. In this case, a composite-edge is defined as the interconnections among multiple ISNs in distinct modules through a *single* RN cluster. It contains the set of all edges whose endpoints are in those ISNs and the RN cluster, respectively. In Figure 2.18, we provide two examples to illustrate composite-edges under the bipartite mode.

The composite-edges under this mode are adopted to construct super-edges under the bipartite mode. Specifically, for each RN cluster, all of its incident super-edges are constructed by using one or more composite-edges under the bipartite mode. Each composite-edge specifies the edges between RNs in the RN cluster and dock nodes in one ISN of each module.

For example, in a 4-ary Fat-Tree topology (see Figure 2.19), each RN (i.e., aggregation switch) in one module is regarded as a dock node and forms an ISN. Such four ISNs are interconnected through two RNs in an RN cluster by a composite-edge under the bipartite mode. For each RN cluster, its four incident super-edges are constructed by selecting one ISN in each module and interconnecting the selected ISNs through an RN cluster using the composite-edge.

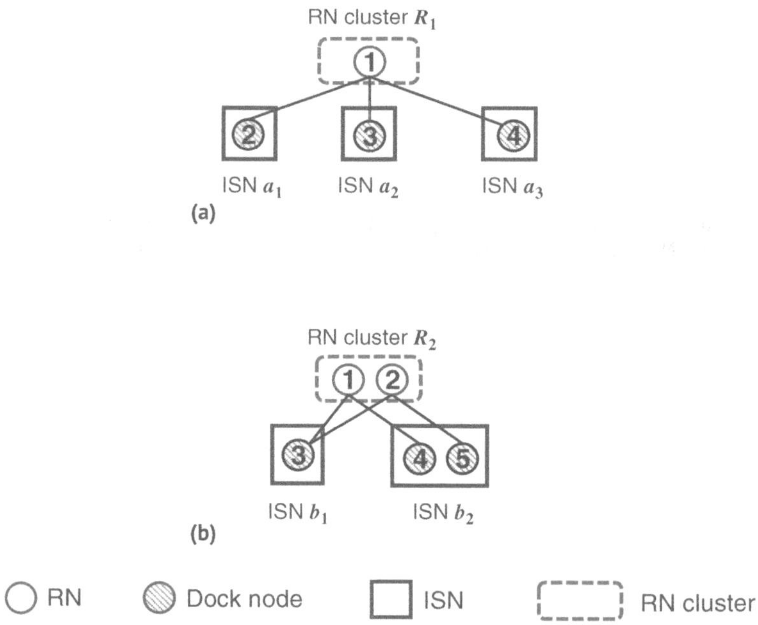

Figure 2.18 An illustration of composite-edges under the bipartite mode. (a) A composite-edge among ISNs a_1, a_2, and a_3 through RN cluster R_1. (b) A composite-edge between ISNs a_1 and a_2 through RN cluster R_2.

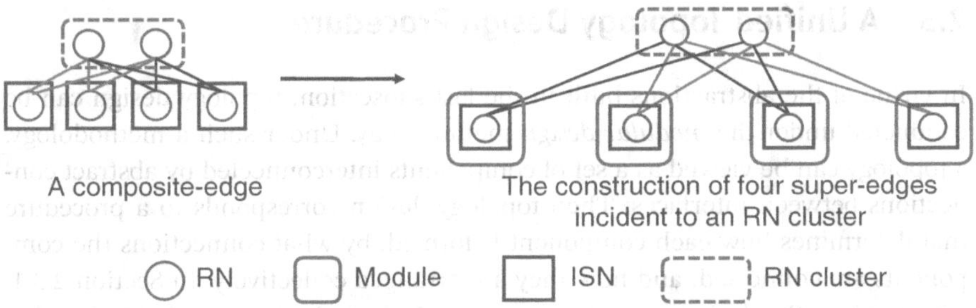

Figure 2.19 An illustration of super-edge construction in a 4-ary Fat-Tree topology.

Remark 2.3 Given m modules that are interconnected through an RN cluster in the skeleton graph, the selected m ISNs in these distinct modules for each composite-edge should be subject to the following two constraints.

- The selected m ISNs (each in a distinct module) should have the same number of dock nodes as required by the composite-edge.
- For each dock node in the selected m ISNs, its outer-degree should be no smaller than the number of its directly-connected RNs in the RN cluster as required by the composite-edge.

In the skeleton graph of a 4-ary Fat-Tree topology, there are two RN clusters. For each RN cluster, its four incident super-edges are all under the bipartite mode. Such super-edges are constructed using one composite-edge under the bipartite mode (see Figure 2.19).

Remark 2.4 In summary, the design of interconnections among distinct components is decomposed into the design of composite-edges, the design of a skeleton graph, and the construction of super-edges in the skeleton graph using the composite-edges. The decoupling of these three parts not only facilitates the independent development of each part using concrete mathematical tools, but also conduces to lowering the complexity of topology design. Specifically, on one hand, the design of the skeleton graph only considers the interconnections among distinct components, abstracting away the details of edges between nodes in distinct components. On the other hand, given the skeleton graph, the construction of each super-edge can reuse the constructed composite-edge. For example, in some network topologies (e.g., a DCell topology and a Fat-Tree topology), the skeleton graph contains modules with the same intra-module design and homogeneous ISNs. In such topologies, only one type of composite-edge is reused to construct the super-edges. Such a way can efficiently reduce the complexity of topology design.

2.3 A Unified Topology Design Procedure

In virtue of the abstractions built in the last subsection, topology design can be conducted under the *modular design* methodology. Under such a methodology, a topology can be viewed as a set of components interconnected by abstract connections between interfaces. Then topology design corresponds to a procedure that determines how each component is formed, by what connections the components are connected, and how they are arranged collectively. In Section 2.3.1, we introduce the design procedure in detail. In fact, the key idea behind such a procedure is decoupling the topology design, which helps to reveal the degrees of freedom in design space and reduce design complexity. The decoupling method behind the design procedure is highlighted in Section 2.3.2. Based on such a decoupling method, the design procedure not only offers a prescriptive model to describe topology design in a unified manner but also provides a constructive approach to generate topologies.

2.3.1 Design Procedure

As shown in Figure 2.20, the design procedure contains four steps, namely *intra-module design*, *connection design*, *inter-module design*, and *finalization*. Following this procedure, topology design is conducted by executing one or more iterations of the former three steps, then applying the finalization step. Such an iterative execution is indicated by the dashed line in Figure 2.20. For ease of understanding, we first consider the case with one iteration to illustrate each step in the procedure.

Step 1: Intra-module design. This step aims to generate the design of *internal graph* for each module. Each internal graph is defined as a connected graph. Each node in the internal graph can be instantiated as an RN, a GN, or an HN. Note that by the definition of modules (see Definition 2.1), it is not allowed to instantiate all nodes as RNs in an internal graph.

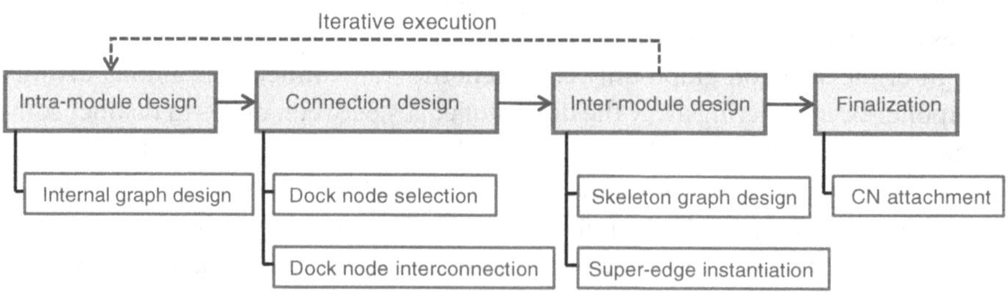

Figure 2.20 An overview of topology design procedure.

Step 2: Connection design. The goal of this step is twofold. One is to design the interfaces of each module. The other is to design the interconnections among interfaces of distinct modules, i.e., composite-edges. To fulfill the goal, this step includes two sub-steps, called *dock node selection* and *dock node interconnection*, respectively. The first sub-step considers how to select dock nodes within each module and group them as intermediate super-nodes (ISNs), i.e., the interfaces of the module. As for the second sub-step, it decides how the designed interfaces of distinct modules are interconnected, i.e., the construction of composite-edges. Depending on the design of composite-edges, this sub-step may require to build RN clusters as communication relays.

Step 3: Inter-module design. This step focuses on the design of interconnections among modules and RN clusters. To this end, two sub-steps are included in this step, namely *skeleton graph design* and *super-edge instantiation*. The former sub-step decides how modules and RN clusters are interconnected via super-edges. The latter sub-step considers how to construct each super-edge in the skeleton graph using the composite-edges designed in Step 2.

Step 4: Finalization (optional). This step considers how to "*directly*" attach CNs to the GNs and HNs in the topology prototype derived from the previous three steps. In this framework, unless otherwise stated, the degree of each CN is assumed to be one and only allowed to be connected to a GN or an HN. Note that such a step can be omitted if CNs are not considered in the topology.

In Figure 2.21, we take a 4-ary Fat-Tree topology as an example to illustrate the procedure.

- *Intra-module design*. The internal graph of each module is a complete bipartite graph (CBG) with two GNs in one node partition and two RNs in the other node partition.[8]
- *Connection design*. Regarding dock node selection, the RNs in each module are selected as dock nodes, each of which forms an ISN. There are two RNs in each module, which implies that each module contains two ISNs. As for dock node interconnection, four ISNs in distinct modules are interconnected through an RN cluster with two RNs, which forms a composite-edge under the bipartite mode. To be specific, each RN in the RN cluster is directly connected to four dock nodes in the four ISNs.
- *Inter-module design*. The skeleton graph is a CBG with two RN clusters in one node partition and four modules in the other node partition. Note that each RN cluster is directly connected to four modules. The four super-edges between the four modules and a common RN cluster are constructed using

8 In a CBG, nodes are partitioned into two disjoint sets, such that any two nodes in the same set are not adjacent to each other. Each set is called a node partition. More details can be found in Section 1.4.6.3.

one composite-edge (see Figure 2.22). For the i-th RN cluster ($i = 1, 2$), it is directly connected to the i-th ISN in each of the four modules using the composite-edge.
- *Finalization.* Each GN is directly connected to two CNs.

Next, we consider the procedure with multiple iterations. In this case, the first three steps are repeated in the following way. At the beginning of the i-th iteration ($i \geq 2$), the topology prototype generated from the ($i - 1$)-th iteration can be treated as one of the candidates for intra-module design in the current iteration.

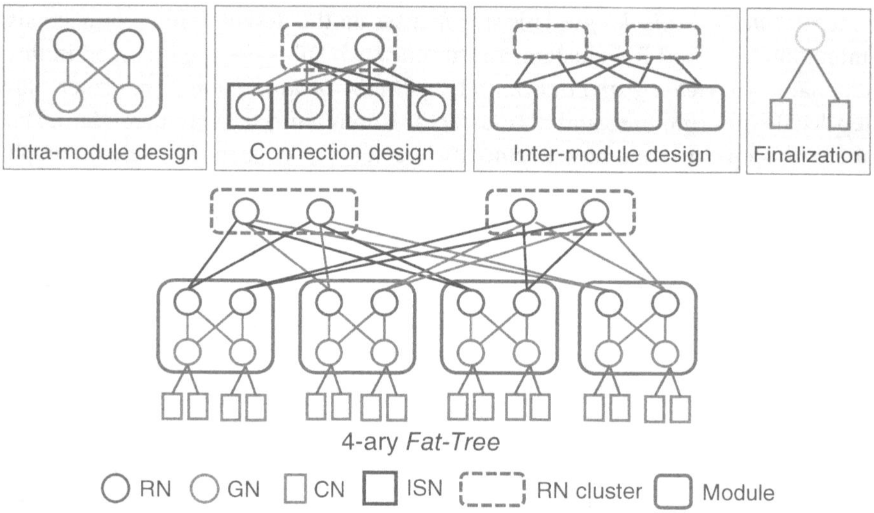

Figure 2.21 An illustration of the design procedure for a 4-ary Fat-Tree topology.

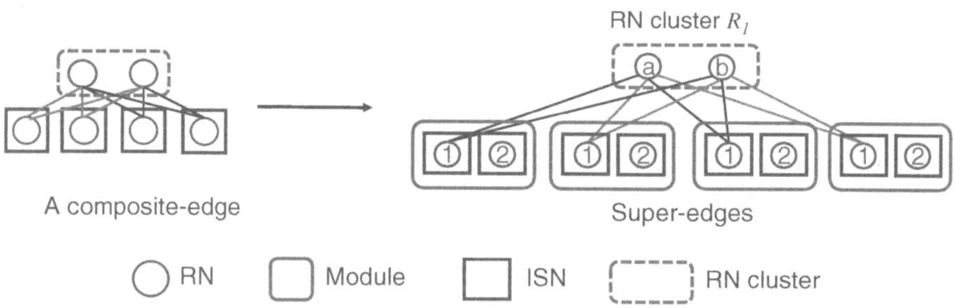

Figure 2.22 An illustration of super-edge construction for a 4-ary Fat-Tree topology. For the first RN cluster R_1, it is directly connected to four modules through four super-edges. These four super-edges are constructed using a composite-edge. To be specific, all RNs labeled with number 1 in each module are directly connected to RNs a and b in RN cluster R_1.

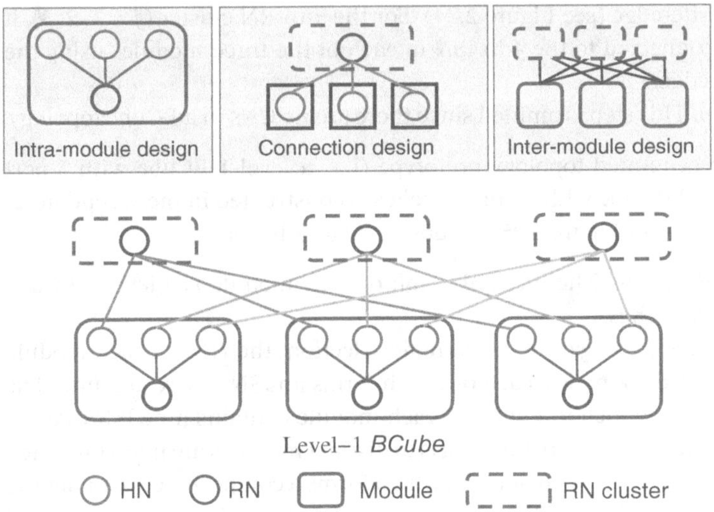

Figure 2.23 The design procedure for a level-1 BCube topology with 3-port switches.

Then the topology prototype and the modules designed in this iteration are both adopted to conduct the connection and inter-module designs. After a number of such iterations, the finalization step is applied to attach CNs to the topology prototype generated from the last iteration. Notably, such a step can be omitted if CNs are not considered in the topology.

For example, the design of a level-2 BCube with 3-port switches can be described by such a procedure with two iterations. As shown in Figure 2.23, the output of the first iteration is a level-1 BCube with 3-port switches. The design steps of a level-1 BCube with 3-port switches are described as follows.

- *Intra-module design.* The internal graph of each module is a CBG with one RN in one node partition and three HNs in the other node partition.
- *Connection design.* Regarding dock node selection, the HNs in each module are selected as dock nodes, each of which forms an ISN. There are three HNs in each module, which implies that each module contains three ISNs. As for dock node interconnection, three ISNs in distinct modules are interconnected through an RN cluster with one RN, which forms a composite-edge under the bipartite mode. To be specific, the RN in the RN cluster is directly connected to three dock nodes in the three ISNs.
- *Inter-module design.* The skeleton graph is a CBG with three RN clusters in one node partition and three modules in the other node partition. Note that each RN cluster is directly connected to three modules. The three super-edges between the three modules and a common RN cluster are constructed using

one composite-edge (see Figure 2.24). For the j-th RN cluster ($j = 1, 2, 3$), it is directly connected to the j-th ISN in each of the three modules using the composite-edge.

- *Finalization.* This step is omitted since there are no CNs in a BCube topology.

Based on the generated topology prototype (i.e., a level-1 BCube with 3-port switches), a level-2 BCube with 3-port switches is constructed in the second iteration, which is shown in Figure 2.25 and described as follows.

- *Intra-module design.* The internal graph of each module is a level-1 BCube with 3-port switches.
- *Connection design.* Regarding dock node selection, the HNs in each module are selected as dock nodes, each of which forms an ISN. There are nine HNs in each module, which implies that each module contains nine ISNs. As for dock node interconnection, three ISNs in distinct modules are interconnected through an RN cluster with one RN, which forms a composite-edge under the bipartite mode. To be specific, the RN in the RN cluster is directly connected to three dock nodes in the three ISNs.
- *Inter-module design.* The skeleton graph is a CBG with nine RN clusters in one node partition and three modules in the other node partition. Note that each RN cluster is directly connected to three modules. The three super-edges between the three modules and a common RN cluster are constructed using one composite-edge (see Figure 2.26). For the k-th RN cluster ($k = 1, 2, ..., 9$), it is directly connected to the k-th ISN in each of the three modules using the composite-edge.
- *Finalization.* This step is omitted since there are no CNs in a BCube topology.

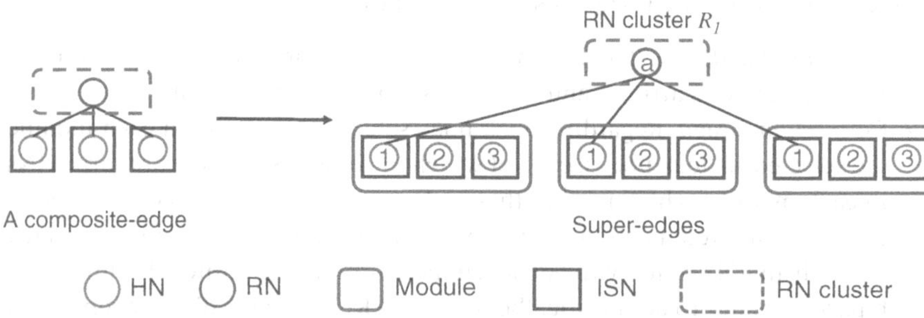

Figure 2.24 An illustration of super-edge construction for a level-1 BCube topology with 3-port switches. For the first RN cluster R_1, it is directly connected to three modules through three super-edges. These three super-edges are constructed using a composite-edge. To be specific, all HNs labeled with number 1 in each module are directly connected to RN a in RN cluster R_1.

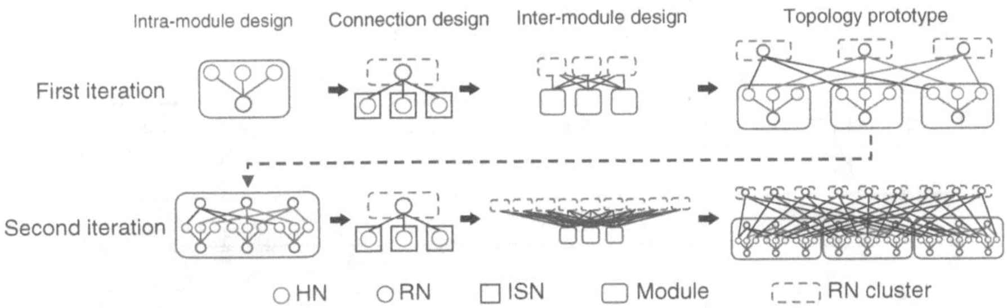

Figure 2.25 The design procedure of a level-2 BCube topology with 3-port switches. The design procedure contains two iterations. The output of the first iteration is a level-1 BCube topology with 3-port switches. By adopting it as the internal graph for intra-module design, a level-2 BCube topology can be obtained from the second iteration. Note that in this case, the finalization step is omitted as a BCube topology contains no CNs.

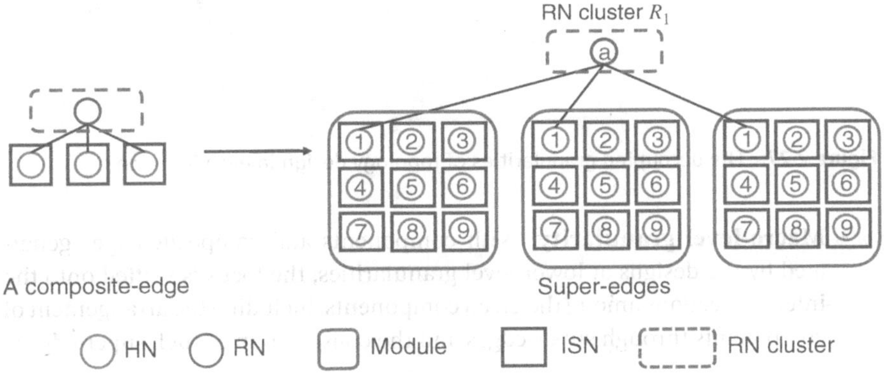

Figure 2.26 An illustration of super-edge construction for a level-2 BCube topology with 3-port switches. For the first RN cluster R_1, it is directly connected to three modules through three super-edges. These three super-edges are constructed using a composite-edge. To be specific, all HNs labeled with number 1 in each module are directly connected to RN a in RN cluster R_1.

2.3.2 Decoupling of Topology Design

Based on the design procedure, in each iteration, the topology design can be decoupled into three levels of design granularity, which is illustrated in Figure 2.27.

- **Micro-level granularity**. The design at this level considers the construction of modules by organizing distinct types of nodes as connected graphs.
- **Medium-level granularity**. Based on the modules constructed at the micro-level granularity, the design at this level is concerned with two parts. One is regarding the *selection* of dock nodes in each module to form interfaces. The other is regarding the *construction* of composite-edges, i.e., interconnections among interfaces of components.

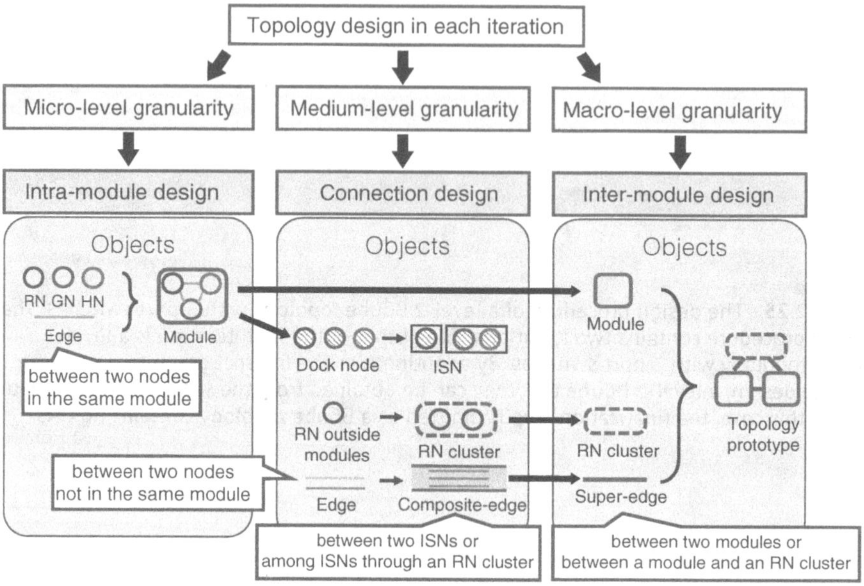

Figure 2.27 The decoupled granularities of topology design in each iteration.

- **Macro-level granularity**. With components and composite-edges generated by the designs at lower-level granularities, the focus is shifted onto the interconnections among the given components, including the arrangement of components through super-edges and the construction of such super-edges.

2.4 Guiding Principles and Concrete Tools

The design procedure elucidates a common series of steps to conduct topology design. Nonetheless, three fundamental questions remain unanswered.

1. How can performance metrics of topology design be characterized?
2. Based on the characterized performance metrics, what are the general rules (i.e., guiding principles) to follow in each design step?
3. How can each step of design be implemented under the guiding principles?

To answer these questions, in Section 2.4.1, we demonstrate a set of key performance metrics based on the previously developed ingredients of topology design. With these performance metrics, the guiding principles for the design steps are introduced in Section 2.4.2. Under the guidance of the proposed principles, a number of concrete tools for each step of design are developed in Section 2.4.3.

2.4.1 Performance Metrics

In general, the performances of large-scale computer networks can be characterized by various metrics (e.g., computing capability, efficiency, and robustness) (Xia et al. 2017). To feature the impacts of topology design on such performances, we consider five principal metrics from the perspective of graph theory,[9] including *scalability*, *cost*, *latency*, *throughput*, and *fault-tolerance*. Their definitions are specified in the following subsubsections.

2.4.1.1 Scalability

For large-scale computer networks, a common goal is to support as more devices with computing function as possible. A trivial way to achieve this goal is to directly connect millions of servers to a switch with millions of ports. However, due to the limitations of production technology and the consideration of cost, commodity network devices have limited communication capacities (e.g., the port number of switches and the number of NICs of servers). For this reason, given the communication capacities of devices with routing function (e.g., the port number of switches and the number of NICs of servers), the design of a topology should be capable of supporting as more devices with computing function as possible. For each network topology, such a capability is also known as its *scalability*. Recall that devices with computing function are modeled as HNs and CNs in the framework. Then the scalability of the network topology is defined in the following way.

Definition 2.4 The scalability S of a network topology is defined as the total number of HNs and CNs in the network.

For example, as shown in Figure 2.28a, there are 16 CNs (servers) and no HNs in a 4-ary Fat-Tree topology. Hence, the scalability of such a topology is 16. In Figure 2.28b, a level-1 DCell with 4-port switches contains 20 servers as HNs, thus its scalability is 20.

Remark 2.5 For a computer network, its scalability can be viewed as a function of the communication capacities of devices with routing function (e.g., the port number of switches and the number of NICs of servers). Intuitively, the improvement of the communication capacities of RNs, GNs, and HNs implies that the network topology can accommodate more devices with computing function (i.e., CNs and HNs). In the following, we take a level-1 DCell topology and a Fat-Tree topology as two examples for illustration. Let parameter p denote the port number of each switch in the two topologies, such that parameter p is a positive

9 In this way, the impacts of topology on network performance can be highlighted independently, instead of being coupled with the other aspects of computer networks (e.g., routing protocols and network traffic patterns).

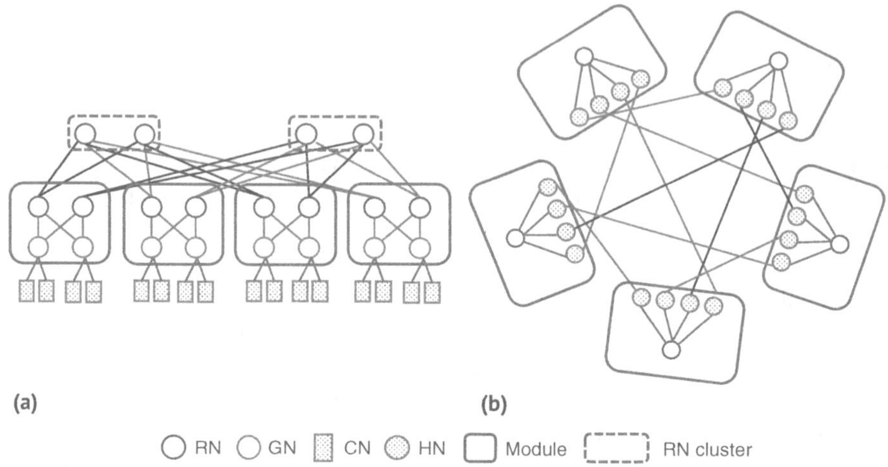

Figure 2.28 Two instances to illustrate the scalability S of different topologies. Nodes that are counted into the scalability are marked with dotted patterns. (a) The scalability S of a 4-ary Fat-Tree topology is 16. (b) The scalability S of a level-1 DCell topology with 4-port switches is 20.

even number. For a level-1 DCell topology, it contains $p^2 + p$ HNs and no CNs, thus its scalability is $p^2 + p$. For a Fat-Tree topology, it consists of $p^3/4$ CNs and no HNs, thus its scalability is $p^3/4$. In comparison, the scalability of the Fat-Tree topology grows faster than that of the DCell topology. In particular, when $p = 4$, the scalability of Fat-Tree is lower than level-1 DCell. However, as the value of parameter p becomes larger than four, Fat-Tree outnumbers level-1 DCell with a higher scalability.

2.4.1.2 Cost

For large-scale computer networks, one of the primary goals for construction is to provide powerful computing capability with as low costs of device acquisition as possible. To this end, we introduce a metric called *apportioned cost* for each device with computing function in a network topology. Given a network topology, it is defined as the *ratio* of the total cost of acquiring all devices *to* the number of devices with computing functions in the network. In this framework, the cost of a topology design is considered as its apportioned cost. More details are given as follows.

Let N_1, N_2, and N_3 denote the number of devices with pure routing function (i.e., GNs and RNs), the number of devices with pure computing function (i.e., CNs), and the number of devices with both routing and computing functions (i.e., HNs), respectively. Meanwhile, we denote the cost of acquiring each device with pure routing function by x_1 and the cost of acquiring each device with pure computing function by x_2. For devices with both routing and computing (e.g., servers in server-centric topologies), each of them can be viewed as the implementation of devices with pure computing functions with added routing components (e.g., network interface cards). In this framework, we consider the cost of acquiring each device with both routing and computing functions as $\epsilon x_1 + x_2$, where ϵ is the ratio of the cost of adding routing components to the cost of each device with pure routing function.[10]

With these notations, the apportioned cost (i.e., the ratio of the total cost of devices to the number of devices with computing function) is characterized as

$$\frac{N_1 \cdot x_1 + N_2 \cdot x_2 + N_3 \cdot (\epsilon x_1 + x_2)}{N_2 + N_3} = \frac{N_1 + \epsilon N_3}{N_2 + N_3} \cdot x_1 + x_2, \tag{2.1}$$

where the values of x_1 and x_2 (i.e., the cost of acquiring individual devices) are usually pre-determined as constants. Note that only the ratio $(N_1 + \epsilon N_3)/(N_2 + N_3)$ is dependent on the topology design. We call such a ratio the *apportionment coefficient* of the network, and formalize it as follows. In Figure 2.29, we provide two examples to demonstrate this definition.[11]

Definition 2.5 Given a topology, its apportionment coefficient is defined as $(N_1 + \epsilon N_3)/(N_2 + N_3)$, in which the value of ϵ is a constant, while parameters N_1, N_2, and N_3 denote the number of GNs and RNs, the number of CNs, and the number of HNs in the topology, respectively.

2.4.1.3 Latency

As a vital metric for large-scale computer networks, communication latency measures the time taken to exchange a given amount of information between devices with computing function in a network. By focusing on the impacts of topology design on communication latency without involving other physical factors (e.g., traffic patterns and routing protocols), we consider the latency from the perspective of graph theory. In the framework, the latency between two devices is characterized by their distance in the topology, which is equal to the hop-count of the shortest path between the two devices plus one. Specifically, given a network topology, we consider its latency as the maximum distance among all pairs of

10 Without loss of generality, we assume that $\epsilon \in (0, 1)$, as the cost of acquiring routing components is often lower than that of a device with pure routing function.

11 In our original INFOCOM paper, Chang et al. (2020), the definition of cost can be viewed as a special case of Definition 2.5 in this monograph. In that case, we focus on network topologies with no HNs (i.e., $N_3 = 0$). Accordingly, the cost is equal to the ratio of the number of nodes with routing function N_1 to the number of CNs N_2.

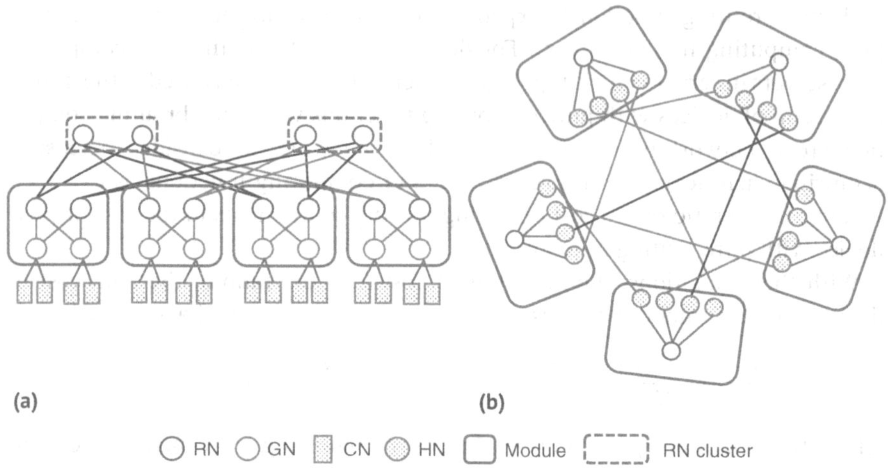

Figure 2.29 An illustration of apportionment coefficient of 4-ary Fat-Tree and level-1 DCell with 4-port switches. In (a), there are 16 CNs, 12 RNs, 8 GNs, and no HNs in a 4-ary Fat-Tree topology. Hence, the apportionment coefficient of such a topology is $(12 + 8)/16 = 1.25$. In (b), a level-1 DCell with 4-port switches contains 20 HNs and 5 RNs, thus its apportionment coefficient is $(20\epsilon + 5)/20 = \epsilon + 0.25$.

nodes with computing function. The following is the formal definition of such a metric.

Definition 2.6 For a network topology, its latency L is defined as the maximum distance between any pair of nodes with computing function (i.e., CNs and HNs) in the topology.

For example, in a 4-ary Fat-Tree topology (see Figure 2.30a), the maximum distance between any two CNs (servers) is 6; i.e., the latency of such a topology is 6. In comparison, in a level-1 DCell shown in Figure 2.30b, the maximum distance between two HNs is 5; i.e., The latency of such a topology is 5.

2.4.1.4 Throughput

Another common goal of large-scale computer networks is to transfer as more data as possible among nodes with computing function (i.e., CNs and HNs) over a fixed time period. Such a capability is usually characterized as *throughput*. In this framework, we focus on the throughput between GNs and HNs. The reason is twofold. On one hand, all data that are sent by a CN must go through its directly-connected node (i.e., a GN or an HN). Consequently, for each CN, the volume of its transferred data to other nodes with computing function (i.e., CNs or HNs) is subject to the throughput between its directly-connected node (i.e., a GN or

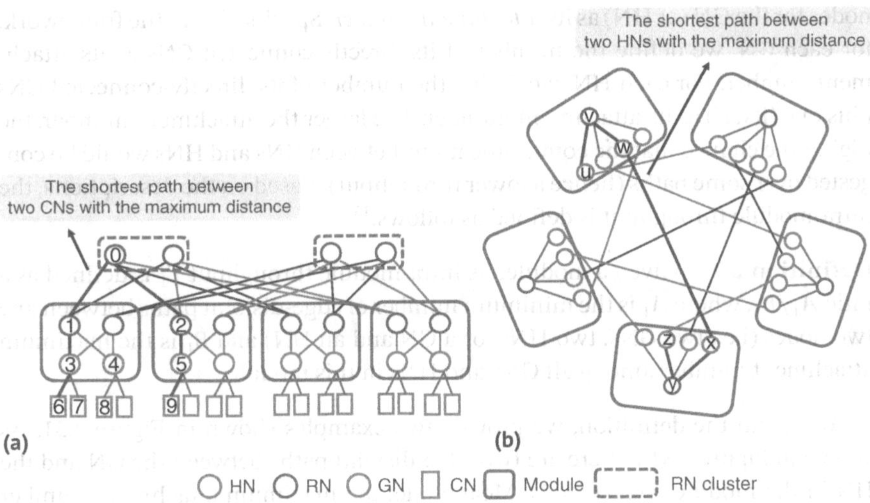

Figure 2.30 Two examples to illustrate the latency L in different topologies. (a) The latency L of a 4-ary Fat-Tree topology is 6. For two CNs attached to the same GN (e.g., CNs 6 and 7), the shortest path between them is $6 - 3 - 7$ with a distance of 2. For two CNs attached to distinct GNs in the same module (e.g., CNs 6 and 8), one of the shortest paths between them is $6 - 3 - 1 - 4 - 8$ with a distance of 4. For two CNs attached to distinct GNs in distinct modules (e.g., CNs 6 and 9), one of the shortest paths between them is $6 - 3 - 1 - 0 - 2 - 5 - 9$ with a distance of 6, which is highlighted by thick lines. (b) The latency L of a level-1 DCell topology with 4-port switches is 5. For two HNs contained in the same module (e.g., HNs u and w), the shortest path between them is $u - v - w$ with a distance of 2. For two HNs contained in distinct modules with the maximum distance (e.g., HNs u and z), one of the shortest paths between them is $u - v - w - x - y - z$ with a distance of 5, which is highlighted by thick lines.

an HN) and that of the other nodes.[12] On the other hand, the throughput between each RN and the other nodes is not considered since RNs have neither computing function nor attached CNs. Note that in the framework, any two GNs, two HNs, or a GN and an HN can be in the same or distinct modules. Accordingly, throughput should be defined in the following two cases.

Intra-Module Throughput. In this case, we focus on the throughput between any two GNs, two HNs, or a GN and an HN in the same module. From the perspective of graph theory, the throughput is jointly featured by two factors. One is the edge-disjoint paths between any two GNs, two HNs, or a GN and an HN. More edge-disjoint paths suggest a higher throughput. The other factor is the number of nodes with computing function whose communications with other nodes must go through a GN (or an HN). For each GN (or HN), we call the number of such

12 Note that the topology is considered as a simple graph with *unweighted* edges here, i.e., the bandwidths of communication links are assumed to be identical here.

nodes for the GN (or HN) as its *attachment number*. Specifically, in the framework, for each GN, we define the number of its directly-connected CNs as its attachment number. For each HN, we define the number of its directly-connected CNs plus one (itself) as its attachment number. The larger the attachment number, the higher the chance that the communications between GNs and HNs would be congested over some paths (hence a lower throughput). Based on the descriptions, the intra-module throughput is defined as follows.[13]

Definition 2.7 Given a module, its intra-module throughput T_1 is defined as a ratio A_1/B_1, where A_1 is the minimum number of edge-disjoint paths between any two nodes (i.e., two GNs, two HNs, or a GN and an HN) and B_1 is the maximum attachment number among all GNs and HNs in this module.

To explain the definition, we provide two examples shown in Figure 2.31. As shown in Figure 2.31a, there are two edge-disjoint paths between the GN and the HN in the module (i.e., $A_1 = 2$). Meanwhile, the maximum attachment number among all GNs and HNs in this module is three ($B_1 = 3$). Hence, the intra-module throughput is 2/3.

The other example (see Figure 2.31b) lies in a 4-ary Fat-Tree topology. In such a topology, each pod corresponds to a module and each edge switch serves as a GN. In each module, there are at least two edge-disjoint paths between two GNs, i.e., $A_1 = 2$. Meanwhile, since each GN is directly connected to two CNs, the maximum attachment number B_1 among GNs is 2. Hence, the intra-module throughput is 1. Intuitively, consider each GN (edge switch) in the modules (pods) of the Fat-Tree topology. For each of its attached CNs, there always exists a path that is edge-disjoint with the paths of the other attached CNs. This demonstrates the non-blocking communication of Fat-Tree topology within each module (Al-Fares et al. 2008).

Inter-module throughput. In this case, we consider the throughput between GNs and HNs in distinct modules. From the perspective of graph theory, the throughput is jointly characterized by two factors. One is the edge-disjoint paths between

---

13 In our original INFOCOM paper, Chang et al. (2020), intra-module throughput T_2 is defined as d_1/s, which follows the similar idea to characterize the intra-module throughput as the definition in this monograph. Specifically, the denominator of d_1/s is equal to the maximum attachment number B_1 among all GNs and HNs (i.e., $B_1 = s$). Meanwhile, the numerators of two definitions (i.e., d_1 and A_1) both represent the minimum number of paths between any two nodes in the same module. In spite of such a similarity, it is still worth pointing out their differences. The definition in our original paper only considers the length-2 paths between two modules. Such a definition applies to the three-layer model in our original paper because the distance between two GNs in the same module of three-layer model is two. In comparison, the definition in this monograph considers all edge-disjoint paths between two nodes in the same module, which applies to more general scenarios.

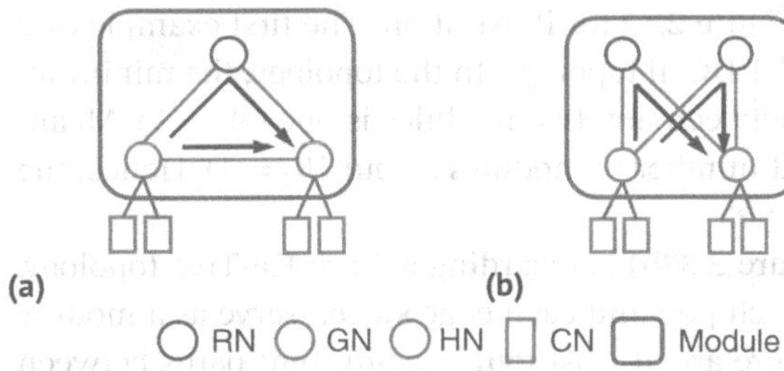

Figure 2.31 Two instances to illustrate intra-module throughput T_1. The thick arrows denote the edge-disjoint paths between a GN and an HN (or between two GNs). (a) The intra-module throughput T_1 of a module with three nodes is 2/3. (b) The intra-module throughput T_1 of a module in a 4-ary Fat-Tree topology is 1.

dock nodes in two distinct modules. More edge-disjoint paths suggests a higher throughput. The other factor is the number of nodes with computing function attached to each module (*aka* its *attachment number*). Each module manages the communications between each of its attached nodes and dock nodes (through which the attached nodes can communicate with nodes attached to other modules). Specifically, in the framework, the attachment number of each module is equal to the total number of HNs in this module and CNs that are directly connected with GNs and HNs in the module. The larger the attachment number, the higher the chance that the communications between dock nodes in distinct modules would be congested over some paths (hence a lower throughput). Formally, we define inter-module throughput of a network topology as follows.[14]

Definition 2.8 For a topology, let A_2 denote the minimum number of edge-disjoint paths between any two modules, such that the edge-disjoint paths between two modules do not traverse through nodes in other modules. Meanwhile, let B_2 denote the maximum attachment number of modules in this topology. Accordingly, the inter-module throughput T_2 of such a topology is defined as a ratio A_2/B_2.

---

14 In our original INFOCOM paper, Chang et al. (2020), inter-module throughput T_2 is defined as $d_2/(\theta_1 k_1(1+c_1)s)$, which follows the similar idea to characterize the inter-module throughput as the definition in this monograph. Specifically, the denominator of $d_2/(\theta_1 k_1(1+c_1)s)$ is equal to the maximum attachment number B_2 among all modules (i.e., $B_2 = \theta_1 k_1(1+c_1)s$). Meanwhile, the numerators of two definitions (i.e., d_2 and A_2) both represent the minimum number of paths between any two modules. In spite of such a similarity, it is still worth pointing out their differences. The definition in our original paper only considers the length-2 paths between two modules. Such a definition applies to the three-layer model in our original paper because the distance between two dock nodes in the three-layer model is two. In comparison, the definition in this monograph considers all edge-disjoint paths between two modules, which applies to more general scenarios.

We provide two examples in Figure 2.32 for illustration. The first example (see Figure 2.32a) is regarding a level-1 DCell topology. In the topology, the minimum number of edge-disjoint paths between any two modules is one ($A_2 = 1$). Meanwhile, the maximum attachment number of modules is four ($B_2 = 4$). Hence, the inter-module throughput is 1/4.

The second example (see Figure 2.32b) is regarding a 4-ary Fat-Tree topology. Recall that in such a topology, each pod and each edge switch serve as a module and a GN, respectively. Then there are at least four edge-disjoint paths between any two modules (i.e., $A_2 = 4$). Meanwhile, the maximum attachment number of modules in this topology is four ($B_2 = 4$). Hence, the inter-module throughput is 1. Intuitively, consider each dock node (aggregation switch) in the modules (pods) of the Fat-Tree topology. For each attached CN (server) of a given module M_1, there always exists an edge-disjoint path between the dock node and another dock node in a different module M_2, over which the CN can communicate with the attached CNs of module M_2. This demonstrates the non-blocking communication of Fat-Tree topology across different modules (Al-Fares et al. 2008).

2.4.1.5 Fault-tolerance

The fault-tolerance reflects the capability of a network to maintain communications in the face of node failures or link failures. Note that the failure of a node results in the disconnections of all links incident to this node, while the failure of a

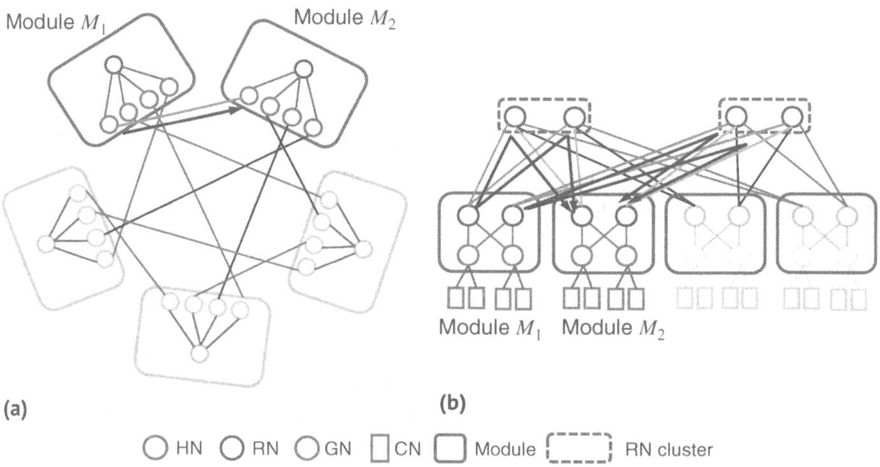

Figure 2.32 Two instances to illustrate the inter-module throughput T_2 in different topologies. The thick arrows denote the edge-disjoint paths between two modules, such that these paths do not traverse through the nodes in other modules. (a) The inter-module throughput T_2 of a level-1 DCell topology with 4-port switches is 1/4. (b) The inter-module throughput T_2 of a 4-ary Fat-Tree topology is 1.

link does not necessarily lead to the failure of its incident nodes. In this framework, we consider the fault-tolerance with respect to the worse case (i.e., node failures).

Among such failures, the failures of dock nodes become our focus as they have significant impacts on the communications among nodes in distinct modules. For each pair of modules, several paths are established to interconnect GNs and HNs in such two modules across their dock nodes. Particularly, to achieve efficient communications (in terms of low latency), we focus on the paths with the *minimum length*. Consider the communications from a source node (a GN or an HN) in module M_1 to a target node (a GN or an HN) in module M_2. Such communications sequentially traverse three sub-paths: first from the source node in module M_1 to a dock node in module M_1, second from the dock node in module M_1 to some dock nodes in module M_2, third from the dock nodes in module M_2 to the target node in module M_2. In this process, the failures of several dock nodes in module M_2 may disconnect such communications from the source node in module M_1 to the target node in module M_2. The recovery of such failures requires to re-establish the communications by using other dock nodes in module M_1 or redirecting the transferred data through other paths with longer distances; in the worst case, the communications may even suffer severe break-downs. Such a re-establishment process may lead to performance degradations of the whole computer network (e.g., increasing the communication delay and causing congestions over some paths). To this end, achieving a better fault-tolerance requires to increase the minimum number of dock nodes in module M_2 whose simultaneous failures would disconnect the communications from any source node in module M_1 to some target node in module M_2. Figure 2.33 shows an example for illustration.

Based on this idea, we define the *fault-tolerance* of a topology as follows.[15]

Definition 2.9 Given a topology, suppose that there are N ordered pairs of modules. For the i-th ordered pair of modules $M_i^{(1)}$ and $M_i^{(2)}$, suppose that the communications between any two nodes (each is a GN or an HN) in the two modules traverse through the paths between such two nodes with the minimum length. Under such a setting, we define the fault-tolerance f_i for the i-th ordered module pair as the minimum number of dock nodes in module $M_i^{(2)}$ whose failures would

15 Note that our original INFOCOM paper Chang et al. (2020) and this monograph follow a similar idea to characterize the fault-tolerance of a network topology. Given a network topology, they both concentrate on the fault-tolerance for each module pair in the network. Specifically, for each pair of modules, their fault-tolerance is characterized as the minimum number of dock nodes in one module whose failures would disconnect the communications between some pairs of nodes (each is a GN or an HN) in the two modules. In spite of such a similarity, it is still worth pointing out their differences. The definition in our original paper only considers the highest fault-tolerance among all module pairs. Such a definition omits the module pairs with inferior fault-tolerances. In comparison, the definition in this monograph involves the fault-tolerances of *all* module pairs, which provides a more comprehensive performance characterization.

disconnect the communications from any GN (or HN) in module $M_i^{(1)}$ to some GN (or HN) in module $M_i^{(2)}$ via a given dock node in module $M_i^{(1)}$. The fault-tolerance F of the given topology is defined as the average over all ordered pairs of modules, i.e., $F = \frac{1}{N} \sum_{i=1}^{N} f_i$.

For ease of understanding, in Figure 2.34, we take a 4-ary Fat-Tree topology and an AB Fat-Tree topology with 4-port switches as examples for illustration. Note that there are four modules (pods) in each of the two topologies, respectively. Accordingly, each topology contains twelve ordered pairs of modules (i.e., $N = 12$).

In the 4-ary Fat-Tree topology (see Figure 2.34a), the connection patterns between each module and RN clusters are all the same. The first dock node in each module (labeled with number 1) is directly connected to RN cluster R_1, while the second dock node in each module (labeled with number 2) is directly connected to RN cluster R_2. Next, we take the ordered module pair (M_1, M_3) as an example to show that the fault-tolerance for each ordered module pair is one. Note that the distance between each pair of GNs in distinct modules is four. The paths with the minimum length first go through a dock node in module M_1 (e.g., dock node 1 in module M_1), then through the dock node labeled with the same number in module M_3 (e.g., dock node 1 in module M_3). Given a dock node in module M_1

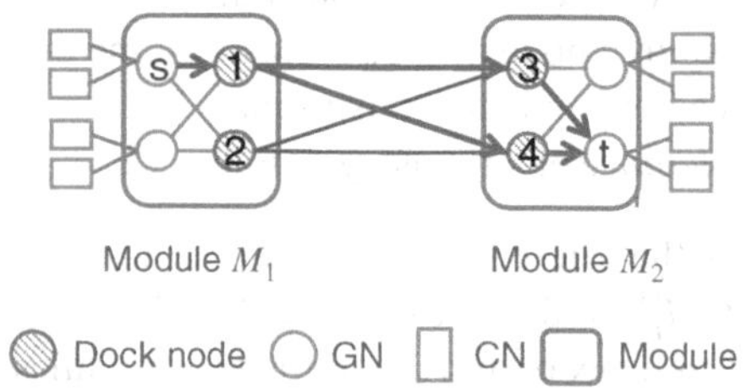

Figure 2.33 An illustration of fault-tolerance for a topology with two modules. For each pair of GNs in distinct modules, the minimum length of paths between them is three. To be specific, the communications from a source node (e.g., GN s) in module M_1 to a target node (e.g., GN t) in module M_2 through such paths sequentially traverse three sub-paths: first from the source node in module M_1 to a dock node in module M_1 (e.g., dock node 1), second from such a dock node in module M_1 to several dock nodes in module M_2 (e.g., dock nodes 3 and 4), third from such dock nodes in module M_2 to the target node in module M_2. An instance of communications from GN s in module M_1 (the source node) to GN t in module M_2 (the target node) is indicated by thick arrows. If one of the dock nodes in module M_2 (e.g., dock node 3) fails, there still exists a path to support the communication (i.e., the path $s - 1 - 4 - t$). However, if the two dock nodes (nodes 3 and 4) in module M_2 both fail, then the communications would break down. In other words, the minimum number of dock nodes in module M_2 whose failures may disconnect the communications is two. Therefore, the fault-tolerance from module M_1 to module M_2 is equal to two.

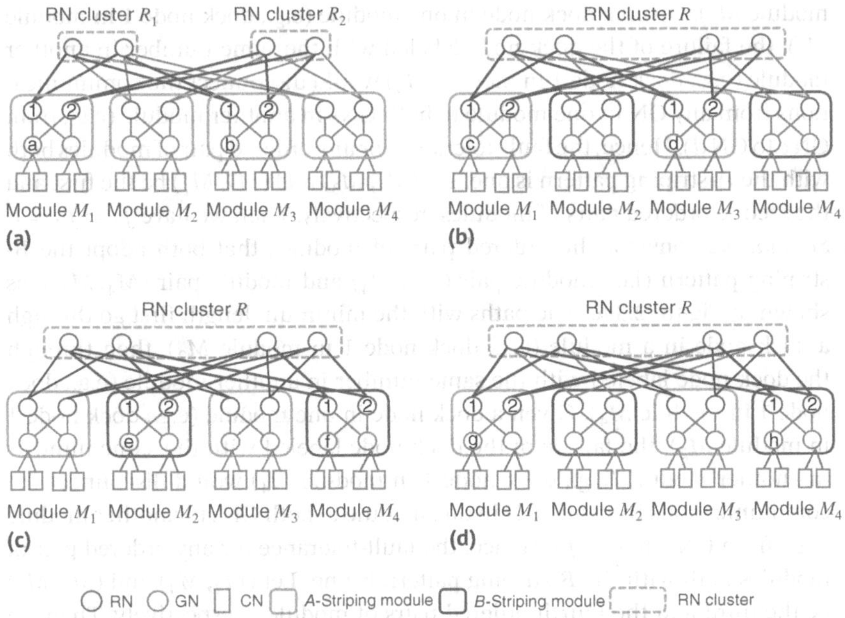

Figure 2.34 Examples to illustrate the fault-tolerance F of different topologies. (a) A 4-ary Fat-Tree topology. (b) An AB Fat-Tree topology with 4-port switches, in which the two A-striping modules are highlighted. (c) An AB Fat-Tree topology with 4-port switches, in which the two B-striping modules are highlighted. (d) An AB Fat-Tree topology with 4-port switches, in which an A-striping module and a B-striping module are highlighted.

(e.g., dock node 1), the failure of one dock node in module M_3 (e.g., dock node 1 in module M_3) would disconnect the communications from any GN in module M_1 to both GNs in module M_3 (e.g., from GN a to GN b). Hence, the fault-tolerance for any ordered pair of modules is one (i.e., $f_i = 1$ for $1 \leq i \leq 12$). By Definition 2.9, the fault-tolerance F of a 4-ary Fat-Tree topology is 1 ($F = 1$).

In the AB Fat-Tree topology (see Figure 2.34b), there are two connection patterns between each module and RN clusters, namely A-striping pattern and B-striping pattern. Modules M_1 and M_3 employ the A-striping pattern, while modules M_2 and M_4 employ the B-striping pattern. The distance between each pair of GNs in distinct modules is four. Next, the fault-tolerances for each ordered pair of modules with distinct types of connection pattern are considered as follows.

- *First*, we consider the ordered pairs of modules that both adopt the A-striping pattern (i.e., module pair (M_1, M_3) and module pair (M_3, M_1)), as shown in Figure 2.34b. The paths with the minimum length first go through a dock node in a module (e.g., dock node 1 in module M_1), then through the dock node labeled with the same number in another module (e.g., dock node 1 in module M_3). Given a dock node in one module (e.g., dock node 1 in module M_1), the failure of the dock node labeled with the same number in another module (e.g., dock node 1 in module M_3) would disconnect the communications from any GN in one module to both GNs in another module (e.g., from GN c to GN d). Hence, the fault-tolerance for any ordered pair of modules both with the A-striping pattern is one. Let (M_1, M_3) and (M_3, M_1) be the first and the second ordered pairs of modules, respectively. Then we have $f_1 = f_2 = 1$.

- *Second*, we consider the ordered pairs of modules that both adopt the B-striping pattern (i.e., module pair (M_2, M_4) and module pair (M_4, M_2)), as shown in Figure 2.34c. The paths with the minimum length first go through a dock node in a module (e.g., dock node 1 in module M_2), then through the dock node labeled with the same number in another module (e.g., dock node 1 in module M_4). Given a dock node in one module (e.g., dock node 1 in module M_2), the failure of the dock node labeled with the same number in another module (e.g., dock node 1 in module M_4) would disconnect the communications from any GN in one module to both GNs in another module (e.g., from GN e to GN f). Hence, the fault-tolerance for any ordered pair of modules both with the B-striping pattern is one. Let (M_2, M_4) and (M_4, M_2) be the third and the fourth ordered pairs of modules, respectively. Then we have $f_3 = f_4 = 1$.
- *Third*, we consider the ordered pairs of modules such that one module adopts the A-striping pattern while the other adopts the B-striping pattern (e.g., modules M_1 and M_4), as shown in Figure 2.34d. The paths with the minimum length first go through a dock node in a module (e.g., dock node 1 in module M_1), then through *any* dock node in another module (e.g., dock node 1 or dock node 2 in module M_4). Given a dock node in one module (e.g., dock node 1 in module M_1), only when two dock nodes in another module (e.g., dock nodes 1 and 2 in module M_4) both fail, the communications from any GN in one module to both GNs in another module (e.g., from GN g to GN h) would be disconnected. Hence, for any ordered pair of modules such that one module adopts the A-striping pattern while the other adopts the B-striping pattern, the fault-tolerance is two. There are eight such ordered pairs of modules. Hence, we have $f_k = 2$ for $5 \leq k \leq 12$.

Therefore, the fault-tolerance F of an AB Fat-Tree topology with 4-port switches is $\frac{1}{N}\sum_{i=1}^{N} f_i = (2 + 2 + 16)/12 = 5/3$. Such a result verifies that AB Fat-Tree achieves a higher fault-tolerance than Fat-Tree (Liu et al. 2013).

2.4.2 Guiding Principles

The aforementioned metrics illuminate a general criterion of topology design; that is, a network topology should be designed with a low latency, a high throughput, a high fault-tolerance, a high scalability, and a low cost. From the perspective of graph theory, the graph of the network topology that contains only nodes with routing function (i.e., RNs, GNs, and HNs) should be subject to a short diameter and a dense connectivity, while ensuring that a large number of nodes are contained in the graph with low degrees. Note that a short diameter implies a low maximum latency among all pairs of GNs and HNs. By Menger's theorem (Diestel 2005), a dense edge-connectivity and a dense vertex-connectivity conduce to a high throughput and a high fault-tolerance, respectively. When it comes to the low-degree requirement, it conforms to the limited communication capacities of commodity network devices (e.g., the port number of switches and the number of NICs of servers). Meanwhile, in a network topology, suppose that the proportion of GNs and HNs and the number of directly-connected CNs with each GN or HN are given. The larger the number of nodes, the larger the number of GNs and HNs. Accordingly, there will be more CNs and HNs contained in the topology. By Definition 2.4, a higher scalability is achieved.

To this end, a possible idea is to adopt graphs with good *expansion* property[16] as the candidates of topology design. Among such graphs, expander graphs are often adopted as they provably strike a decent balance among these performance requirements (Valadarsky et al. 2016). Such graphs have been applied to design efficient network topologies in various fields, such as parallel computing and data center networks (Hoory et al. 2006).

However, a straightforward adoption of expander graphs as the whole topology may lead to the following issues. On one hand, the network topology based on expander graphs is less structured as the construction usually requires randomized pairing between network devices. This characteristic brings great challenges in the operation and management of the network. On the other hand, most adopted expander graphs are regular graphs in which each node has the same degree. In other words, they require devices in a network to be homogeneous (e.g., the switches should have the same number of ports). This makes it nontrivial to generalize the topology design to support communications among heterogeneous devices.

Inspired by recent research which affirms that "*the lower the diameter of the graph and (more importantly) the larger its scale, the better the expansion property guarantees*" (Dinitz et al. 2020), we adopt an incremental way to construct graphs with good expansion property. The key idea is to construct a network topology by continuously increasing the number of nodes while restricting its diameter. Such an idea is well aligned with the unified design procedure. In other words, starting from the design of internal graphs for modules, a network topology with good

16 Intuitively, good expansion property is often used to describe large-scale graphs with dense connectivity properties. Such graphs also have a short diameter (*cf.* Section 1.5.1).

expansion property can be built progressively by interconnecting modules with well-devised connections. In this process, to guarantee the good expansion property, the design procedure should be aimed to restrict the diameter of the topology.

Accordingly, the guiding principle for each design step is specified as follows.

1. **Intra-module design**. The candidates for internal graphs are Ramanujan graphs (*cf.* Section 1.5.3) with low constant diameters.[17] The reason is twofold. On one hand, Ramanujan graphs are graphs with asymptotically optimal expansion property and thus strike a decent balance among the key performance metrics. On the other hand, internal graphs with low constant diameters conduce to restricting the distance between nodes in the same module.
2. **Connection design**. In this framework, the goal is to achieve efficient communications between nodes with computing function. To realize such a goal, a low latency needs to be considered in the process of topology design. By Definition 2.6, this is equivalent to restricting the distances between GNs and HNs. For any such two nodes (i.e., two GNs, two HNs, or a GN and an HN) in distinct modules, their distance includes the following two measures. One is the distance between each GN (or HN) and the dock nodes in the same module. The other is the distance between dock nodes in distinct modules. Accordingly, the principle for dock node selection is to select the dock nodes that are close to GNs and HNs in the same module. Meanwhile, the principle for dock node interconnection is to restrict the distance between dock nodes in distinct modules.
3. **Inter-module design**. This design step considers how to restrict the distance between two nodes (i.e., two GNs, two HNs, or a GN and an HN) in distinct modules at the macro-level granularity. Since GNs and HNs are only contained in modules but not in RN clusters, the principle is to restrict the distance between each pair of *modules* in the skeleton graph. To this end, in the skeleton graph, the shortest paths between each pair of modules should achieve the minimum feasible distance under each atomic mode. By using super-edges under the plain mode, the minimum feasible distance between two modules is *one*, which is achieved when such two modules are interconnected by a super-edge under the plain mode (see Figure 2.35a). By using super-edges under the bipartite mode, the minimum feasible distance between two modules is *two*, which is achieved when such two modules are interconnected through one or more RN clusters as one-hop communication relays (see Figure 2.35b). In summary, two modules should be interconnected either by a super-edge under the plain mode, through one or more RN clusters as one-hop communication relays, or both of them.

---

17 We call strictly Ramanujan graphs and bipartite Ramanujan graphs as "Ramanujan graphs".

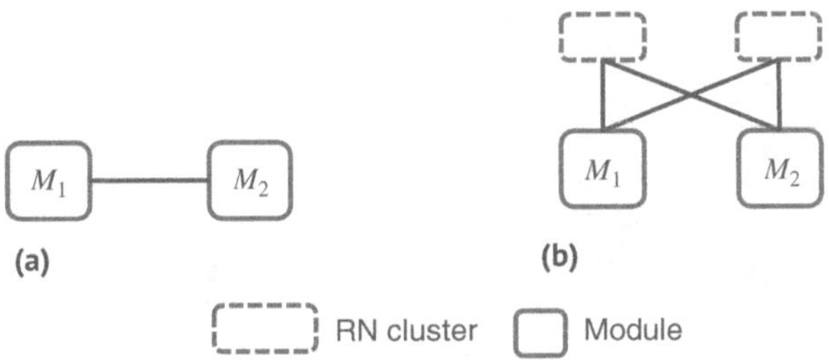

Figure 2.35 Two examples to illustrate the paths between two modules with the minimum feasible distance under the two atomic modes. (a) Modules M_1 and M_2 are directly interconnected by a super-edge under the plain mode. (b) Modules M_1 and M_2 are interconnected using super-edges under the bipartite mode which go through two RN clusters as one-hop communication relays.

4. **Finalization.** In this step, CNs are attached to GNs and HNs, which mainly affects the scale of the graph. To maximize the scale of the graph, it is required to use all unoccupied degrees of GNs and HNs (e.g., unoccupied ports of switches and unoccupied NICs of servers with routing function) to attach CNs.

2.4.3 Concrete Tools

Based on the guiding principles, a set of concrete tools are developed for each design step, as specified in the following parts, respectively.

2.4.3.1 Intra-module Design

In this step, four members of the family of Ramanujan graph are considered as the candidates for internal graph design.

Candidate 1: *Complete graph (CG)*. CG is a graph where each pair of nodes are connected via a single edge. In general, a CG with n nodes is denoted by K_n which exists for any integer n with $n \geq 2$. As shown in Section 1.5.3, CG is a Ramanujan graph with a low constant diameter which equals one. Hence, CG is selected as a candidate of internal graph.

There are two ways for the node instantiation in CG. One way is to instantiate all nodes as GNs or HNs. The other way is to instantiate part of nodes as RNs and other nodes as GNs or HNs. Either way, the distance between two nodes (i.e., two GNs, two HNs, or a GN and an HN) is equal to one. Nonetheless, since CNs can only be attached to GNs or HNs but not to RNs, the former way (i.e., instantiating all nodes as GNs or HNs) is preferred as it conduces to a higher scalability. Figure 2.36 shows an instance of CG and its corresponding internal graph.

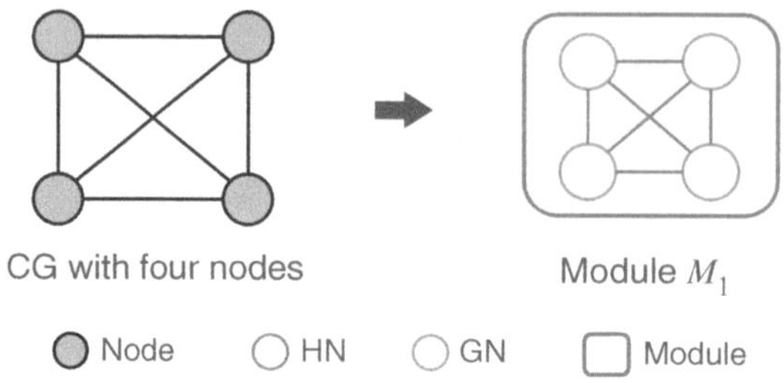

Figure 2.36 An instance of a complete graph (CG) K_4 and its corresponding internal graph.

Candidate 2: *Complete bipartite graph (CBG)*. CBG is a bipartite graph, where each node in one node partition is adjacent to all nodes in the other node partition. Suppose there are m and n nodes in the two node partitions, respectively. Then such a CBG is denoted by $K_{m,n}$. As shown in Section 1.5.3, CBG is a bipartite Ramanujan graph with a low constant diameter which equals two. Hence, CBG is selected as a candidate for internal graph.

There are two ways for the node instantiation in CBG as illustrated in Figure 2.37. One way is to instantiate all nodes as GNs or HNs. The other way is to instantiate each node in one node partition as a GN (or an HN) and each node in the other node partition as an RN. Since the latter way is a special case of the following block incidence graph, in this framework, we only consider the former way (i.e., instantiating all nodes as GNs or HNs) when refer to CBG as the internal graph.

Candidate 3: *Block incidence graph (BIG)*. The definition and property of such a candidate have been provided in Section 1.7.1 of Chapter 1.

As a Ramanujan graph with a low constant diameter, BIG is selected as a candidate of internal graph. Note that the distance between each pair of element nodes is two. Therefore, to restrict the distance between two GNs/HNs (i.e., two GNs, two HNs, or a GN and an HN), in this framework, each element node is instantiated as a GN (or an HN) and each block node is instantiated as an RN (see Figure 2.38). In this way, the distance between any two such nodes is restricted to two.

Given the node degrees r and k, Lemma 1.6 in Section 1.7.1 of Chapter 1 reveals a tradeoff between the total number of GNs and HNs (positively proportional to the value of θ by Definition 1.23) and the number of edge-disjoint paths λ between two element nodes (i.e., two GNs, two HNs, or a GN and an HN). By Equation (1.50), a larger value of θ leads to a smaller value of λ. In other words, given fixed node degrees r and k, a BIG with more GNs and HNs will have fewer edge-disjoint paths between each pair of element nodes (i.e., two GNs, two HNs, or a GN and an HN).

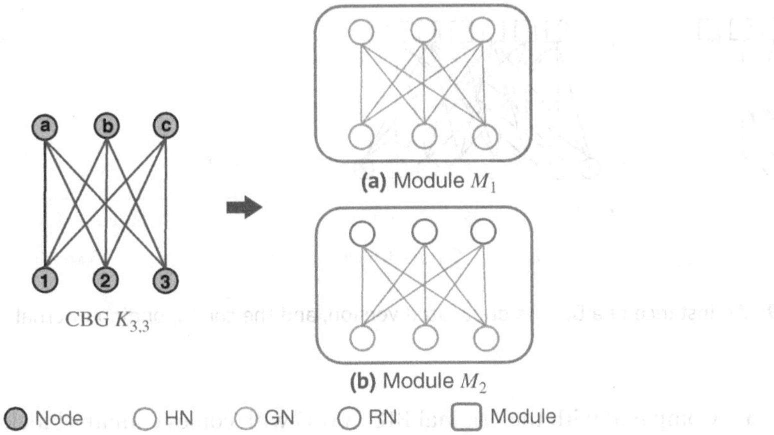

Figure 2.37 An instance of a complete bipartite graph (CBG) $K_{3,3}$ and its corresponding internal graph. (a) All nodes in such a CBG are instantiated as GNs and HNs (module M1). (b) Each node in one node partition is instantiated as an HN and each node in the other node partition is instantiated as an RN (module M2). The latter instantiation method is a special case of block incidence graph and is not considered when refer to CBG.

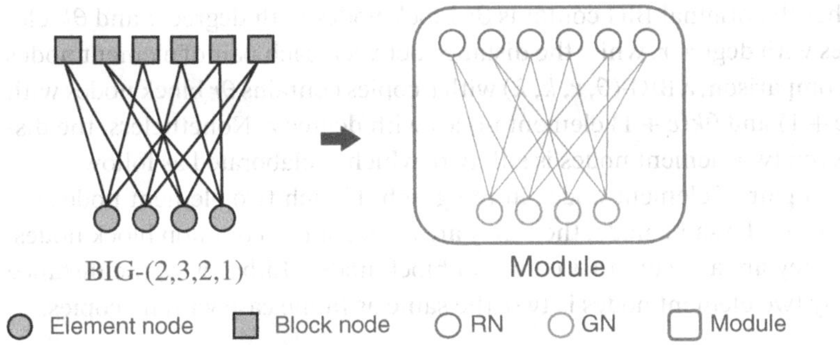

Figure 2.38 An illustration of a BIG-(2, 3, 2, 1) with each element node as a GN (or an HN) and each block node as an RN. The distance between each pair of GNs is two.

Next, we introduce an operation over BIGs called *duplicating*. Under such an operation, given a BIG-(θ, r, k, λ), its element nodes are duplicated c times ($c \geq 0$), while each group of copied element nodes (called a *copy* of element nodes) are connected with block nodes in the same way as the original element nodes. We call such a graph as a BIG-(θ, r, k, λ) with c copies. Note that if $c = 0$, the resulting BIG with copies is exactly the same as the original BIG. Figure 2.39 shows a BIG-(2, 3, 2, 1) with one copy and its corresponding internal graph after node instantiation.

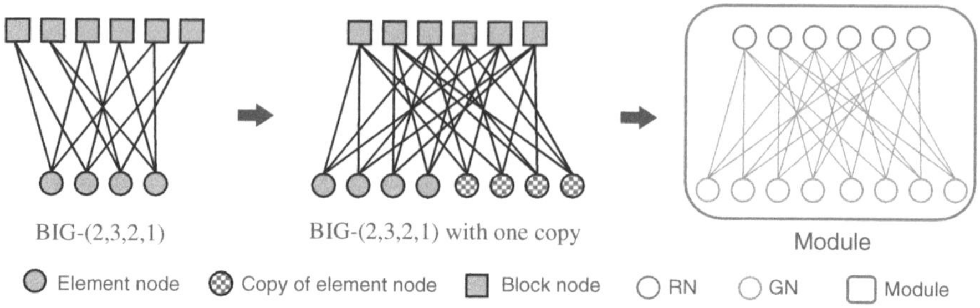

Figure 2.39 An instance of a BIG, its duplicated version, and the corresponding internal graph.

Remark 2.6 Compared with the original BIG, a BIG with copies contains block nodes with higher degrees and more element nodes (i.e., GNs and HNs) without increasing the distance between each pair of element nodes. Such a duplicating operation can be adopted to increase the number of GNs and HNs (element nodes) by occupying more degrees of RNs (block nodes). This point is explained with more details in the following.

Recall that the original BIG contains θr block nodes with degree k and θk element nodes with degree r, while the distance between each pair of element nodes is two. In comparison, a BIG-(θ, r, k, λ) with c copies contains θr block nodes with degree $k(c+1)$ and $\theta k(c+1)$ element nodes with degree r. Nonetheless, the distance between two element nodes is still two, which is elaborated as follows.

Consider a pair of element nodes in the graph. If such two element nodes are duplicated from the same node, then they are adjacent to r common block nodes; otherwise, they are adjacent to λ common block nodes. Either way, the distance between any two element nodes is two, the same as in the case with no copies.

Candidate 4: *Transversal design graph (TDG)*. The definition and property of such a candidate have been provided in Section 1.7.2 of Chapter 1.

As a Ramanujan graph with a low constant diameter, TDG is selected as another candidate of internal graph. Note that in a TDG, the distance between two element nodes and the distance between two block nodes are both two or both four. Accordingly, there are two ways of node instantiation. The first way is to instantiate each block node as a GN (or an HN) and each element node as an RN. The second way is to instantiate each element node as a GN (or an HN) and each block node as an RN. In this framework, the second way (see Figure 2.40) is employed due to the following two considerations.

- Given a TDG with fixed node degrees, the larger the number of GNs and HNs, the higher the scalability.

- Given a GN (or an HN), the larger the proportion of GNs and HNs within distance two, the lower the latency.

Based on these considerations, we first explain why the first way of node instantiation is *not* adopted in this framework. In such a way, block nodes are instantiated as GNs or HNs, and the number of GNs and HNs is μv^2. Recall that the degree of each element node is μv and the degree of each block node is n, denoted by constants d_1 and d_2, respectively. To achieve a high scalability (i.e., μv^2), given fixed node degree d_1, the value of v should be increased. Meanwhile, by Lemma 1.10, given a GN (or an HN), the proportion of GNs and HNs within distance two is

$$\frac{\mu n(v-1)+\mu-1}{\mu v^2}=\frac{d_1 d_2-1}{d_1 v}-\frac{d_2-1}{v^2}, \tag{2.2}$$

which is denoted by y. Then we have

$$\begin{aligned}
\frac{dy}{dv} &= -\frac{d_1 d_2-1}{d_1 v^2}+\frac{2(d_2-1)}{v^3}\\
&= \frac{1}{v^2}\cdot\left[-\frac{d_1 d_2-1}{d_1}+\frac{2(d_2-1)}{v}\right]\\
&\le \frac{1}{v^2}\cdot\left[-\frac{d_1 d_2-1}{d_1}+\frac{2(d_2-1)}{2}\right]\\
&= \frac{1}{v^2}\cdot\frac{(d_2-1)d_1-(d_1 d_2-1)}{d_1}\\
&= \frac{1}{v^2}\cdot\frac{1-d_1}{d_1}<0,
\end{aligned} \tag{2.3}$$

which implies that the value of y is decreasing as the value of v increases. Accordingly, given fixed node degrees (i.e., d_1 and d_2), the larger the number of GNs and HNs (i.e., $d_1 v$), the smaller the proportion of GNs and HNs within distance two

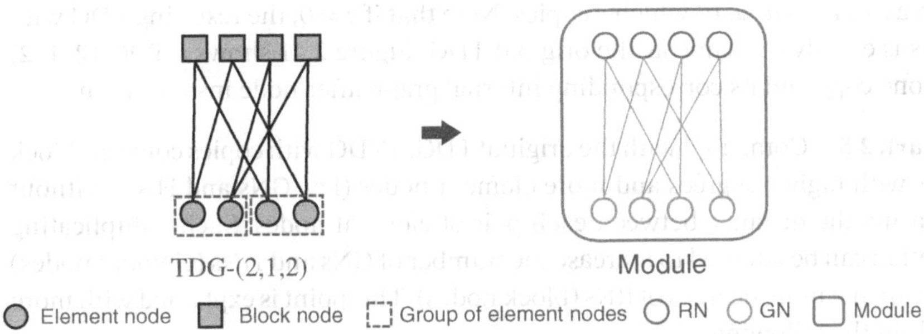

Figure 2.40 An illustration of a TDG-(2, 1, 2) with each element node as a GN (or an HN) and each block node as an RN.

(i.e., y). In other words, a high scalability and a low latency can *not* be achieved at the same time. Hence, such a way is not considered.

Next, we explain why the second way is adopted. In such a way, element nodes are instantiated as GNs or HNs, and the number of GNs and HNs is nv. Recall that the degree of each element node is μv and the degree of each block node is n, denoted by constants d_1 and d_2, respectively. To achieve a high scalability (i.e., $d_2 v$), given fixed node degree d_2, the value of v should be increased. Meanwhile, by Lemma 1.10, given a GN (or an HN), the proportion of GNs and HNs within distance two is

$$\frac{(n-1)v}{nv} = \frac{d_2 - 1}{d_2}, \tag{2.4}$$

which implies that the proportion of GNs and HNs within distance two remains unchanged with the increase of the number of GNs and HNs. In other words, the scalability can be improved without increasing the latency. Hence, the second way is adopted in this framework.

Remark 2.7 Under the second way of node instantiation, given fixed node degrees (i.e., n for block nodes and μv for element nodes), there exists a trade-off between the number of GNs and HNs (i.e., nv) and the number of length-2 edge-disjoint paths (i.e., μ) between any two element nodes (i.e., two GNs, two HNs, or a GN and an HN). Specifically, the larger the value of μ, the larger the number of length-2 edge-disjoint paths between two element nodes (i.e., two GNs, two HNs, or a GN and an HN). In this case, the value of v is decreased as the degree of element nodes is fixed as μv, leading to a smaller number of GNs and HNs.

Next, we introduce the duplicating operation over TDGs. Under such an operation, given a TDG-(n, μ, v), its element nodes are duplicated c times ($c \geq 0$), while each group of copied element nodes (called a *copy* of element nodes) are connected with block nodes in the same way as the original element nodes. We call such a graph as a TDG-(n, μ, v) which c copies. Note that if $c = 0$, the resulting TDG with copies is exactly the same as the original TDG. Figure 2.41 shows a TDG-(2, 1, 2) with one copy and its corresponding internal graph after node instantiation.

Remark 2.8 Compared with the original TDG, a TDG with copies contains block nodes with higher degrees and more element nodes (i.e., GNs and HNs) without increasing the distance between each pair of element nodes. Such a duplicating operation can be adopted to increase the number of GNs and HNs (element nodes) by occupying more degrees of RNs (block nodes). This point is explained with more details in the following.

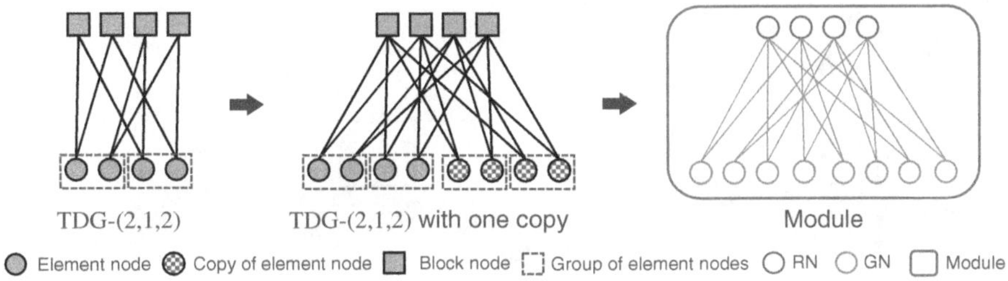

Figure 2.41 An instance of a TDG, its duplicated version, and the corresponding internal graph.

Recall that the original TDG contains μv^2 block nodes with degree n and nv element nodes with degree μv, which the distance between each pair of element nodes is two or four. In comparison, a TDG-(n, μ, v) with c copies contains μv^2 block nodes with degree $n(c+1)$ and $nv(c+1)$ element nodes with degree μv. Nonetheless, the distance between any two element nodes is still two or four, which is elaborated as follows.

Consider a pair of element nodes in the graph. If such two element nodes are duplicated from the same node, then they are adjacent to μv common block nodes, which implies that the distance between them is two. Otherwise, the distance between them is the same as the distance between the original pairs of element nodes (i.e., two or four). In summary, the distance between any two element nodes is two or four, the same as in the case with no copies.

2.4.3.2 Connection Design

This step includes two sub-steps, namely *dock node selection* and *dock node interconnection*.

Sub-step 1: Dock node selection. In general, the principle of this sub-step is to select dock nodes that are close to GNs and HNs in the same module. To be precise, the concept of *intermediate super-node* is introduced as follows.

Definition 2.10 An intermediate super-node, denoted by an ISN-(x, w, l), is a subset of dock nodes in the same module, such that

- x, w, l are positive integers;
- there are x dock nodes in the ISN;
- for any GN or HN in the same module that contains such an ISN, there are w dock nodes in the ISN within a distance of l from such a GN or an HN. If $l = 1$, then the ISN is denoted as an ISN-(x, w).

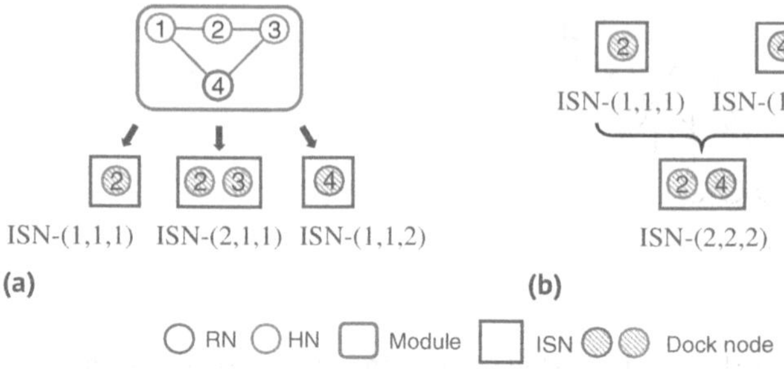

Figure 2.42 An illustration of intermediate super-nodes (ISNs) and their superposition. (a) Three examples of ISN in a given module. In the first example, node 2 is selected as the only dock node that forms an ISN-(1, 1, 1) ($x = 1$). The distance between dock node 2 and any other HN (node 1 or 3) is one, which implies that $w = 1$ and $l = 1$. In the second example, nodes 2 and 3 are selected as dock nodes to form an ISN-(2, 1, 1) ($x = 2$). For node 1 (an HN), its distance is one from dock node 2 and two from dock node 3. Hence, only dock node 2 is within a distance of one from any other HN (node 1) in the same module, which implies that $w = 1$ and $l = 1$. In the third example, node 4 is selected as the only dock node that forms an ISN-(1, 1, 2) ($x = 1$). Note that the distance between dock node 4 and node 2 (HN) is two, while the distance between dock node 4 and other HNs (node 1 or node 3) is one. Hence, for any HN (node 1 or 2 or 3) in the same module, node 4 is within a distance of two, which implies that $w = 1$ and $l = 2$. (b) The two ISNs in sub-figure (a), ISN-(1, 1, 1) (with node 2 as the dock node) and ISN-(1, 1, 2) (with node 4 as the dock node), can be superposed as an ISN-(2, 2, 2). For any HN in the module (node 1 or 2 or 3), nodes 2 and 4 are within a distance of two, which implies that $w = 2$ and $l = 2$.

In Figure 2.42, we present some ISNs with different parameters as examples.

By this definition, ISNs can be superposed as a new ISN. To be specific, given two ISNs *without* common dock nodes that are denoted by ISN-(x_1, w_1, l_1) and ISN-(x_2, w_2, l_2) with $l_1 \leq l_2$, the dock nodes in these two ISNs can be grouped to form an ISN-($x_1 + x_2, w_1 + w_2, l_2$). We provide an example in Figure 2.42b. Such a superposition can be demonstrated by the definition of ISN. First, there are $x_1 + x_2$ dock nodes in the new ISN. Second, for any GN (or HN) in the same module, there are $w_1 + w_2$ dock nodes in the new ISN within a distance of l_2 from such a GN (or an HN). These $w_1 + w_2$ dock nodes include w_1 dock nodes in the original ISN-(x_1, w_1, l_1) and w_2 dock nodes in the original ISN-(x_2, w_2, l_2).

Remark 2.9 Recall that the guiding principle for connection design requires a short distance between each GN (or HN) and the dock nodes in the same module. Based on the concept of ISN-(x, w, l), such a requirement can be satisfied by restricting the parameter l. Meanwhile, given the values of x and l, the larger value of w, the more dock nodes each GN or HN can employ to communicate with GNs and HNs in other modules within the given distance l.

Under the guiding principle for connection design, we focus on how to construct ISNs-(x, w, l) with a small value of l. Recall that in the design procedure, the first three steps can be iteratively applied. In the first iteration, the intra-module design of each module adopts any of the internal graphs in Section 2.4.3.1. Given such an internal graph, we consider constructing ISNs with $l = 1$ or $l = 2$. In the i-th iteration ($i > 1$), the intra-module design of each module can either employ any of the internal graphs in Section 2.4.3.1 or adopt the topology design generated from the $(i - 1)$-th iteration. In the former case, we consider constructing ISNs with $l = 1$ or $l = 2$. However, we do not explicitly restrict the value of l in the latter case, because for ISNs-(x, w, l) in each module, the value of l is at most the diameter of the topology design generated from the $(i - 1)$-th iteration. In the following discussions, we provide approaches and analysis for the selection of ISNs with $l = 1$ or $l = 2$.

- *Complete graph (CG).* Given a CG, to achieve a high scalability, the internal graph is constructed by instantiating each node as a GN or an HN. By the following theorem, any group of x nodes in the CG can form an ISN-(x, x). An illustrative example is given in Figure 2.43. The proof of the following theorem is provided in Section A.2.1.

  Theorem 2.1 *In a CG with m nodes, each node is supposed to be a GN or an HN. Then any group of s nodes (for $1 \leq s \leq m$) form an ISN-(x, w, l) with $x = w = s$ and $l = 1$* (i.e., *the minimum feasible value of l*).

- *Complete bipartite graph (CBG).* Given a CBG, the internal graph is constructed by instantiating each node as a GN or an HN. The dock nodes in such an internal graph form an ISN-(x, w, l) with its parameters subject to the following theorem. The proof of this theorem is provided in Section A.2.2.

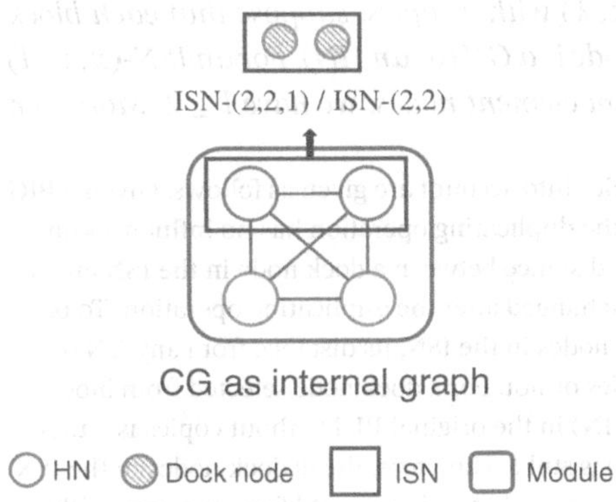

Figure 2.43 An example of ISN selection from nodes in a CG. In the example, any two HNs form an ISN-(2, 2, 1) (or an ISN-(2, 2)).

Theorem 2.2 *In a $CBGK_{m,n}$, each node is instantiated as a GN or an HN. Then any node forms an ISN-$(1, 1, 2)$ (i.e., $x = w = 1$ and $l = 2$). Moreover, in the case with $2 \leq x \leq m + n$, an ISN-(x, w, l) can be formed with $l = 1$, in which $w \leq \min\left\{m, n, \left\lfloor \frac{x}{2} \right\rfloor\right\}$.*

When $x > 1$, to construct ISNs-(x, w, l) with the minimum value of l (l=1) and the maximum value of w $\left(w = \min\left\{m, n, \left\lfloor \frac{x}{2} \right\rfloor\right\}\right)$, it requires that each node partition of CBG contains at most half of dock nodes in the ISN unless the number of nodes in one node partition is smaller than $\left\lfloor \frac{x}{2} \right\rfloor$ and all nodes in this node partition are in the ISN. An illustrative example is given in Figure 2.44.

- *Block incidence graph (BIG).* Given a BIG with c copies ($c \geq 0$),[18] the internal graph is constructed by instantiating each element node and each block node as a GN (or an HN) and an RN, respectively. The dock nodes are selected only from element nodes, only from block nodes, or both. Particularly, the ISNs selected from both element nodes and block nodes can be equivalently induced by the superposition of ISNs (each of which is selected from either element nodes or block nodes). To be specific, given an ISN-(x_1, w_1, l_1) selected only from element nodes and an ISN-(x_2, w_2, l_2) selected only from block nodes, the combination of dock nodes in these two ISNs form an ISN-$(x_1 + x_2, w_1 + w_2, l)$, where $l = \max\{l_1, l_2\}$. Next, we focus on two basic cases where ISNs are selected only from element nodes and only from block nodes, respectively.
  - For any group of x element nodes, they can form an ISN-$(x, x, 2)$, as demonstrated by the following theorem and exemplified in Figure 2.45. The following theorem is proved in Section A.2.3.

Theorem 2.3 *In a BIG-(θ, r, k, λ) with c copies, suppose that each block node is an RN and each element node is a GN (or an HN). For an ISN-(x, w, l) whose dock nodes are selected from element nodes, we have $l \geq 2$. Moreover,*

---

18 The reasons to directly take BIGs with c copies into account are given as follows. Given a BIG and an ISN-(x, w, l) selected from such a BIG, the duplicating operation has no influences on the parameters of the ISN-(x, w, l), because the distance between a dock node in the ISN and a GN (or an HN) in the same module remains unchanged after the duplicating operation. To be specific, for a dock node selected from element nodes in the ISN, its distance from any GN (or HN) is two no matter whether the BIG has copies or not. For a dock node selected from block nodes in the ISN, if its distance from a GN (or HN) in the original BIG without copies is l, then its distance from the copy of such a GN (or HN) is still l. Therefore, the w dock nodes in the ISN that are at a distance of l from some GN (or HN) are still at a distance of l from the copy of the original GN (or HN). This implies that the parameters of such an ISN-(x, w, l) are kept unchanged in the BIG with copies. The analysis also applies to BIGs with no copies, as they can be viewed as a special case when $c = 0$.

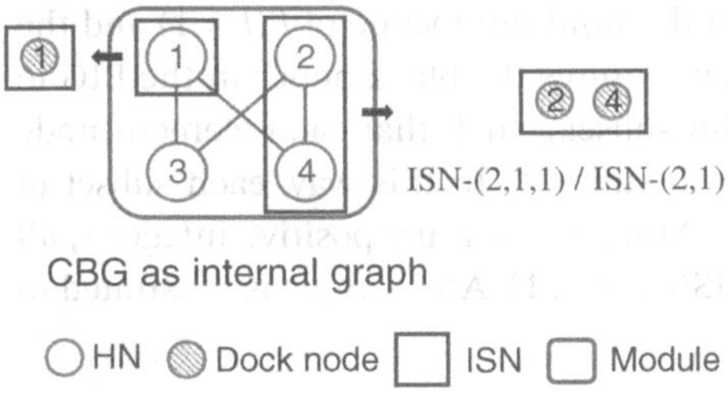

Figure 2.44 An example of ISN selection from nodes in a CBG. In the example, one HN 1 forms an ISN-(1, 1, 2), while two HNs 2 and 4 form an ISN-(2, 1, 1) (or an ISN-(2, 1)).

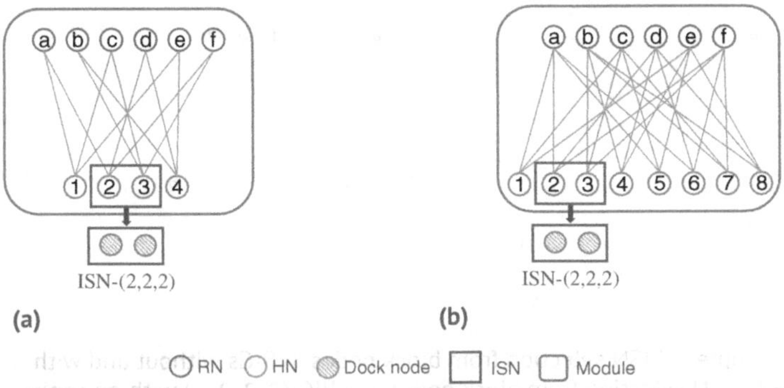

Figure 2.45 Two examples of ISN selection from element nodes in BIGs without and with copies, respectively. (a) ISN selection from element nodes in a BIG-(2, 3, 2, 1) with no copies. In this case, element nodes 2 and 3 form an ISN-(2, 2, 2) since for each of the remaining HNs (i.e., element nodes 1 and 4), its distance from either element node 2 or element node 3 is two. (b) ISN selection from element nodes in a BIG-(2, 3, 2, 1) with one copy. In this case, element nodes 2 and 3 form an ISN-(2, 2, 2) since for each of the remaining HNs (i.e., element nodes {1, 4, 5, 6, 7, 8}), its distance from either element node 2 or element node 3 is two.

any group of s element nodes (for $1 \le s \le \Theta k(c+1)$) form an ISN-$(x, w, l)$ with $x = w = s$ and $l = 2$, which achieves the minimum feasible value of l.

- If all dock nodes are selected from the block nodes, then they form an ISN-(x, w, l) with its parameters subject to the following theorem. The following theorem is proved in Section A.2.4.

Theorem 2.4 *In a BIG-(Θ, r, k, λ) with c copies, suppose that each block node is an RN and each element node is a GN (or an HN). Then by selecting dock nodes from the block nodes, an ISN-(x, w, l) can be formed with $l = 1$, in which $x \ge \Theta w$.*

To construct ISNs-(x, w, l) with the minimum value of l ($l = 1$) and the maximum value of w ($w = x/\theta$), it requires the block nodes in the BIG to be partitioned into several disjoint subsets, such that each element node is adjacent to a block node in each subset.[19] In this way, each subset of block nodes form an ISN-$(\theta, 1, 1)$. Moreover, for any positive integer s, all block nodes in s subsets form an ISN-$(\theta s, s, 1)$. An example is illustrated in Figure 2.46.

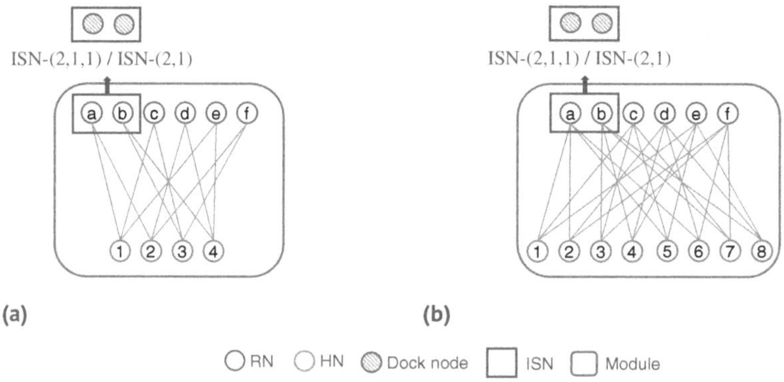

Figure 2.46 Two examples of ISN selection from block nodes in BIGs without and with copies, respectively. (a) ISN selection from block nodes in a BIG-(2, 3, 2, 1) with no copies. The six block nodes in the BIG can be partitioned into three pairs (i.e., $\{a, b\}$, $\{c, d\}$, and $\{e, f\}$), such that each pair of block nodes forms an ISN-(2, 1, 1). In this case, for element nodes (HNs) 1 and 2, block node a is at a distance of one; for element nodes (HNs) 3 and 4, block node b is at a distance of one. Hence, block nodes a and b form an ISN-(2, 1, 1). Consider all block nodes in the three subsets (i.e., $s = 3$). Such block nodes form an ISN-(6, 3, 1). To be specific, for element node 1 (an HN), block nodes a, c, and e are at a distance of one; for element node 2 (an HN), block nodes a, d, and f are at a distance of one; for element node 3 (an HN), block nodes b, c, and f are at a distance of one; for element node 4 (an HN), block nodes b, d, and e are at a distance of one. (b) ISN selection from block nodes in a BIG-(2, 3, 2, 1) with one copy. The six block nodes in the BIG can be partitioned into three pairs (i.e., $\{a, b\}$, $\{c, d\}$, and $\{e, f\}$), such that each pair of block nodes forms an ISN-(2, 1, 1). In this case, for element nodes (HNs) 1, 2, 5, and 6, block node a is at a distance of one; for element nodes (HNs) 3, 4, 7, and 8, block node b is at a distance of one. Hence, block nodes a and b form an ISN-(2, 1, 1). Consider all block nodes in the three subsets (i.e., $s = 3$). Such block nodes form an ISN-(6, 3, 1). To be specific, for element nodes 1 and 5 (HNs), block nodes a, c, and e are at a distance of one; for element nodes 2 and 6 (HNs), block nodes a, d, and f are at a distance of one; for element nodes 3 and 7 (HNs), block nodes b, c, and f are at a distance of one; for element nodes 4 and 8 (HNs), block nodes b, d, and e are at a distance of one.

---

19 Such a BIG corresponds to a CBD or a BIBD where blocks can be partitioned into several subsets, such that each subset of blocks contains each element exactly once. Such a property of CBD or BIBD is also known as *resolvable* property (*cf.* Section 1.6.2).

- *Transversal design graph (TDG).* Given a TDG with c copies ($c \geq 0$),[20] the internal graph is constructed by instantiating each element node and each block node as a GN (or an HN) and an RN, respectively. The dock nodes are selected only from element nodes, only from block nodes, or both. Particularly, the ISNs selected from both element nodes and block nodes can be equivalently induced by the superposition of ISNs (each of which is selected from either element nodes or block nodes). To be specific, given an ISN-(x_1, w_1, l_1) selected only from element nodes and an ISN-(x_2, w_2, l_2) selected only from block nodes, the combination of dock nodes in these two ISNs form an ISN-$(x_1 + x_2, w_1 + w_2, l)$, where $l = \max\{l_1, l_2\}$. Next, we focus on two basic cases where ISNs are selected only from element nodes and only from block nodes, respectively.
  - If all dock nodes are selected from element nodes, then the formed ISNs-(x, w, l) satisfy the following theorem, as illustrated in Figure 2.47. The following theorem is proved in Section A.2.5.

    Theorem 2.5 *In a TDG-(n, μ, v) with c copies, suppose that each block node is an RN and each element node is a GN or an HN. For an ISN-(x, w, l) whose dock nodes are selected from the element nodes, we have $l \geq 2$. In the case of $l = 2$, it holds that $w \leq x - 1$.*

    A construction method of ISNs-(x, w, l) with $w = x - 1$ and $l - 2$ is provided as follows. Recall that element nodes are partitioned into n groups. Given a positive integer with $x \leq n$, a subset of x element nodes in distinct groups form an ISN-$(x, x - 1, 2)$. An example is provided in Figure 2.47.
  - If all dock nodes are selected from the block nodes, then they form the ISNs-(x, w, l) subject to the following theorem. The proof of the following theorem is provided in Section A.2.6.

    Theorem 2.6 *In a TDG-(n, μ, v) with c copies, suppose that each block node is an RN and each element node is a GN or an HN. Then by selecting*

20 The reasons to directly take TDGs with c copies into account are given as follows. Given a TDG and an ISN-(x, w, l) selected from such a TDG, the duplicating operation has no influence on the parameters of the ISN-(x, w, l), because the distance between a dock node in the ISN and a GN (or an HN) in the same module remains unchanged after the duplicating operation. To be specific, for a dock node selected from element nodes in the ISN, its distance from any GN (or an HN) is two or four no matter whether the TDG has copies or not. For a dock node selected from block nodes in the ISN, if its distance from a GN (or an HN) in the original TDG without copies is l, then its distance from the copy of such a GN (or an HN) is still l. Therefore, the w dock nodes in the ISN that are at a distance of l from some GN (or HN) are still at a distance of l from the copy of the original GN (or HN). This implies that the parameters of the ISN-(x, w, l) are kept unchanged in the TDG with copies. The analysis also applies to TDGs with no copies, as they can be viewed as a special case when $c = 0$.

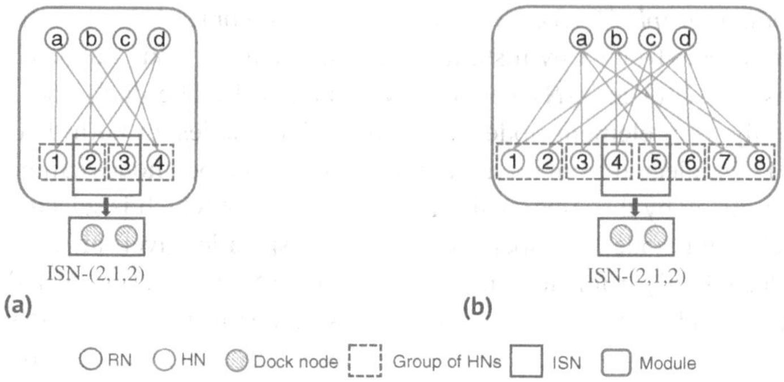

Figure 2.47 Two examples of ISN selection from element nodes in TDGs without and with copies. (a) ISN selection from element nodes in a TDG-(2, 1, 2) with no copies. In this case, element nodes 2 and 3 are contained in distinct groups. For element nodes (HNs) 3 and 4, element node 2 is at a distance of two; for element nodes (HNs) 1 and 2, element node 3 is at a distance of two. Hence, element nodes 2 and 3 form an ISN-(2, 1, 2). (b) ISN selection from element nodes in a TDG-(2, 1, 2) with one copy. In this case, element nodes 4 and 5 are contained in distinct groups. For element nodes (HNs) 1, 2, 5, and 6, element node 4 is at a distance of two; for element nodes (HNs) 3, 4, 7, and 8, element node 5 is at a distance of two. Hence, element nodes 4 and 5 form an ISN-(2, 1, 2).

dock nodes from the block nodes, an ISN-(x, w, l) can be formed with $l = 1$, in which $x \geq vw$.

To construct ISNs-(x, w, l) with the minimum value of l ($l = 1$) and the maximum value of w ($w = x/v$), it requires the block nodes in the TDG to be partitioned into several subsets, such that each element node is adjacent to a block node in each subset.[21] Following this way, each subset of block nodes form an ISN-(v, 1, 1). Moreover, for any positive integer s, all block nodes in s subsets form an ISN-(vs, s, 1). An example is illustrated in Figure 2.48.

Sub-step 2: Dock node interconnection. This sub-step considers how to construct composite-edges among ISNs in distinct modules. In general, ISNs may be heterogeneous in terms of the number of dock nodes. Consequently, the design of composite-edges would involve the optimization for the combination of these heterogeneous ISNs, leading to a high design complexity for topology design. In this framework, to reduce design complexity, we assume that all ISNs contain the same number of dock nodes, i.e., each ISN-(x, w, l) has the same parameter x.

21 Such a TDG corresponds to a TD where blocks can be partitioned into several subsets, such that each subset of blocks contains each element exactly once. In fact, such a property of TD is also known as *resolvable* property (*cf.* Section 1.6.3.3).

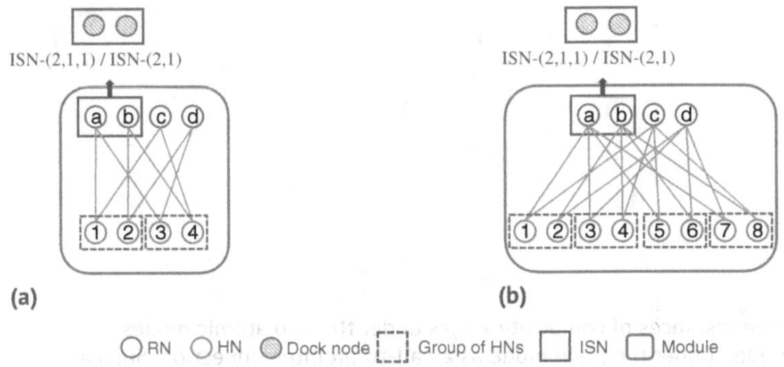

Figure 2.48 Two examples of ISN selection from block nodes in TDGs without and with copies. (a) ISN selection from block nodes in a TDG-(2, 1, 2) with no copies. The four block nodes in the BIG can be partitioned into two pairs (i.e., $\{a, b\}$ and $\{c, d\}$), such that each pair of block nodes forms an ISN-(2, 1, 1). In this case, for element nodes 1 and 3 (HNs), block node a is at a distance of one; for element nodes 2 and 4 (HNs), block node b is at a distance of one. Hence, block nodes a and b form an ISN-(2, 1, 1). Consider all block nodes in the two subsets (i.e., $s = 2$). Such block nodes form an ISN-(4, 2, 1). To be specific, for element node 1 (an HN), block nodes a and c are at a distance of one; for element node 2 (an HN), block nodes b and d are at a distance of one; for element node 3 (an HN), block nodes a and d are at a distance of one; for element node 4 (an HN), block nodes b and c are at a distance of one. (b) ISN selection from block nodes in a TDG-(2, 1, 2) with one copy. The four block nodes in the TDG can be partitioned into two pairs (i.e., $\{a, b\}$ and $\{c, d\}$), such that each pair of block nodes forms an ISN-(2, 1, 1). In this case, for element nodes 1, 3, 5, and 7 (HNs), block node a is at a distance of one; for element nodes 2, 4, 6, and 8 (HNs), block node b is at a distance of one. Hence, block nodes a and b form an ISN-(2, 1, 1). Consider all block nodes in the two subsets (i.e., $s = 2$). Such block nodes form an ISN-(4, 2, 1). To be specific, for element nodes 1 and 5 (HNs), block nodes a and c are at a distance of one; for element nodes 2 and 6 (HNs), block nodes b and d are at a distance of one; for element nodes 3 and 7 (HNs), block nodes a and d are at a distance of one; for element nodes 4 and 8 (HNs), block nodes b and c are at a distance of one.

Next, we discuss the design of composite-edges under the two atomic modes, respectively. For composite-edges under the plain mode, the minimum distance between two dock nodes is one. For composite-edges under the bipartite mode, the minimum distance between two dock nodes is two (through one RN in RN clusters). To restrict the distance between dock nodes in distinct ISNs, we provide two design schemes for composite-edges to achieve the minimum distance under the two atomic modes, respectively. Figure 2.49 shows some instances of composite-edges under the two modes.

- *Composite-edge under the plain mode*. In this case, the distance between each pair of dock nodes in distinct ISNs is restricted to one. In other words, each dock node in one ISN-(x, w, l) should be adjacent to all x dock nodes in another ISN-(x, w, l) (*aka* an *all-to-all interconnection*). Figure 2.49a shows

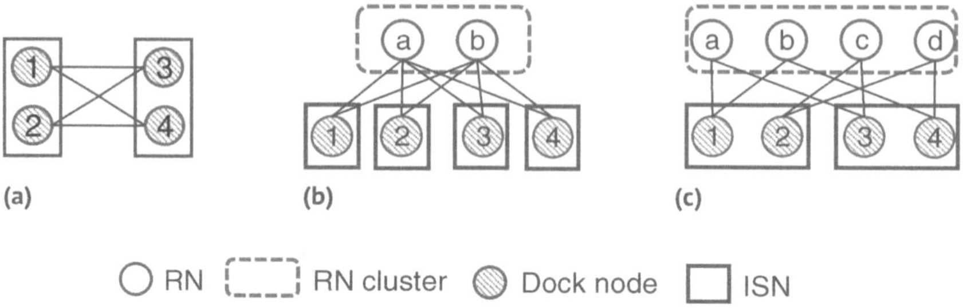

Figure 2.49 Some instances of composite-edges under the two atomic modes. (a) A composite-edge under the plain mode as an all-to-all interconnection. Such a composite-edge contains edges $\{e_{13}, e_{14}, e_{23}, e_{24}\}$. In this case, each ISN-(x, w, l) contains two dock nodes (i.e., $x = 2$). (b) A composite-edge under the bipartite mode as a TD-(n, μ, x) interconnection with $n = 4$, $\mu = 2$, and $x = 1$. Such a composite-edge contains edges $\{e_{1a}, e_{1b}, e_{2a}, e_{2b}, e_{3a}, e_{3b}, e_{4a}, e_{4b}\}$. In this case, each ISN-(x, w, l) contains one dock node (i.e., $x = 1$). (c) A composite-edge under the bipartite mode as a TD-(n, μ, x) interconnection with $n = 2$, $\mu = 1$, and $x = 2$. Such a composite-edge contains edges $\{e_{1a}, e_{1b}, e_{2c}, e_{2d}, e_{3a}, e_{3c}, e_{4b}, e_{4d}\}$. In this case, each ISN-(x, w, l) contains two dock nodes (i.e., $x = 2$). The corresponding TD-(2, 1, 2) contains four elements $\{1, 2, 3, 4\}$ and four blocks $\{a, b, c, d\}$. The four elements are partitioned into two groups $\{1, 2\}$ and $\{3, 4\}$. The blocks a, b, c, d are denoted by $\{1, 3\}$, $\{1, 4\}$, $\{2, 3\}$, and $\{2, 4\}$, respectively.

an example of an all-to-all interconnection between two ISNs-(x, w, l) with $x = 2$.

- *Composite-edge under the bipartite mode.* In this case, the distance between each pair of dock nodes in distinct ISNs is restricted to two. In other words, the composite-edge among n ISNs-(x, w, l) through a single RN cluster should guarantee that each pair of dock nodes in distinct ISNs are connected to the same number of common RNs in the RN cluster. Such a composite-edge is called a *TD-(n, μ, x) interconnection*, which is defined as follows.

Definition 2.11 A TD-(n, μ, x) interconnection with $n \geq 2$ and $\mu, x \geq 1$ is defined as a biregular bipartite graph that satisfies the following properties.

- One node partition includes nx dock nodes with degree μx in n distinct ISNs-(x, w, l), while the other node partition includes μx^2 RNs with degree n in an RN cluster.
- Every pair of dock nodes in distinct ISNs are connected to the same number of μ common RNs.

In a TD-$(n, \mu, 1)$ interconnection (i.e., $x = 1$), each ISN contains one dock node. Then dock nodes in such ISNs are connected to all μ RNs in the RN cluster of the TD-$(n, \mu, 1)$ interconnection (see Figure 2.49b). When each ISN contains more dock nodes (i.e., $x \geq 2$), the TD-(n, μ, x) interconnection can be induced by a transversal design TD-(n, μ, x). To be specific, in such a composite-edge, dock nodes are regarded as elements, each ISN is regarded as a group of elements, and RNs in the RN cluster are regarded as blocks. If a block contains an element in the TD, then the corresponding block node and element node are directly connected in the composite-edge. Figure 2.49c shows an example of a TD-(2, 1, 2) interconnection between two ISNs-(x, w, l) with $x = 2$.

Based on the existence conditions of TD, we have the following lemma for the existence of TD-(n, μ, x) interconnections, which is proved in Section A.2.7.

Lemma 2.1 *Regarding the existence of TD-(n, μ, x) interconnections, three sufficient conditions are given as follows, respectively.*

- *If $x = 1$, there exists a TD-(n, μ, x) interconnection for any integer $n \geq 2$.*
- *If x is a prime power, there exists a TD-(n, μ, x) interconnection for any integer $2 \leq n \leq x + 1$.*
- *If $n = 2$ or $n = 3$, there exists a TD-(n, μ, x) interconnection for any integer $x \geq 1$.*

2.4.3.3 Inter-module Design

The inter-module design involves two sub-steps, namely skeleton graph design and super-edge instantiation.

Regarding skeleton graph design, according to the guiding principle, the shortest paths between each pair of modules should achieve their minimum feasible distances. Next, we consider CG and BIG as the candidates of skeleton graph. The reasons for their adoption are elaborated in two cases where super-edges under the plain mode and the bipartite mode are adopted, respectively.[22]

- By using super-edges under the plain mode, the minimum distance between each pair of modules is one. This is achieved when CG is adopted as the skeleton graph. In such a graph, the distance between any two nodes is one.
- By using super-edges under the bipartite mode, the minimum distance between each pair of modules is two. This is achieved when BIG is adopted as the skeleton graph. In such a graph, the distance between each pair of element nodes is two.

Note that TDG is not considered as a candidate because the distance between a pair of element nodes or a pair of block nodes may be larger than two (*cf.* Lemma 1.10).

22 In this framework, skeleton graph design is considered under only one of the two atomic modes. As for the case where the design involves both atomic modes, it is left as a future work.

Regarding super-edge instantiation, in general, super-edges can be instantiated as the combination of various composite-edges. However, such a general design requires the optimization for the selection of composite-edges and their appropriate combination to instantiate super-edges in the skeleton graph. The optimization would lead to sophisticated topology design in practice. In this framework, to reduce design complexity, it is assumed that the instantiation of super-edges in a topology uses a common composite-edge.

Next, two candidates of skeleton graph and their corresponding super-edge instantiations are introduced as follows.

Candidate 1: CG. When the skeleton graph is a CG, each pair of nodes in the graph are directly connected. In such a graph, all nodes are instantiated as modules. Note that RN clusters are not considered because in this framework, they are assumed to be disconnected with each other. Figure 2.50 illustrates an instance of CG and its corresponding skeleton graph.

By the definition of CG, any two modules are directly connected by a super-edge under the plain mode. To instantiate such a super-edge, a number of t ISNs-(x, w, l) are selected in each module. Each selected ISN is connected to a distinct ISN in another module as a composite-edge. In this way, a super-edge is constructed as a combination of t paralleled composite-edges such that no composite-edges share the same ISNs. Such a super-edge is called a *t-parallelized all-to-all interconnection*.

In Figure 2.51, we present two examples of all-to-all interconnections where each ISN-(x, w, l) contains two dock nodes (i.e., $x = 2$). Specifically, based on the all-to-all interconnections in Figure 2.51a, two t-parallelized all-to-all interconnections with $t = 1$ and $t = 2$ can be constructed accordingly (see Figure 2.51b,c, respectively).

Candidate 2: BIG. When the skeleton graph is a BIG-(θ, r, k, λ), any two element nodes are connected through λ common block nodes at a distance of two. By instantiating element nodes as modules and block nodes as RN clusters, there are λ length-2 paths between each pair of modules. Each path goes through a single

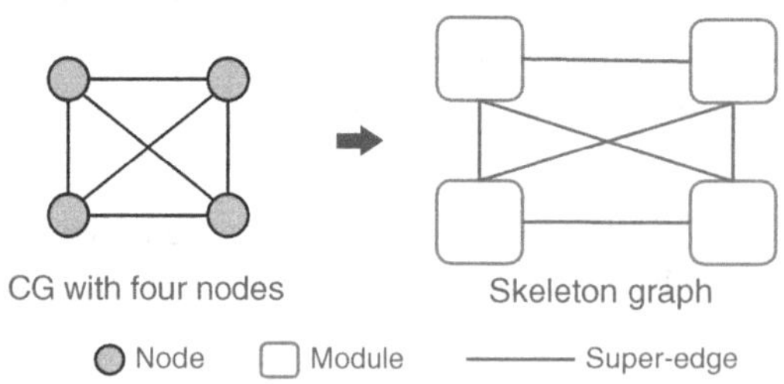

Figure 2.50 An instance of a CG and its corresponding skeleton graph.

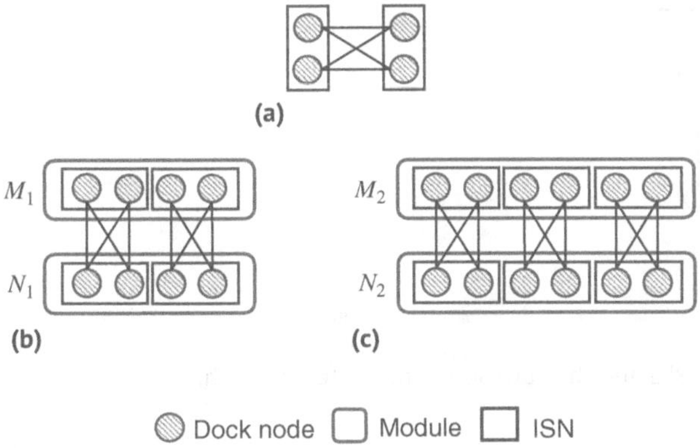

Figure 2.51 An illustration of t-parallelized all-to-all interconnections. (a) An all-to-all interconnection as the composite-edge under the plain mode ($x = 2$). (b) A 2-parallelized all-to-all interconnection as the super-edge (i.e., $t = 2$). Such a super-edge is constructed based on the composite-edge shown in sub-figure (a). (c) A 3-parallelized all-to-all interconnection as the super-edge (i.e., $t = 3$). Such a super-edge is constructed based on the composite-edge shown in sub-figure (a).

RN cluster (i.e., a block node). An instance of BIG and the corresponding skeleton graph are shown in Figure 2.52.

In the BIG-(θ, r, k, λ), each RN cluster is directly connected to k modules through k super-edges between the RN cluster and a module. To instantiate such k super-edges between modules and a common RN cluster, a number of t ISNs-(x, w, l) are selected in each of the k modules. In each module, one of its t ISNs-(x, w, l) is selected, such that these k ISNs in distinct k modules are connected to a common RN cluster. Such interconnections are viewed as a composite-edge under the bipartite mode, i.e., a TD-(n, μ, x) interconnection.[23] Recall that n ISNs are interconnected through such a TD-(n, μ, x) interconnection, thus it is required that $k = n$. In this way, the super-edges are constructed as a combination of t composite-edges in a parallelized fashion, aka a t-parallelized TD-(n, μ, x) interconnection.

In Figure 2.53, we show two examples of t-parallelized TD-(n, μ, x) interconnections in the case of $n = 2$, $\mu = 1$, and $x = 2$. Each ISN-(x, w, l) contains two dock nodes (i.e., $x = 2$). Figure 2.53a shows an example of TD-(2, 1, 2) interconnections as the composite-edge between two ISNs. Based on the composite-edge, Figure 2.53b,c shows two examples of t-parallelized TD-(2, 1, 2) interconnections with $t = 1$ and $t = 2$, respectively.

23 For the t ISNs in the same module, their interconnections with the RN clusters are assumed to be identical.

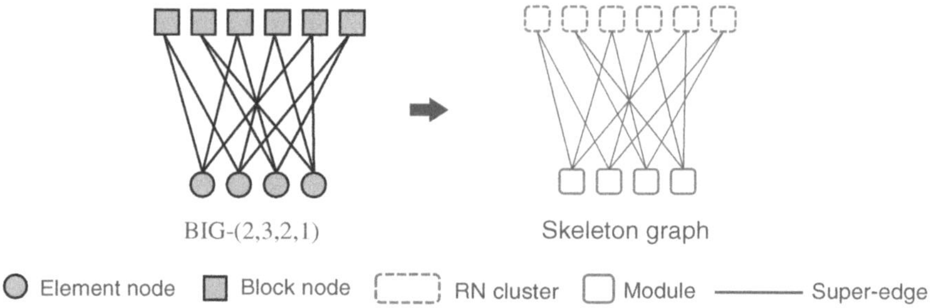

Figure 2.52 An instance of a BIG and the corresponding skeleton graph.

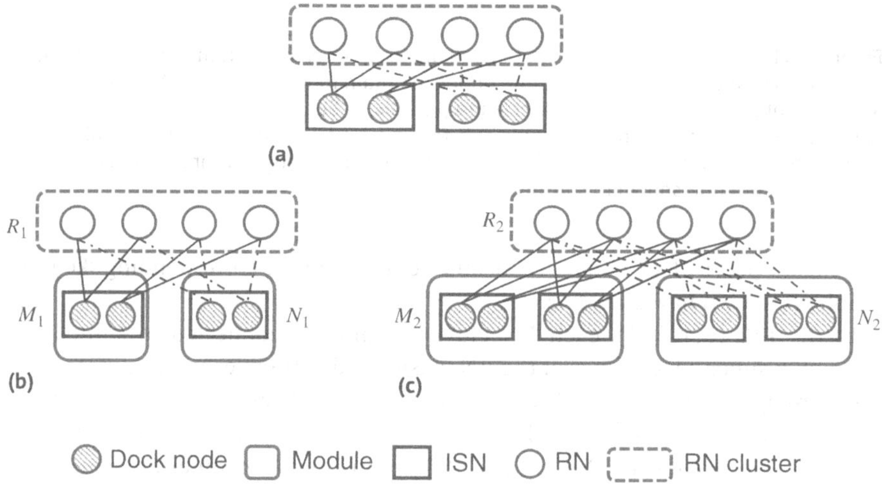

Figure 2.53 An illustration of t-parallelized TD-(n, μ, x) interconnections. (a) A TD-(2, 1, 2) interconnection as the composite-edge under the bipartite mode (i.e., $n = 2$, $\mu = 1$, and $x = 2$). (b) A 1-parallelized TD-(2, 1, 2) interconnection as the super-edges (i.e., $t = 1$). Such super-edges are constructed based on the composite-edge shown in sub-figure (a). (c) A 2-parallelized TD-(2, 1, 2) interconnection as the super-edges (i.e., $t = 2$). Such super-edges are constructed based on the composite-edge shown in sub-figure (a).

Duplicating operation. Under such an operation, given a BIG-(θ, r, k, λ) as the skeleton graph, all modules (i.e., element nodes) are duplicated c times ($c \geq 0$), while the dock nodes in each group of copied modules (called a *copy* of the original modules) are connected with the RNs in RN clusters in the same way as the dock nodes in the original modules. Note that if $c = 0$, the resulting BIG with copies is exactly the same as the original BIG. Figure 2.54 shows a BIG-(2, 3, 2, 1) with one copy and its mapped skeleton graph.

Remark 2.10 Compared with the original BIG, a BIG with copies contains more element nodes (i.e., modules) without increasing the distance between each pair

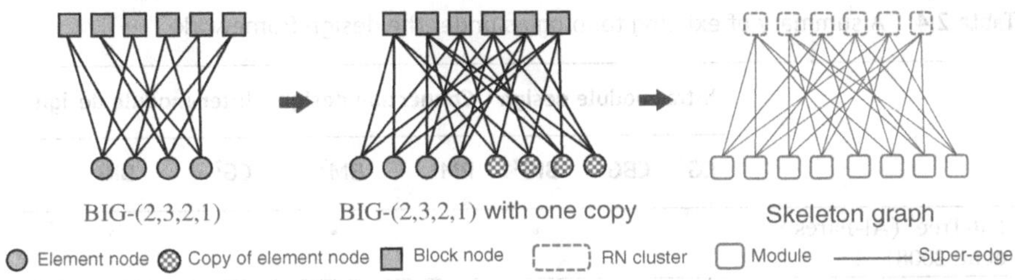

Figure 2.54 An instance of a BIG, its duplicated version, and the corresponding skeleton graph.

of element nodes, while each block node (i.e., RN cluster) is directly connected to more modules. Such a duplicating operation can be adopted to increase the number of modules (equivalently, more GNs and HNs) for a higher scalability by leveraging more degrees of RNs in RN clusters. This point is explained with more details in the following.

Recall that the original BIG contains θk element nodes (i.e., modules), while the distance between each pair of element nodes is two. By instantiating super-edges as a t-parallelized TD-(n, μ, x) interconnection, the degree of each RN in RN clusters is nt (or kt since we have $n = k$). In comparison, a BIG-(θ, r, k, λ) with c copies contains $\theta k(c + 1)$ element nodes (i.e., modules). By instantiating super-edges as a t-parallelized TD-(n, μ, x) interconnection, the degree of each RN in RN clusters is $nt(c + 1)$ (or $kt(c + 1)$ since we have $n = k$). Recall that HNs are contained in modules (HNs) and CNs are attached to some nodes in modules (i.e., GNs and HNs). Since a BIG with copies contains more modules, its induced topology contains more nodes with computing function (i.e., HNs and CNs), leading to a higher scalability.

Next, we explain why the distance between two element nodes remains unchanged. Consider a pair of element nodes in the graph. If such two element nodes are duplicated from the same node, the they are adjacent to r common block nodes; otherwise, they are adjacent to λ common block nodes. Either way, the distance between any two element nodes is two, the same as in the case with no copies.

2.5 Revisit of Existing Topologies

By following the design procedure with the concrete tools, a great number of existing network topologies can be reverse-engineered, as summarized in Table 2.4. In the following subsections, we illustrate the corresponding designs in each step of the procedure of eight representative existing network topologies, respectively.

Table 2.4 A summary of existing topologies under the design framework.

	Intra-module design			**Connection design**		**Inter-module design**	
	CG[2]	**CBG[2]**	**BIG[2]**	**PM[3]**	**BM[3]**	**CG[2]**	**BIG[2]**
Fat-Tree[1] (Al-Fares et al. 2008)			•		•		•
Aspen Trees (Walraed-Sullivan et al. 2013)			•		•		•
AB Fat-Tree (Liu et al. 2013)			•		•		•
Quasi Fat-Tree (Zahavi et al. 2014)			•		•		•
Subways (Liu et al. 2015)			•		•		•
Diamond (Sun et al. 2014)		•			•		•
DCCube (Azizi et al. 2017)		•		•		•	
HyperX (Ahn et al. 2009)	•			•		•	
Dragonfly (Kim et al. 2008)	•			•		•	
Dragonfly+ (Shpiner et al. 2017)			•	•		•	
DCell (Guo et al. 2008)			•	•		•	
FiConn (Li et al. 2009)			•	•		•	
BCube (Guo et al. 2009)			•		•		•
HCN & BCN (Guo et al. 2011)			•	•		•	

1 The designs of Fat-Tree also apply to its variants in the family of folded Clos networks, e.g., Facebook's "four-post" (Farrington and Andreyev 2013) and Google's Jupiter (Singh et al. 2015).

2 CG for Complete Graph; CBG for Complete Bipartite Graph; BIG for Bipartite Incidence Graph.

3 PM for Plain Mode; BM for Bipartite Mode.

2.5.1 Fat-Tree

Fat-Tree is a switch-centric topology, where switches are represented by GNs and RNs, while servers are represented by CNs. Next, given a three-layer p-ary Fat-Tree where the port number of switches p is an even integer, its design in each step is illustrated as follows.

- **Intra-module design**: The internal graph is a BIG-$(1, p/2, p/2, p/2)$, where element nodes and block nodes are viewed as GNs and RNs, respectively.
- **Connection design**: In each module, all RNs are selected as dock nodes. Each dock node serves as an intermediate super-node, i.e., an ISN-(1, 1). Then the dock node interconnection is constructed under the bipartite mode. To be specific, each composite-edge among p ISNs is designed as a TD-$(p, p/2, 1)$ interconnection through an RN cluster.
- **Inter-module design**: The skeleton graph is a BIG-$(1, p/2, p, p/2)$. Such a graph contains p element nodes and $p/2$ block nodes that correspond to modules and RN clusters, respectively. For any p modules, they are connected with a common RN cluster via p super-edges. Such super-edges form a 1-parallelized TD-$(p, p/2, 1)$ interconnection.
- **Finalization**: Each GN is directly connected with $p/2$ CNs.

Based on the descriptions, we show the design of a 4-ary Fat-Tree under the design framework in Figure 2.55.

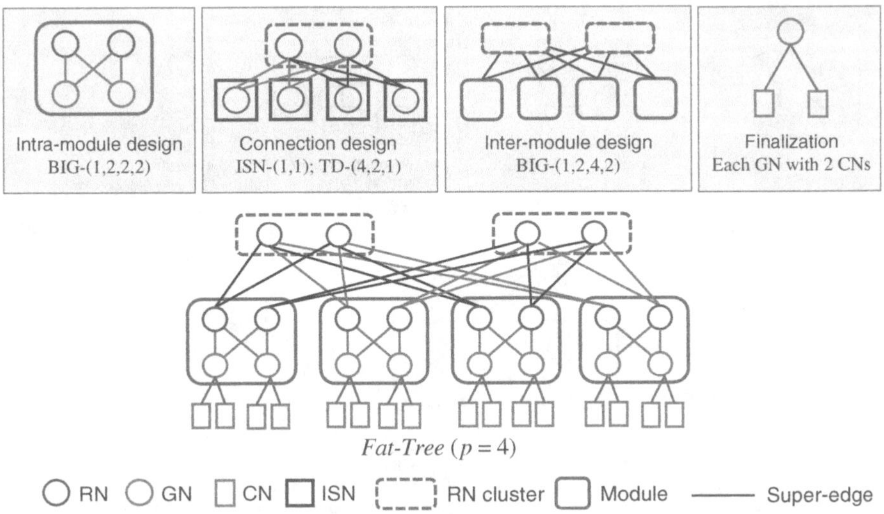

Figure 2.55 The design of a 4-ary Fat-Tree under the design framework.

2.5.2 Aspen Trees

Aspen Trees are switch-centric topologies, where switches are represented by GNs and RNs, while servers are represented by CNs. Next, given a three-layer p-ary Aspen Tree where the port number of switches p is an even integer, its design in each step is illustrated as follows.

- **Intra-module design**: The internal graph is a BIG-$(1, p/2, p/2, p/2)$, where element nodes and block nodes are viewed as GNs and RNs, respectively.
- **Connection design**: In each module, all block nodes (RNs) are selected as dock nodes. Each dock node serves as an intermediate super-node, i.e., an ISN-(1, 1). Then the dock node interconnection is constructed under the bipartite mode. Given c_3 as a positive factor of $p/2$, each composite-edge is designed as a TD-$(p/c_3, p/2, 1)$ interconnection among p/c_3 ISNs through an RN cluster.
- **Inter-module design**: The skeleton graph is a BIG-$(1, p/2c_3, p/c_3, p/2c_3)$. Such a graph contains p/c_3 element nodes and $p/2c_3$ block nodes that correspond to modules and RN clusters, respectively. For any p/c_3 modules, they are connected with a common RN cluster via p/c_3 super-edges. Such super-edges form a c_3-parallelized TD-$(p/c_3, p/2, 1)$ interconnection.
- **Finalization**: Each GN is directly connected with $p/2$ CNs.

Based on these descriptions, we show the design of a 3-layer 4-ary Aspen Tree with $c_3 = 2$ under the design framework in Figure 2.56.

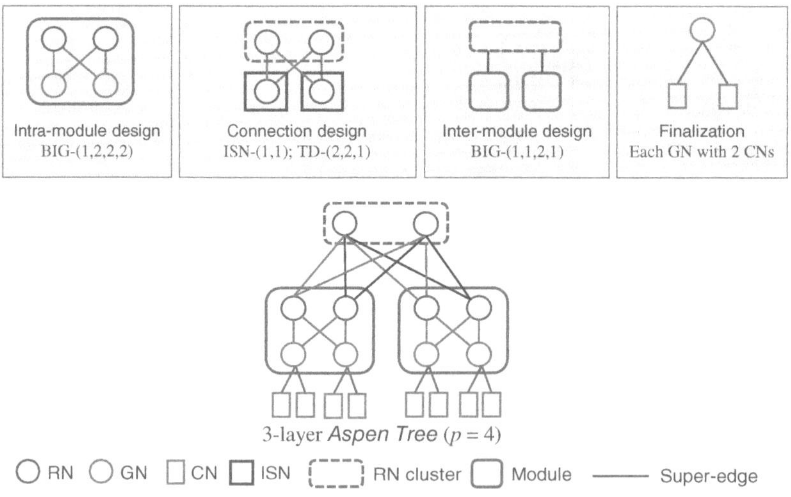

Figure 2.56 The design of a 3-layer 4-ary Aspen Tree with $c_3 = 2$ under the design framework.

The design of higher-layer Aspen Trees: For an l-layer Aspen Tree ($l \geq 4$), its intra-module design is conducted by removing half of the modules in an $(l-1)$-layer Aspen Tree. As for the connection design, the RNs in the $(l-1)$-th layer are selected as dock nodes. Each dock node serves as an ISN-$(1, 1, l-2)$. Given c_l as a positive factor of $p/2$, each composite-edge is designed as a TD-$(p/c_l, p/2, 1)$ interconnection among p/c_l ISNs through an RN cluster. Regarding the inter-module design, the skeleton graph is a BIG-$(1, p/2c_l, p/c_l, p/2c_l)$. For any p modules, they are connected with a common RN cluster via p super-edges. Such super-edges are instantiated as a c_l-parallelized TD-$(p/c_l, p/2, 1)$ interconnection. In this way, an L-layer Aspen Tree ($L \geq 4$) can be constructed by iteratively applying the procedure for $L-2$ times. Finally, each GN is directly connected with $p/2$ CNs.

In Figure 2.57, we show the design of a 4-layer 4-ary Aspen Tree with $c_4 = 2$ under the design framework.

2.5.3 AB Fat-Tree

AB Fat-Tree is a switch-centric topology, where switches are represented by GNs and RNs, while servers are represented by CNs. Next, given a three-layer p-ary AB Fat-Tree where the port number of switches p is an even integer, its design in each step is illustrated as follows.

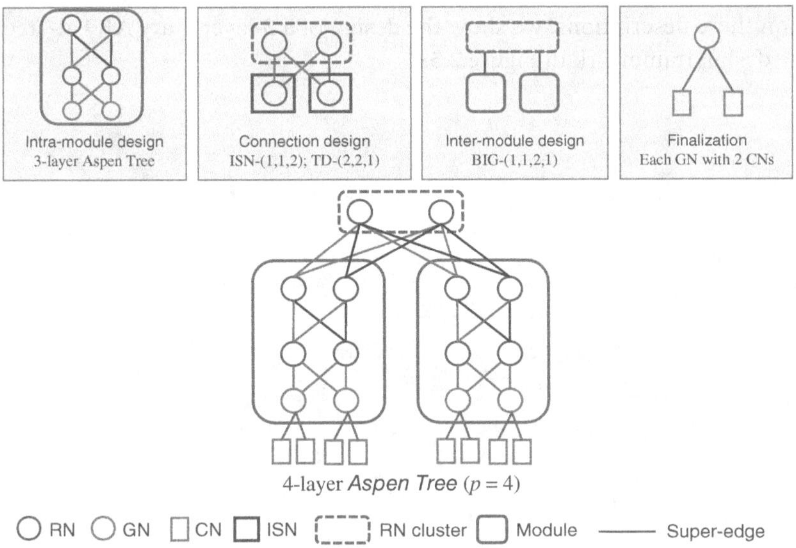

Figure 2.57 The design of a 4-layer 4-ary Aspen Tree with $c_4 = 2$ under the design framework.

- **Intra-module design**: The internal graph is a BIG-(1, $p/2$, $p/2$, $p/2$), where element nodes and block nodes are viewed as GNs and RNs, respectively.
- **Connection design**: In each module, all block nodes (RNs) are selected as dock nodes. Each group of $p/2$ RNs serves as an intermediate super-node, i.e., an ISN-($p/2$, $p/2$). Then the dock node interconnection is constructed under the bipartite mode. The composite-edge between any two ISNs is designed as a TD-(2, 1, $p/2$) interconnection through an RN cluster.
- **Inter-module design**: The skeleton graph is a BIG-(1, 1, 2, 1) with $p/2 - 1$ copies. Such a graph contains p element nodes and one block node that correspond to modules and an RN cluster, respectively. The p modules are split into $p/2$ groups, each of which contains two modules. For any two modules in the same group, they are connected with a common RN cluster via two super-edges. Such super-edges form a 1-parallelized TD-(2, 1, $p/2$) interconnection between the two modules. Note that for an AB Fat-Tree, its skeleton graph is a BIG *with c copies*. The reason for adopting the duplicating operation is given as follows. In the case with $c = 0$ (i.e., no copies), the whole topology only contains two modules and each RN in RN clusters has two occupied degrees (i.e., two ports of each switch are utilized). In comparison, when $c > 0$, more ports of each switch are utilized and the scalability is improved simultaneously. Particularly, by setting $c = p/2 - 1$, the ports of each switch (corresponding to an RN in RN clusters) are fully utilized.
- **Finalization**: Each GN is directly connected with $p/2$ CNs.

Based on these descriptions, we show the design of a 3-layer 4-ary AB Fat-Tree under the design framework in Figure 2.58.

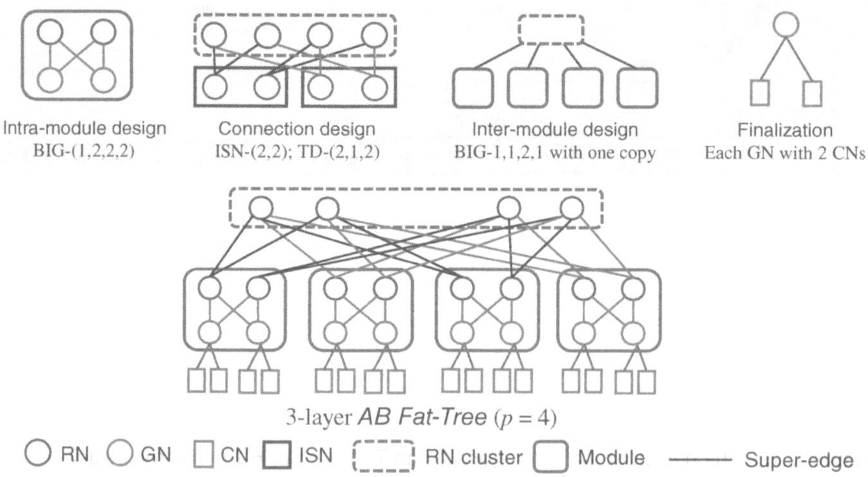

Figure 2.58 The design of a 3-layer 4-ary AB Fat-Tree under the design framework.

The design of higher-layer AB Fat-Tree: For an l-layer AB Fat-Tree ($l \geq 4$), its intra-module design is conducted by removing half of the modules in an $(l-1)$-layer AB Fat-Tree. Regarding the connection design, the $(p/2)^{(l-2)}$ RNs in the $(l-1)$-th layer are selected as dock nodes, which are partitioned into $(p/2)^{(l-3)}$ ISNs-$(p/2, p/2, l-2)$. Then each composite-edge is designed as a TD-$(2, 1, p/2)$ interconnection between two ISNs through an RN cluster. As for the inter-module design, the skeleton graph is a BIG-$(1, (p/2)^{(l-3)}, 2, (p/2)^{(l-3)})$ with $p/2-1$ copies. Such a graph contains p element nodes and $(p/2)^{(l-3)}$ block nodes that are regarded as modules and RN clusters, respectively. The p modules are partitioned into $p/2$ groups, each of which contains two modules. For any two modules in the same group, they are connected with a common RN cluster via two super-edges. Such super-edges form a 1-parallelized TD-$(2, 1, p/2)$ interconnection between the two modules. Following this way, an L-layer AB Fat-Tree ($L \geq 4$) can be constructed by iteratively applying the procedure for $L-2$ times. As the final step, each GN is directly connected with $p/2$ GNs.

In Figure 2.59, we show the design of a 4-layer 4-ary AB Fat-Tree under the design framework.

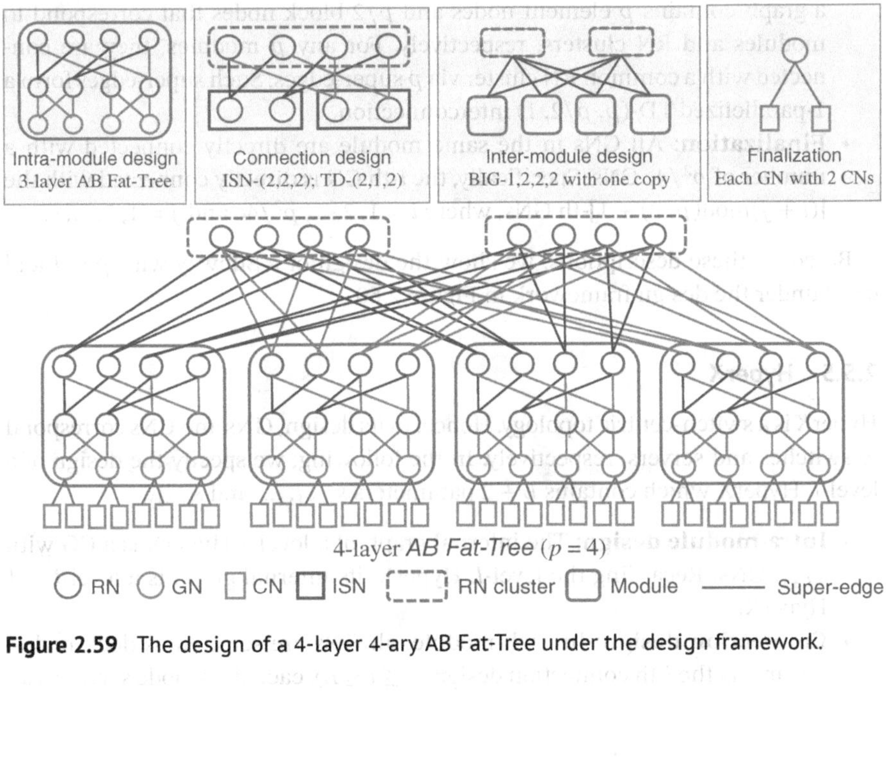

Figure 2.59 The design of a 4-layer 4-ary AB Fat-Tree under the design framework.

2.5.4 Subway

As a modification of Fat-Tree topology, Subway is a switch-centric topology. Particularly, in the topology, each server is equipped with multiple NICs and directly connected with several servers for better fault-tolerance and load balancing performances. In the design of Subway, each server does not execute routing function and is considered a CN that is attached to multiple GNs. Next, we take the design demonstrated in Liu et al. (2015) as an example for illustration. In the example, the port number of each switch is denoted by an positive even integer p. Meanwhile, the number of NICs on each server is denoted by parameter c which is a factor of $p/2$.

- **Intra-module design**: For each module, the internal graph is designed as a BIG-$(1, p/2, p/2, p/2)$, where element nodes and block nodes are viewed as GNs and RNs, respectively.
- **Connection design**: In each module, all RNs are selected as dock nodes. Each dock node serves as an intermediate super-node, i.e., an ISN-(1, 1). Then the dock node interconnection is constructed under the bipartite mode. The composite-edge among any p ISNs is designed as a TD-$(p, p/2, 1)$ interconnection through an RN cluster.
- **Inter-module design**: The skeleton graph is a BIG-$(1, p/2, p, p/2)$. Such a graph contains p element nodes and $p/2$ block nodes that correspond to modules and RN clusters, respectively. For any p modules, they are connected with a common RN cluster via p super-edges. Such super-edges form a 1-parallelized TD-$(p, p/2, 1)$ interconnection.
- **Finalization**: All GNs in the same module are directly connected with a number of $p^2/4c$ CNs. Specifically, the i-th CN is directly connected with the $[(i+j)\text{mod}(p/2)+1]$-th GNs, where $i = 1, 2, \ldots, p^2/4c$ and $j = 1, 2, \ldots, c$.

Based on these descriptions, we show the design of a Subway with $p = 4$ and $c = 2$ under the design framework in Figure 2.60.

2.5.5 HyperX

HyperX is a switch-centric topology. Hence, in its design, GNs and CNs correspond to switches and servers, respectively. In the following, we specify the design of a level-L HyperX which contains $L + 2$ parameters $s_0, s_1, \ldots,$ and $s_{(L+1)}$.

- **Intra-module design**: The internal graph of a level-1 HyperX is a CG with $s_{(L+1)}$ GNs. Regarding the level-L HyperX, its internal graph is a level-$L-1$ HyperX.
- **Connection design**: In each module, all GNs are selected as dock nodes. Regarding the i-th connection design $(1 \le i \le L)$, each dock node serves as an

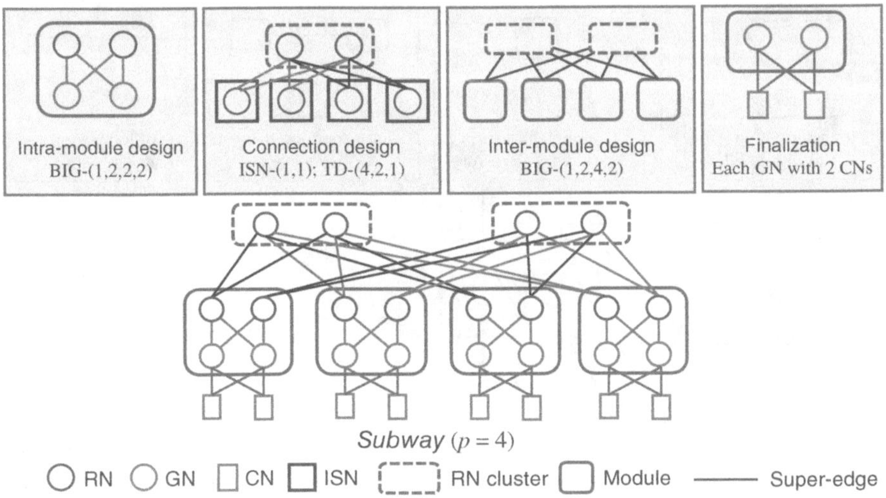

Figure 2.60 The design of a Subway with $p = 4$ and $c = 2$ under the design framework.

intermediate super-node, i.e., an ISN-(1, 1, i). Then the dock node interconnection is constructed under the plain mode. Each composite-edge between any two ISNs is designed as an all-to-all interconnection.

- **Inter-module design**: Regarding the i-th inter-module design ($1 \leq i \leq L$), its skeleton graph is a CG with $s_{(L-i+1)}$ nodes. Each node corresponds to a module. Each super-edge between two modules is instantiated as a t-parallelized all-to-all interconnection, where $t = \prod_{j=L+2-i}^{L+1} s_j$.
- **Finalization**: Each GN is directly connected with a number of s_0 CNs.

In Figure 2.61, we show the design of a level-1 HyperX with $s_0 = 1$, $s_1 = 3$, and $s_2 = 4$ under the design framework.

2.5.6 Dragonfly

Dragonfly is a switch-centric topology, where GNs and CNs correspond to p-port switches and servers, respectively. The design of a Dragonfly is specified as follows.

- **Intra-module design**: The internal graph of a Dragonfly is a CG with s_0 GNs.
- **Connection design**: In each module, all GNs are selected as dock nodes. Each dock node serves as an intermediate super-node, i.e., an ISN-(x, x). Then the dock node interconnection is constructed under the plain mode. Each composite-edge between any two ISNs is designed as an all-to-all interconnection.

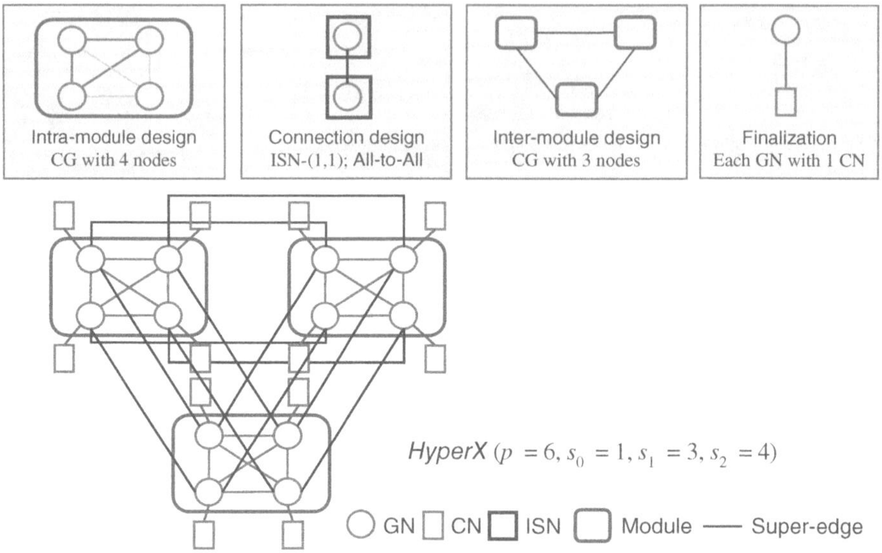

Figure 2.61 The design of a level-1 HyperX with 6-port switches under the design framework.

- **Inter-module design**: The skeleton graph is a CG with s_2 nodes. Each node corresponds to a module. Each super-edge between two modules is instantiated as a t-parallelized all-to-all interconnection, where $t = \dfrac{s_0(p+1-s_1-s_0)}{(s_2-1)x^2}$.
- **Finalization**: Each GN is directly connected with a number of s_1 CNs.

Note that in these designs, it is required that $(s_2-1)x^2$ should be a factor of $s_0(p+1-s_1-s_0)$. In Figure 2.62, we show the design of a Dragonfly topology with $x = 1$, $s_0 = 4$, $s_1 = 1$, and $s_2 = 2$ under the design framework.

2.5.7 DCell

DCell is a server-centric topology. Hence, in its design, HNs correspond to servers while RNs correspond to switches. Next, we describe the design of a level-1 DCell with p-port switches step by step.

- **Intra-module design**: The internal graph (i.e., a level-0 DCell) is a BIG-(1, 1, p, 1) where element nodes and block nodes are viewed as HNs and RNs, respectively.
- **Connection design**: In each module, all HNs are selected as dock nodes. Each dock node serves as an ISN-(1, 1, 2). Then dock node interconnection is

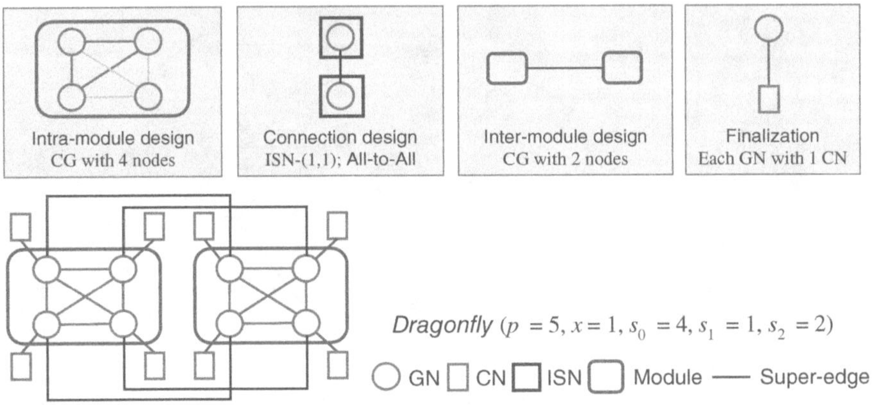

Figure 2.62 The design of a Dragonfly topology with 5-port switches under the design framework.

constructed under the plain mode. Specifically, each composite-edge between any two ISNs is designed as an all-to-all interconnection.

- **Inter-module design**: The skeleton graph is a CG with $p + 1$ nodes. Each node corresponds to a module. Each super-edge between two modules is instantiated as a 1-parallelized all-to-all interconnection.
- **Finalization**: This step is omitted as the topology contains no CNs.

In Figure 2.63, we show the design of a level-1 DCell with 2-port switches under the design framework.

The design of higher-level DCells: In general, for a level-i DCell ($i \geq 2$), level-($i - 1$) DCells are regarded as modules. Regarding the connection design, in each module, each HN serves as an ISN-(1, 1, l_i), where $l_1 = 2$ and $l_i = 2l_{i-1} + 1$. The composite-edge between two ISNs is still an all-to-all interconnection. As for the inter-module design, the skeleton graph is a CG with m_i nodes, where $m_0 = 1$ and $m_i = p \cdot m_{i-1} + 1$. Each node corresponds to a module. Each super-edge between two modules is instantiated as a 1-parallelized all-to-all interconnection. Following this way, a level-L DCell ($L \geq 2$) can be constructed by repeating the second and the third steps (i.e., connection design and inter-module design) L times. The finalization step is neglected as no CNs should be attached.

The design of a level-2 DCell with 2-port switches under the design framework is illustrated in Figure 2.64.

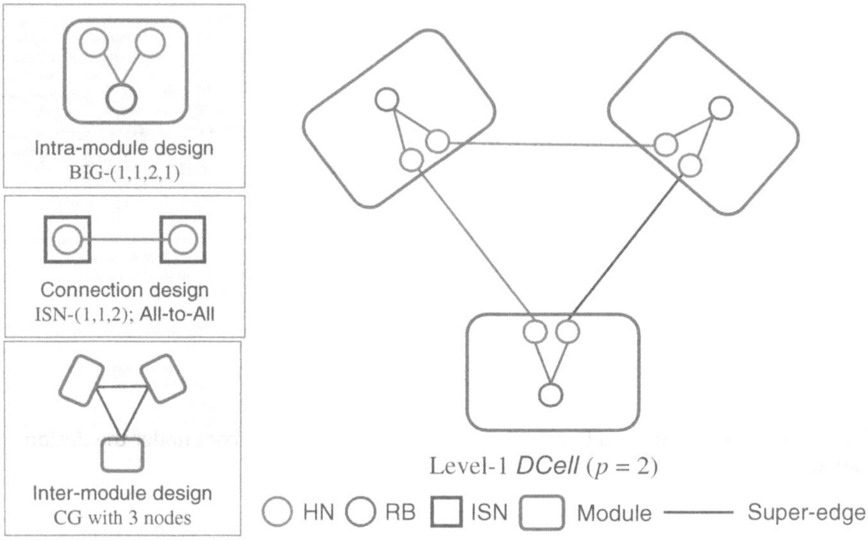

Figure 2.63 The design of a level-1 DCell with 2-port switches under the design framework.

2.5.8 BCube

BCube is a server-centric topology. In its design, HNs correspond to servers while RNs correspond to switches. Next, we demonstrate the design of a level-1 BCube with p-port switches step by step.

- **Intra-module design**: The internal graph (i.e., a level-0 BCube) is a BIG-$(1, 1, p, 1)$ in which element nodes and block nodes are regarded as HNs and RNs, respectively.
- **Connection design**: In each module, all HNs are selected as dock nodes. Each dock node serves as an intermediate super-node, i.e., an ISN-$(1, 1, 2)$. Then dock node interconnection is constructed under the bipartite mode. The composite-edge among any p ISNs through an RN cluster is designed as a TD-$(p, 1, 1)$ interconnection.
- **Inter-module design**: The skeleton graph is a BIG-$(1, p, p, p)$ where element nodes and block nodes correspond to modules and RN clusters, respectively. For any p modules, they are connected with a common RN cluster via p super-edges. Such super-edges are instantiated as a 1-parallelized TD-$(p, 1, 1)$ interconnection.
- **Finalization**: This step is omitted as the topology contains no CNs.

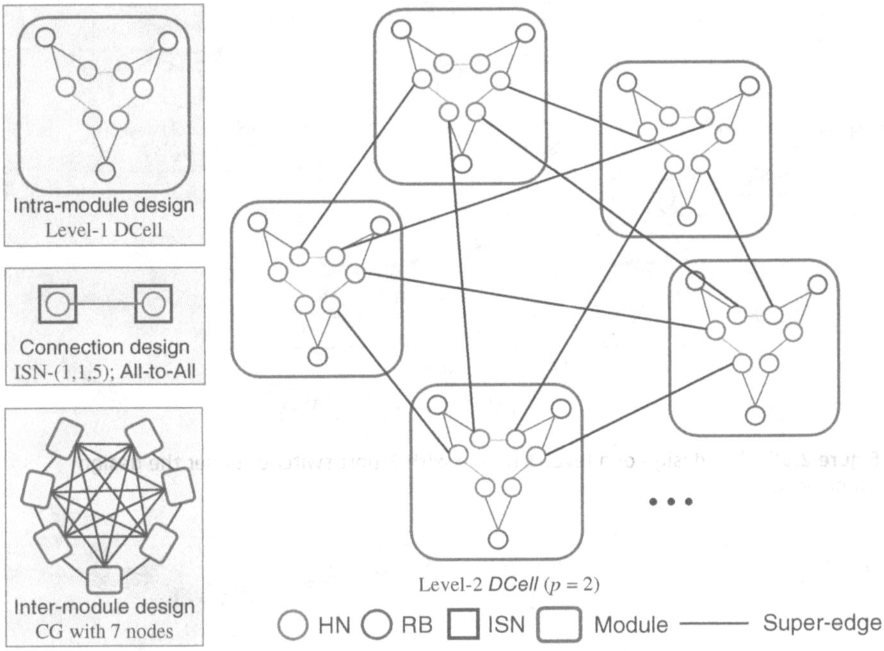

Figure 2.64 The design of a level-2 DCell with 2-port switches under the design framework.

In Figure 2.65, we present the design of a level-1 BCube with 3-port switches under the design framework.

The design of higher-level BCubes: In general, the design of a level-i BCube ($i \geq 2$) takes level-($i - 1$) BCubes as modules. As for the connection design, each HN in a module serves as an ISN-(1, 1, 2i) and a TD-(p, 1, 1) interconnection serves as a composite-edge between such two ISNs. Regarding the inter-module design, the skeleton graph of such a topology is a BIG-(1, p^i, p, p^i). For any p modules, they are connected with a common RN cluster via p super-edges. Such super-edges are instantiated as a 1-parallelized TD-(p, 1, 1) interconnection between the p modules. In this way, a level-L BCube ($L \geq 2$) can be constructed by iteratively applying the second and third steps (i.e., connection design and inter-module design) L times. The finalization step is neglected as no CNs should be attached.

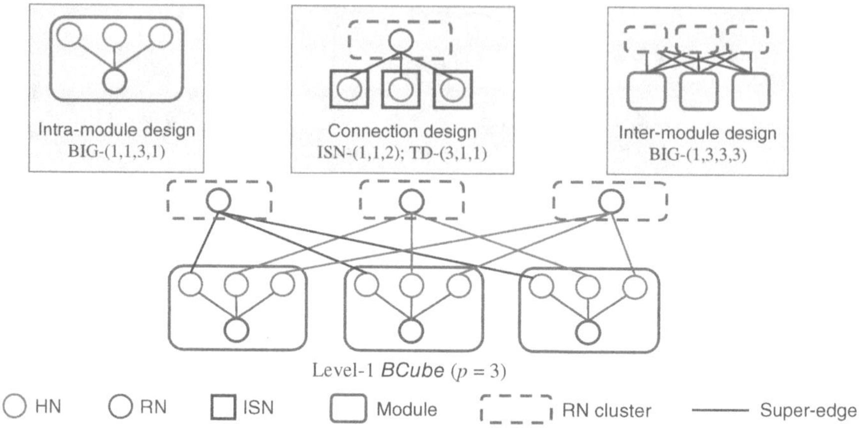

Figure 2.65 The design of a level-1 BCube with 3-port switches under the design framework.

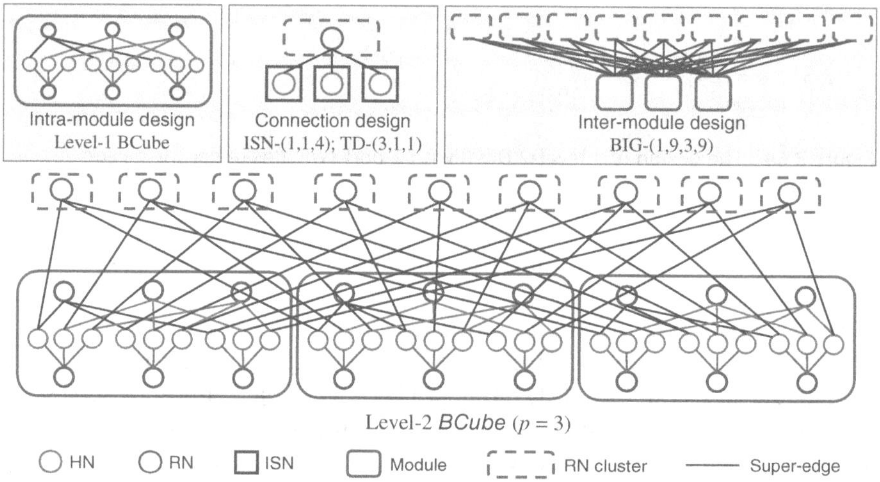

Figure 2.66 The design of a level-2 BCube with 3-port switches under the design framework.

The design of a level-2 BCube with 3-port switches under the design framework is illustrated in Figure 2.66.

3

Evolution of Classical Topologies

Guided by the design procedure, network topologies can be derived under different choices of design options and parameter settings in each step. Such topology designs would lead to various performance trade-offs that deserve detailed investigations. Depending on whether topology design should be subject to pre-given constraints, the analysis can be conducted under two general design schemes in practice. One scheme is to start from a family of existing topologies that share common design options in some design step (e.g., intra-module design based on the same internal graph), then fix such options and decide the options for the other design steps. This scheme is well suited for scenarios where the design goal is to improve existing topologies in an incremental fashion (e.g., to expand or modify an operating data center network). The other scheme is to conduct a *clean-slate* design by fully exploring feasible combinations of design options over the topology design space. This scheme applies to the early-stage blueprint generation before a network is constructed.

In this chapter, we focus on analysis under the first scheme over different topology families. Particularly, for ease of illustration, we present an analysis on the families of network topologies that follow the design procedure with one iteration.[1] In each family, the topologies share the same design options in partial design steps. Our goal is to proceed the analysis with such fixed options and investigate *feasible* combinations of design options in the remaining design steps. In this process, we illustrate the impacts of different options and parameters on performance metrics, justify the performance trade-offs achieved by existing topologies, and quest for their possible modifications. For each topology family, such an analysis procedure is called the family's *evolution*.

In the following sections, three topology families are considered, including the family of switch-centric tree-like topologies (Al-Fares et al. 2008; Walraed-Sullivan et al. 2013; Liu et al. 2013), the family of switch-centric HyperX-like topologies

1 With similar analysis, our results can be extended to the topology designs that follow the design procedure with multiple iterations.

(Ahn et al. 2009; Cheng and Chuang 1994; Kim et al. 2007), and the family of server-centric topologies (Guo et al. 2008, 2009). In Section 3.1, we analyze the evolutions of switch-centric topology families, including the tree-like family (see Section 3.1.1) and the HyperX-like family (see Section 3.1.2). For these two families, the internal graphs for intra-module design are fixed as particular kinds of block incidence graph (BIG) and complete graph (CG), respectively. Then in Section 3.2, we switch the focus to the family of server-centric topologies. In this family, the intra-module design is commonly conducted using BIGs with parameter settings different from tree-like family, while connection design is performed with fixed means of dock node selection.

3.1 Evolution of Switch-centric Topologies

In this section, we consider two families of *switch-centric* topologies that are widely adopted in existing large-scale computer networks, namely *tree-like* topologies and *HyperX-like* topologies. In tree-like topologies such as Fat-Tree (Al-Fares et al. 2008) and Aspen Tree (Walraed-Sullivan et al. 2013), BIGs are adopted as the internal graphs for intra-module design. In HyperX-like topologies such as YARC (Scott et al. 2006) and hypercube (Cheng and Chuang 1994), CGs are commonly employed in the intra-module design. With fixed options of internal graphs, we illuminate the design space and the evolutions of the two topology families in the following subsections, respectively. At the end of each subsection, we provide performance comparisons among topology designs that are derived in the evolution of the corresponding topology family.

3.1.1 Evolution of Tree-like Topologies

For tree-like topologies, the port number of each switch is denoted by $p = 2q$, where q is an integer greater than one (i.e., p is an even number greater than two). Throughout the evolution of the tree-like topology family, the intra-module design and part of the connection design remain unchanged. Typically, the design procedure of tree-like topologies can be described as follows.

- Step 1: *Intra-module design.* A BIG-$(1, q, q, q)$ is employed as the internal graph for each module, in which element nodes and block nodes are instantiated as GNs and RNs, respectively. An example with the internal graph fixed as a BIG-$(1, q, q, q)$ when $q = 2$ is presented in Figure 3.1.
- Step 2: *Connection design.*
  - *Dock node selection.* In each module, only RNs are selected as dock nodes. For each dock node, its inner-degree and outer-degree are both equal to q. Such dock nodes are grouped as ISNs-(x, w, l).

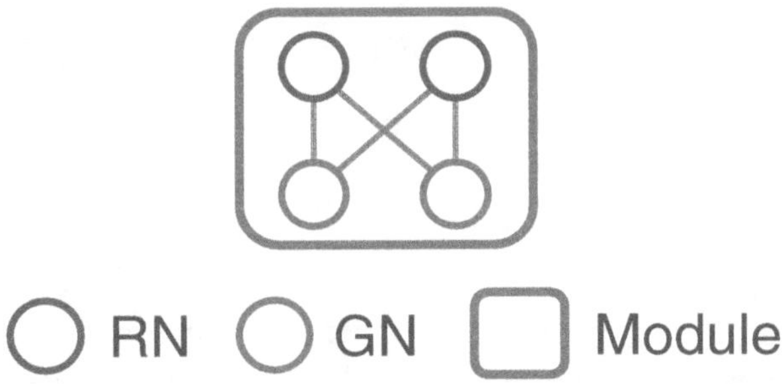

Figure 3.1 An example with the internal graph as a BIG-$(1, q, q, q)$ when $q = 2$.

  - *Dock node interconnection.* Dock nodes in distinct modules are all connected under the bipartite mode. Specifically, every n ISNs-(x, w, l) are interconnected by a composite edge that is constructed as a TD-(n, μ, x) interconnection through a single RN cluster.
- Step 3: *Inter-module design.* With the connection design, a BIG with copies is employed as the skeleton graph to interconnect modules.
- Step 4: *Finalization.* Each GN in the modules is directly connected with q CNs.

Based on the design procedure, tree-like topologies can be derived and modified by adopting different parameter settings in the second and the third steps.[2] A roadmap of the evolution of tree-like topologies is presented in Figure 3.2.

To provide a guideline for the parameter determination, we discuss the impact of parameter settings on performance metrics as follows. First, the latency and the intra-module throughput are investigated by the following theorem, whose proof is provided in Section A.3.1.

Theorem 3.1 *Following the design procedure with one iteration, tree-like topologies have a constant latency of six and a constant intra-module throughput of one.*

Next, we discuss the design options and derive feasible settings for the selection and the interconnection of dock nodes, respectively.

- *Dock node selection.* Each module employs a BIG-$(1, q, q, q)$ as the internal graph. The graph consists of q RNs (block nodes). All of them are selected as dock nodes and partitioned into q/x ISNs-(x, w, l), where x is a factor of q with $1 \leq x \leq q$. Note that such a BIG-$(1, q, q, q)$ is essentially a complete bipartite graph (CBG) and each GN is adjacent to all RNs. Then given an ISN-(x, w, l), for each GN in the same module, all of the x RNs in the ISN are

2 A few examples can be found in Sections 2.5.1–2.5.3.

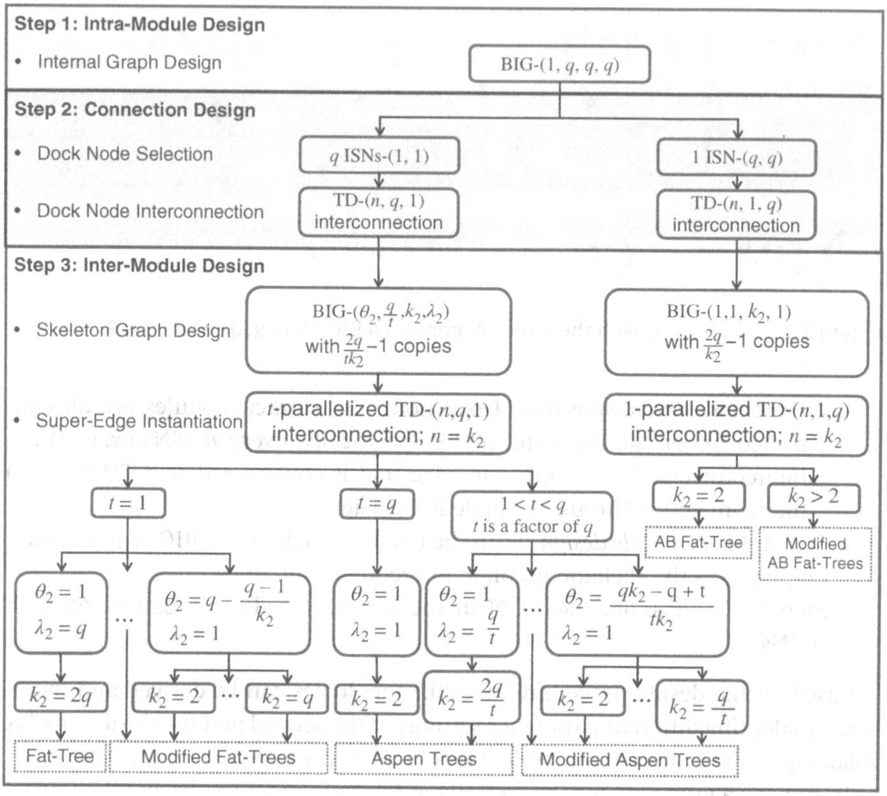

Figure 3.2 A roadmap of the evolution of tree-like topologies. The finalization step is omitted since in tree-like topologies, each GN in the modules is directly connected with q CNs (servers).

within a distance of *one* from it. By Definition 2.10, the ISN-(x, w, l) satisfies that $w = x$ and $l = 1$. Figure 3.3 shows the dock node selections in two extreme cases where $x = 1$ and $x = q$, respectively, given the internal graph as a BIG-(1, 2, 2, 2) with $q = 2$.

- *Dock node interconnection*. Based on the above dock node selection, every n ISNs-($x, x, 1$) (each as an ISN-(x, x)) are interconnected through a TD-(n, μ, x) interconnection. Under the TD-(n, μ, x) interconnection, for any two ISNs-(x, x) in distinct modules, each dock node in one ISN-(x, x) is connected with all x dock nodes in the other ISN-(x, x) through a single RN cluster. In this way, the two ISNs can communicate with each other unless *all* of the x dock nodes in one ISN fail simultaneously. Specifically, consider the case where the two ISNs belong to two distinct modules. In this case, to

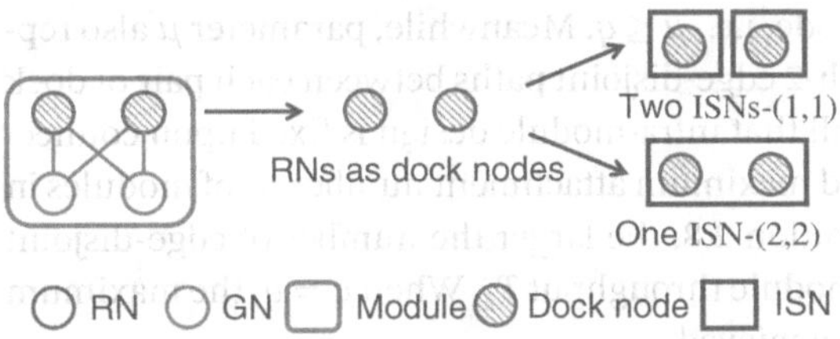

Figure 3.3 An illustration of dock node selection. In the given module, The RNs in the module are both selected as dock nodes. Accordingly, there are two ways of dock node selection. In one way, the two RNs are grouped as two ISNs-(1, 1). In the other way, they are grouped as one ISN-(2, 2).

achieve efficient communications, GNs in the two modules can communicate through any pair of dock nodes in such two ISNs over the shortest path between them. Each path equally has a length of two. Under this setting, any two GNs in the two modules can communicate with each other unless *all* of the x dock nodes in one ISN fail simultaneously. This implies that the minimum number of dock nodes in each module whose failures disconnect communications between any pair of GNs equals x. By Definition 2.9, this indicates that the fault-tolerance of tree-like topologies is equal to x. In other words, the larger the value of parameter x, the higher the fault-tolerance of a tree-like topology.

Recall that x is a factor of q with $1 \leq x \leq q$. Particularly, when $x = 1$, there exists a TD-(n, μ, x) interconnection for any $n \geq 2$ and $\mu \geq 1$. However, when $1 < x \leq q$, such a statement does not necessarily hold. In other words, given the value of x, the value selections of parameters μ and n may not guarantee the existence of TD-(n, μ, x) interconnection with a high fault-tolerance. In this monograph, we focus on two extreme cases with $x = 1$ and $x = q$, respectively. As for the cases in which $1 < x < q$, it is left to interested readers to conduct further analysis.

Next, given that $x = 1$ and $x = q$, we carry out two branches of analysis (see Figure 3.2 for an overview) in Sections 3.1.1.1 and 3.1.1.2, respectively.

3.1.1.1 Branch 1: q ISNs-$(1, 1)$ in Each Module

Parameter settings of connection design. Along with this branch (i.e., given $x = 1$), in connection design, dock node selection in each module is fixed as q ISNs-$(1, 1)$ in each module ($x = 1$). Regarding dock node interconnection, the parameter settings of the corresponding TD-$(n, \mu, 1)$ interconnection are discussed as follows.

- *Parameter* μ. Under a TD-$(n, \mu, 1)$ interconnection, each dock node is adjacent to μ RNs in each RN cluster. To this end, the value of μ is subject to the outer-degree of each dock node, i.e., $\mu \leq q$. Meanwhile, parameter μ also represents the number of length-2 edge-disjoint paths between each pair of dock nodes in distinct ISNs. Recall that intra-module design is fixed upon connection design and hence a fixed maximum attachment number B_2 of modules in the topology. Then by Definition 2.8, the larger the number of edge-disjoint paths, the higher the inter-module throughput T_2. When $\mu = q$, the maximum inter-module throughput is achieved.

- *Parameter n*. In general, for any integer $n \geq 2$, there exists a TD-$(n, \mu, 1)$ interconnection in which n ISNs are interconnected through an RN cluster. Under such an interconnection, the RN cluster contains μ RNs. For each RN in the RN cluster, it is adjacent to n dock nodes in distinct ISNs. Since the degree of each RN is p $(p = 2q)$, we always have $n \leq 2q$.

Given the above analysis, we focus on the case in which a TD-$(n, \mu, 1)$ interconnection with $\mu = q$ and $2 \leq n \leq 2q$ is employed as the composite edge for dock node interconnection. Note that in such a composite edge, n ISNs-$(1, 1)$ are connected through a single RN cluster with q RNs. Figure 3.4 shows an example of a TD-$(n, q, 1)$ interconnection with $n = 2$ and $q = 2$.

Parameter settings of inter-module design. Since the dock nodes are interconnected under the bipartite mode, for inter-module design, a BIG-$(\theta_2, r_2, k_2, \lambda_2)$ with c_2 copies $(c_2 \geq 0)$ is employed as the skeleton graph. In such a graph, element nodes and block nodes are instantiated as modules and RN clusters, respectively. Particularly, using the duplicating operation, modules are partitioned into $c_2 + 1$ groups, such that each group of modules and all RN clusters form a BIG-$(\theta_2, r_2, k_2, \lambda_2)$. An example with the skeleton graph as a BIG-$(1, 1, 2, 1)$ with one copy (i.e., $c_2 = 1$) is presented in Figure 3.5a.

For a group of modules, each RN cluster is directly connected with k_2 modules in the group via a number of k_2 super-edges. Such super-edges are constructed using a t-parallelized TD-$(n, q, 1)$ interconnection where $n = k_2$. In Figure 3.5b, we show an example in which super-edges in a skeleton graph (see Figure 3.5a) are constructed using a 2-parallelized TD-$(2, 2, 1)$ interconnection (i.e., $t = 2$, $q = 2$, and $n = k_2 = 2$).

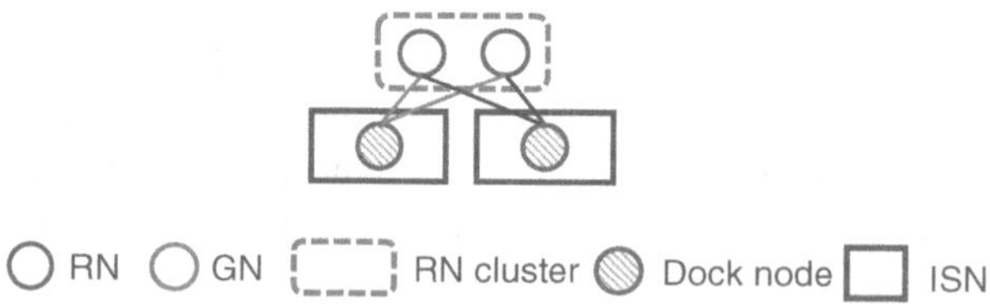

Figure 3.4 An example of a TD-$(n, q, 1)$ interconnection with $n = 2$ and $q = 2$. Under this interconnection, two ISNs-$(1, 1)$ are connected through a single RN cluster with two RNs.

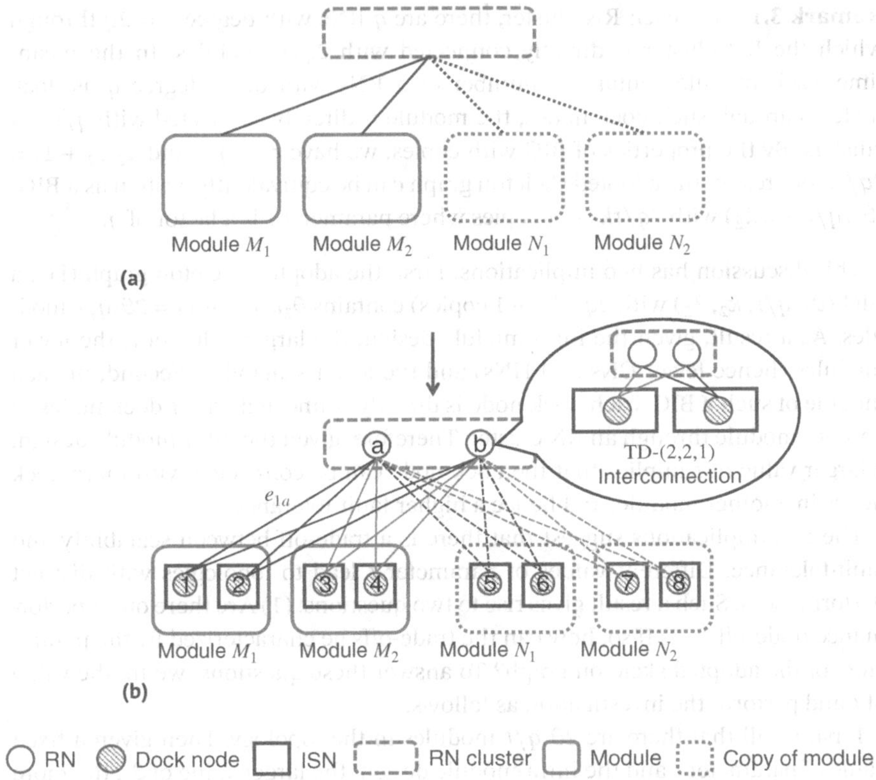

Figure 3.5 An illustration of skeleton graph design and super-edge instantiation. (a) A BIG-(1, 1, 2, 1) with one copy (i.e., $c_2 = 1$) as the skeleton graph. Such a skeleton graph contains two groups of modules. One group consists of modules M_1 and M_2. The other group consists of modules N_1 and N_2. (b) The two super-edges between an RN cluster and a group of modules are instantiated as a 2-parallelized TD-(2, 2, 1) interconnection. To be specific, the super-edge between module M_1 and an RN cluster consists of four edges $\{e_{1a}, e_{1b}, e_{2a}, e_{2b}\}$, while the super-edge between module M_2 and the RN cluster consists of another four edges $\{e_{3a}, e_{3b}, e_{4a}, e_{4b}\}$. These two super-edges form a 2-parallelized TD-(2, 2, 1) interconnection. The connection pattern with the RN cluster for module N_1 (resp. module N_2) is the same as that for module M_1 (resp. module M_2).

Under a t-parallelized TD-(n, q, 1) interconnection, each super-edge between an RN cluster and a module consists of tq edges. Each edge is incident to an RN in the RN cluster and a dock node in the module. In other words, each super-edge occupies tq degrees of RNs in the RN cluster and tq outer-degrees of dock nodes in its corresponding module. Accordingly, we have the following remark.

Remark 3.1 For each RN cluster, there are q RNs with degree $p = 2q$ through which the RN cluster is directly connected with $2q/t$ modules. In the meantime, each module contains a number of q RNs with outer-degree q as dock nodes. Through such dock nodes, the module is directly connected with q/t RN clusters. By the properties of BIG with copies, we have $r_2 = q/t$ and $k_2(c_2 + 1) = 2q/t$. As a result, the adopted skeleton graph can be equivalently written as a BIG-$(\theta_2, q/t, k_2, \lambda_2)$ with $2q/tk_2 - 1$ copies where parameter t is a factor of q.

This discussion has two implications. First, the adopted skeleton graph (i.e., a BIG-$(\theta_2, q/t, k_2, \lambda_2)$ with $2q/tk_2 - 1$ copies) contains $\theta_2 k_2(c_2 + 1) = 2\theta_2 q/t$ modules. As a result, given the intra-module design, the larger value of t, the fewer modules (hence fewer CNs and HNs) and the lower scalability. Second, in each module of such a BIG, each dock node is directly connected with t dock nodes in another module through an RN cluster. Therefore, given the intra-module design, a larger value of t implies that the dock node can be connected with more dock nodes in another module, and hence a higher fault-tolerance.

The two implications suggest that there is a trade-off between scalability and fault-tolerance. Different values of parameter t lead to topologies with distinct performances. Such a result gives rise to two questions: (1) Are there other performance trade-offs? (2) If so, how can the trade-offs be characterized by the parameters of the adopted skeleton graph? To answer these questions, we fix the value of t and perform the investigation as follows.

First, recall that there are $2\theta_2 q/t$ modules in the topology. Then given a fixed value of parameter t and the intra-module design, the larger value of θ_2, the more modules (hence more CNs and HNs) and the higher scalability.

Second, given the skeleton graph as a BIG-$(\theta_2, q/t, k_2, \lambda_2)$ with $2q/tk_2 - 1$ copies, there are at least λ_2 length-2 paths between any two modules. Each path goes through one RN cluster as the communication relay. Therefore, given the intra-module design and connection design, the larger value of λ_2, the more paths and the higher inter-module throughput. By Lemma 1.6, we have

$$\lambda_2 = \frac{r_2(k_2 - 1)}{\theta_2 k_2 - 1} = \frac{q(k_2 - 1)}{t(\theta_2 k_2 - 1)}. \tag{3.1}$$

Equation (3.1) implies that given fixed values of parameters k_2 and t, the value of parameter λ_2 (i.e., the number of length-2 paths between any two modules) is negatively correlated with the value of parameter θ_2 (positively proportional to the total number of modules). In other words, given the intra-module design and connection design, there is a trade-off between inter-module throughput and scalability. Below we consider three cases, including the case where $t = 1$, the case where $t = q$, and the case where t is a factor of q and $1 < t < q$.

Case 1: $t = 1$.

In this case, equation (3.1) can be equivalently rewritten as

$$\lambda_2 = \frac{q(k_2 - 1)}{\theta_2 k_2 - 1}. \tag{3.2}$$

Recall that $\theta_2, \lambda_2 \geq 1$. Based on equation (3.2), we analyze three sub-cases, following which a number of tree-like topologies with different performance trade-offs can be induced, as illustrated in Figure 3.6.

- Sub-case 1: $\theta_2 = 1$ and $\lambda_2 = q$, in which a higher throughput is achieved at the cost of a lower scalability.
- Sub-case 2: $\theta_2 = q - (q-1)/k_2$ and $\lambda_2 = 1$, in which a higher scalability is achieved at the cost of a lower throughput.
- Sub-case 3: $1 < \theta_2 < q - (q-1)/k_2$ and $1 < \lambda_2 < q$.

Further discussions are given as follows.

Sub-case 1: $\theta_2 = 1$ and $\lambda_2 = q$. In this sub-case, a BIG-$(1, q, k_2, q)$ with $2q/k_2 - 1$ copies is employed as the skeleton graph. Such a graph is isomorphic to a BIG-$(1, q, 2q, q)$ with no copies for any value k_2. Hence, the skeleton graph can be equivalently regarded as a BIG-$(1, q, 2q, q)$. Under such settings, given the value of parameter q with $p = 2q$, the adopted options for the design steps are given as follows.

- *Intra-module design.* The internal graph is a BIG-$(1, q, q, q)$ with no copies.
- *Connection design.* In each module, all RNs are selected as dock nodes. Each dock node forms an intermediate super-node, i.e., an ISN-$(1, 1)$. The dock node interconnection is constructed under the bipartite mode. To be specific,

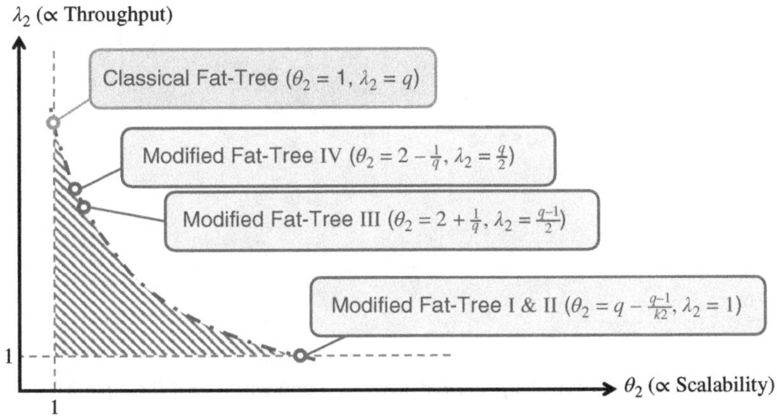

Figure 3.6 The trade-off between scalability and throughput in terms of parameters θ_2 and λ_2.

each composite-edge among $2q$ ISNs is designed as a TD-$(2q, q, 1)$ interconnection through an RN cluster.

- *Inter-module design.* The skeleton graph is given as a BIG-$(1, q, 2q, q)$. For all of the $2q$ modules, they are connected with a common RN cluster via $2q$ super-edges. Such super-edges form a 1-parallelized TD-$(2q, q, 1)$ interconnection.
- *Finalization.* Each GN is directly connected with q CNs.

In fact, the resulting topology based on these designs is a p-ary Fat-Tree topology. For ease of understanding, we show an instance of such a topology with $p = 4$ and $q = 2$ in Figure 3.7. The performances of the topology are analyzed as follows.[3]

- *Scalability.* Given the value of parameter q, the skeleton graph is a BIG-$(1, q, 2q, q)$ that contains $2q$ modules. For each module, its internal graph is fixed as a BIG-$(1, q, q, q)$ that contains q GNs. Each GN is directly connected with q CNs. Hence, the scalability S of the topology (i.e., the total number of attached CNs in the topology) is given by

$$S = 2q \cdot q \cdot q = 2q^3. \tag{3.3}$$

- *Inter-module throughput.* Provided that the skeleton graph is designed as a BIG-$(1, q, 2q, q)$, each pair of modules are connected through q length-2

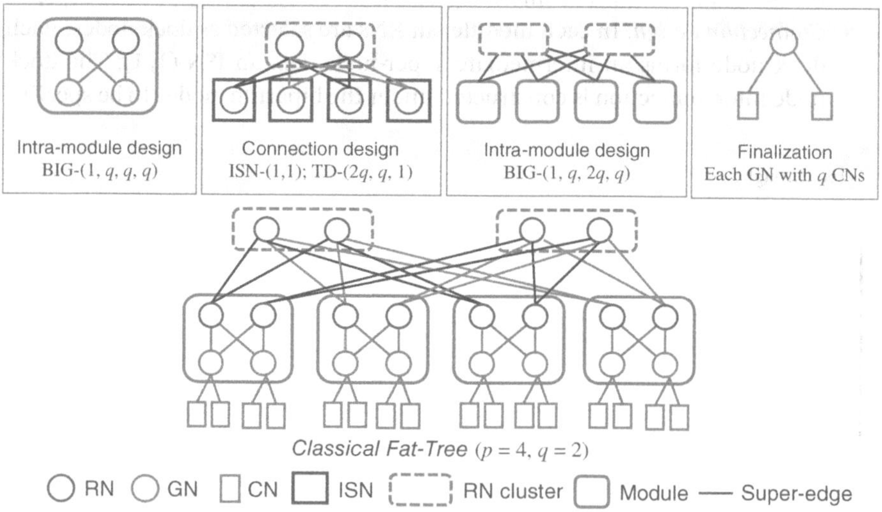

Figure 3.7 The complete design of a p-ary Fat-Tree topology with $p = 4$ (i.e., $q = 2$).

3 By Theorem 3.1, for tree-like topology designs with one iteration, the latency L and the intra-module throughput T_1 are always equal to six and one, respectively. Unless otherwise stated, they are omitted in the subsequent analysis in the rest of this subsection.

paths over super-edges. By the property of 1-parallelized TD-$(2q, q, 1)$ interconnection, each length-2 path of super-edges between two modules consists of q edge-disjoint paths between dock nodes in such two modules. Hence, the minimum number of edge-disjoint paths A_2 between the dock nodes in any two modules is $q \cdot q = q^2$. Meanwhile, the internal graph of each module is fixed as a BIG-$(1, q, q, q)$ that contains q GNs. Each GN is directly connected with q CNs. Hence, the attachment number B_2 of each module is $q \cdot q = q^2$. By Definition 2.8, the inter-module throughput is given by

$$T_2 = \frac{A_2}{B_2} = \frac{q^2}{q^2} = 1. \tag{3.4}$$

- *Fault-tolerance.* Let N denote the number of distinct pairs of modules. Note that each group of modules are connected with a common RN cluster via $2q$ super-edges using a 1-parallelized TD-$(2q, q, 1)$ interconnection. Therefore, for each pair of modules, any dock node in one module is connected with all dock nodes of one ISN (i.e., $t = 1$) in the other module. Since each ISN contains one dock node, for the i-th pair of modules, each dock node in one module is connected with one dock node in the other module through an RN cluster. Hence, given a dock node in module M_i^1, if the dock node in module M_i^2 fails, the communication between some pair of GNs in the two modules would break down (i.e., $f_i = 1$). By Definition 2.9, the fault-tolerance F of the topology is equal to

$$F = \frac{1}{N}\sum_{i=1}^{N} f_i = \frac{1}{N}\sum_{i=1}^{N} 1 = 1. \tag{3.5}$$

Sub-case 2: $\theta_2 = q - (q-1)/k_2$ and $\lambda_2 = 1$. In this sub-case, a BIG-$(q - (q-1)/k_2, q, k_2, 1)$ with c_2 copies ($c_2 = 2q/k_2 - 1$) is employed as the skeleton graph of the topology.

Next, let's consider the value of the undetermined parameter k_2 of the skeleton graph. First, recall that the number of copies c_2 is a nonnegative integer that equals $2q/k_2 - 1$. Then parameter k_2 needs to be a factor of $2q$ with $k_2 \leq 2q$. Second, by Definition 1.23, it follows that $k_2 \geq 2$ and thus $\theta_2 = q - (q-1)/k_2 \geq q - (q-1)/2 = (q+1)/2 > 1$. Third, given that $\theta_2 > 1$, a BIG-$(\theta_2, r_2, k_2, \lambda_2)$ is an incidence graph of a BIBD-$(\theta_2 k_2, \theta_2 r_2, r_2, k_2, \lambda_2)$ that satisfies $k_2 \leq r_2$ (cf. Lemma 1.1 in Section 1.6.2). In other words, $k_2 \leq r_2$ is a necessary condition of the existence of BIG-$(\theta_2, r_2, k_2, \lambda_2)$. Based on the discussion in Section 2.4.3.1, such a necessary condition also holds for BIGs with copies. Hence, given a BIG-$(q - (q-1)/k_2, q, k_2, 1)$ with c_2 copies, we have $k_2 \leq q$. In summary, the value of parameter k_2 is supposed to be a factor of $2q$ with $2 \leq k_2 \leq q$.

Given these settings, we focus on two extreme conditions where $k_2 = 2$ and $k_2 = q$, respectively.[4]

- *Condition 1: $k_2 = 2$.* In this case, the skeleton graph is designed as a BIG-$((q+1)/2, q, 2, 1)$ with $q-1$ copies. Under such settings, given the value of parameter q, the adopted options for the design steps are given as follows.
  - *Intra-module design.* The internal graph for each module is fixed as a BIG-$(1, q, q, q)$.
  - *Connection design.* In each module, all RNs are selected as dock nodes. Each dock node serves as an intermediate super-node, i.e., an ISN-(1, 1). Then the dock node interconnection is constructed under the bipartite mode. To be specific, each composite-edge among two ISNs is designed as a TD-$(2, q, 1)$ interconnection through an RN cluster.
  - *Inter-module design.* The skeleton graph is given as a BIG-$((q+1)/2, q, 2, 1)$ with $q-1$ copies. For each copy of modules, they are connected with a common RN cluster via two super-edges. Such two super-edges are constructed using a 1-parallelized TD-$(2, q, 1)$ interconnection.
  - *Finalization.* Each GN is directly connected with q CNs.

We call the resulting topology based on these designs a *modified Fat-Tree I*. An instance of such a topology with $p = 4$ and $q = 2$ is presented in Figure 3.8. The performances of the topology are analyzed as follows.

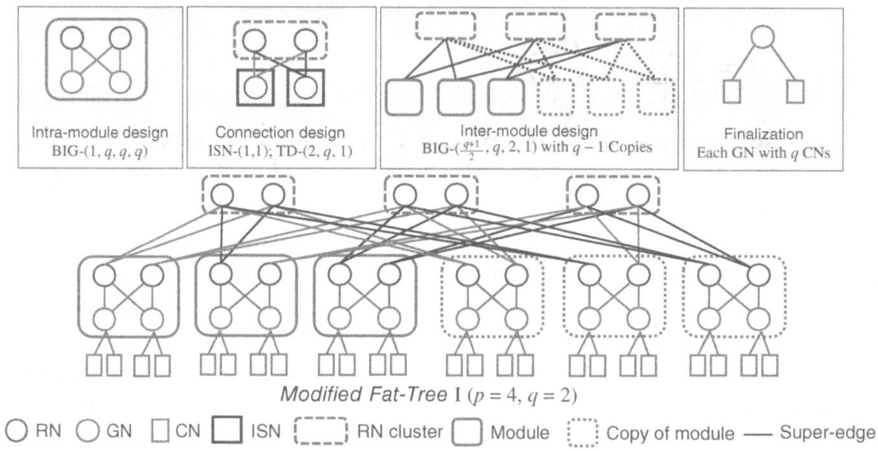

Figure 3.8 The complete design of a modified Fat-Tree I topology with $p = 4$ and $q = 2$.

---

4 For cases with $2 < k_2 < q$, further analysis is required on the value selection of parameter q to guarantee the existence of BIG-$(q - (q-1)/k_2, q, k_2, 1)$ with a high fault-tolerance. Such analysis can be considered as an interesting direction for future work.

- *Scalability.* Given the value of parameter q, the skeleton graph is fixed as a BIG-($(q+1)/2, q, 2, 1$) with $q-1$ copies that contain $2 \cdot (q+1)/2 \cdot q = q^2 + q$ modules. For each module, its internal graph is fixed as a BIG-($1, q, q, q$) that contains q GNs. Each GN is directly connected with q CNs. Hence, the scalability S of the topology (i.e., the total number of attached CNs in the topology) is given by

$$S = (q^2 + q) \cdot q \cdot q = q^4 + q^3. \tag{3.6}$$

- *Inter-module throughput.* Recall that the skeleton graph is fixed as a BIG-($(q+1)/2, q, 2, 1$) with $q-1$ copies. In such a graph, each pair of modules are connected with at least one length-2 path over super-edges under the bipartite mode. By the property of 1-parallelized TD-($2, q, 1$) interconnection, each length-2 path of super-edges between two modules consists of q edge-disjoint paths. Hence, the minimum number of edge-disjoint paths A_2 between any two modules is $1 \cdot q = q$. Meanwhile, the internal graph of each module is fixed as a BIG-($1, q, q, q$) that contains q GNs. Each GN is directly connected with q CNs. Hence, the attachment number B_2 of each module is $q \cdot q = q^2$. By Definition 2.8, the inter-module throughput is given by

$$T_2 = \frac{A_2}{B_2} = \frac{q}{q^2} = \frac{1}{q}. \tag{3.7}$$

- *Fault-tolerance.* Let N denote the number of distinct ordered pairs of modules. In the topology, each group of modules are connected with a common RN cluster via $2q$ super-edges using a 1-parallelized TD-($2q, q, 1$) interconnection. Therefore, for each pair of modules, any dock node in one module is connected with all dock nodes of one ISN (i.e., $t = 1$) in the other module. Since each ISN contains one dock node, for the i-th pair of modules, each dock node in one module is connected with one dock node in the other module through an RN cluster. Hence, given a dock node in module M_i^1, if the dock node in module M_i^2 fails, the communication between some pair of GNs in the two modules would break down (i.e., $f_i = 1$). By Definition 2.9, the fault-tolerance F of the topology is equal to

$$F = \frac{1}{N} \sum_{i=1}^{N} f_i = \frac{1}{N} \sum_{i=1}^{N} 1 = 1. \tag{3.8}$$

- *Condition 2:* $k_2 = q$. The skeleton graph of the topology is given by a BIG-($q-1+1/q, q, q, 1$) with one copy. By Lemma 1.6, such a skeleton graph exists if $q = 2$ or $q-1$ is a prime power. Meanwhile, it can be induced by a projective BIBD-($l^2+l+1, l^2+l+1, l+1, l+1, 1$) with $q = l+1$. Under such settings, given the value of parameter q, the adopted options of the design steps are given as follows.

- *Intra-module design*. The internal graph is a BIG-$(1, q, q, q)$.
- *Connection design*. In each module, all RNs are selected as dock nodes. Each dock node forms an intermediate super-node, i.e., an ISN-$(1, 1)$. Then the dock node interconnection is constructed under the bipartite mode. To be specific, each composite-edge among q ISNs is designed as a TD-$(q, q, 1)$ interconnection through an RN cluster.
- *Inter-module design*. The skeleton graph is given as a BIG-$(q-1+1/q, q, q, 1)$ with one copy. For each copy of modules, they are connected with a common RN cluster via q super-edges. Such q super-edges are constructed using a 1-parallelized TD-$(q, q, 1)$ interconnection.
- *Finalization*. Each GN is directly connected with q CNs.

We call the resulting topology based on these designs a *modified Fat-Tree II*. An instance of such a topology with $p = 4$ and $q = 2$ is presented in Figure 3.9. The performances of the topology are analyzed as follows.

- *Scalability*. Given the value of parameter q, the skeleton graph is fixed as a BIG-$(q-1+1/q, q, q, 1)$ with one copy that contains $2 \cdot (q-1+1/q) \cdot q = 2q^2 - 2q + 2$ modules. For each module, its internal graph is fixed as a BIG-$(1, q, q, q)$ that contains q GNs. Each GN is directly connected with q CNs. Hence, the scalability S of the topology (i.e., the total number of CNs contained in the topology) is given by

$$S = (2q^2 - 2q + 2) \cdot q \cdot q = 2q^4 - 2q^3 + 2q^2. \tag{3.9}$$

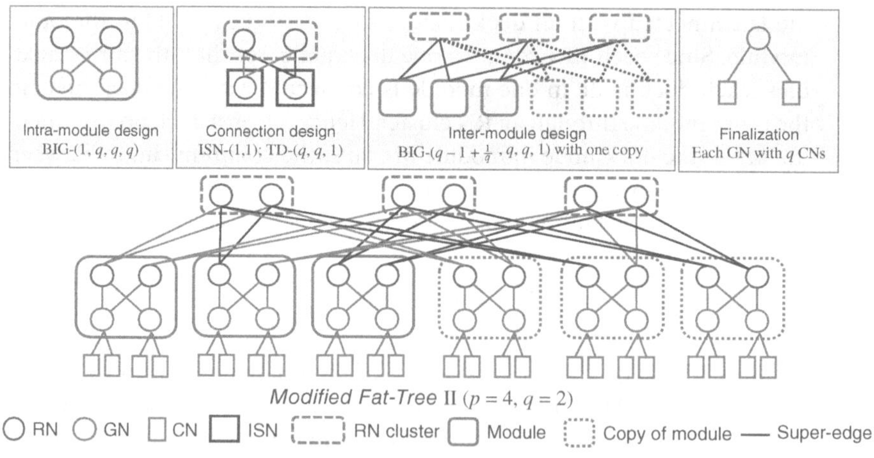

Figure 3.9 The complete design of a modified Fat-Tree II topology with $p = 4$ and $q = 2$.

- *Inter-module throughput.* Recall that the skeleton graph of the topology is fixed as a BIG-$(q-1+1/q, q, q, 1)$ with one copy. For each pair of modules, they are connected with at least one length-2 path over super-edges under the bipartite mode. By the property of 1-parallelized TD-$(q, q, 1)$ interconnection, each length-2 path over super-edges between two modules consists of q edge-disjoint paths. Hence, the minimum number of edge-disjoint paths A_2 between any two modules is $1 \cdot q = q$. Meanwhile, the internal graph of each module is fixed as a BIG-$(1, q, q, q)$ that contains q GNs. For each GN, it is directly connected with q CNs. Hence, the attachment number B_2 of each module is $q \cdot q = q^2$. By Definition 2.8, the inter-module throughput T_2 of the topology is equal to

$$T_2 = \frac{A_2}{B_2} = \frac{q}{q^2} = \frac{1}{q}. \tag{3.10}$$

- *Fault-tolerance.* Let N denote the number of distinct pairs of modules. Note that each group of modules are connected with a common RN cluster via $2q$ super-edges by using a 1-parallelized TD-$(2q, q, 1)$ interconnection. Therefore, for each pair of modules, any dock node in one module is connected with all dock nodes of one ISN (i.e., $t = 1$) in the other module. Since each ISN contains one dock node, for the i-th pair of modules, each dock node in one module is connected with one dock node in the other module through an RN cluster. Hence, given a dock node in module M_i^1, if the dock node in module M_i^2 fails, the communication between some pair of GNs in the two modules would break down (i.e., $f_i = 1$). By Definition 2.9, the fault-tolerance F of the topology is

$$F = \frac{1}{N}\sum_{i=1}^{N} f_i = \frac{1}{N}\sum_{i=1}^{N} 1 = 1. \tag{3.11}$$

Sub-case 3: $1 < \theta_2 < q - (q-1)/k_2$ and $1 < \lambda_2 < q$. In this sub-case, there exist feasible settings of parameters θ_2 and λ_2 to guarantee the existence of BIGs as the skeleton graph. Two examples are illustrated as follows.

- *Example 1.* Provided that q is an odd number with $3 \le q \le 211$, there exists a BIG-$(2 + 1/q, q, q, (q-1)/2)$ (i.e., $\theta_2 = 2 + 1/q$, $\lambda_2 = (q-1)/2$, and $k_2 = q$) induced by a Hadamard BIBD. Under such settings, given the value of parameter q, the adopted options for the design steps are given as follows.
  - *Intra-module design.* The internal graph is a BIG-$(1, q, q, q)$.
  - *Connection design.* In each module, all RNs are selected as dock nodes. Each dock node serves as an intermediate super-node, i.e., an ISN-$(1, 1)$. Then the dock node interconnection is constructed under the bipartite mode. To be specific, each composite-edge among q ISNs is designed as a TD-$(q, q, 1)$ interconnection through an RN cluster.

- *Inter-module design.* The skeleton graph is given as a BIG-$(2 + 1/q, q, q, (q - 1)/2)$ with one copy. For each copy of modules, they are connected with a common RN cluster via q super-edges. Such q super-edges are constructed using a 1-parallelized TD-$(q, q, 1)$ interconnection.
- *Finalization.* Each GN is directly connected with q CNs.

We call the resulting topology based on these designs a *modified Fat-Tree III*. An instance of such a topology with $p = 6$ and $q = 3$ is presented in Figure 3.10. The performances of the topology are analyzed as follows.

- *Scalability.* Given the value of parameter q, the skeleton graph is fixed as a BIG-$(2 + 1/q, q, q, (q - 1)/2)$ with one copy that contains $2 \cdot (2 + 1/q) \cdot q = 4q + 2$ modules. For each module, its internal graph is fixed as a BIG-$(1, q, q, q)$ that contains q GNs. Each GN is directly connected with q CNs. Hence, the scalability S of the topology (i.e., the total number of CNs contained in the topology) is given by

$$S = (4q + 2) \cdot q \cdot q = 4q^3 + 2q^2. \tag{3.12}$$

- *Inter-module throughput.* Recall that the skeleton graph is given by a BIG-$(2 + 1/q, q, q, (q - 1)/2)$ with one copy. For each pair of modules, they are

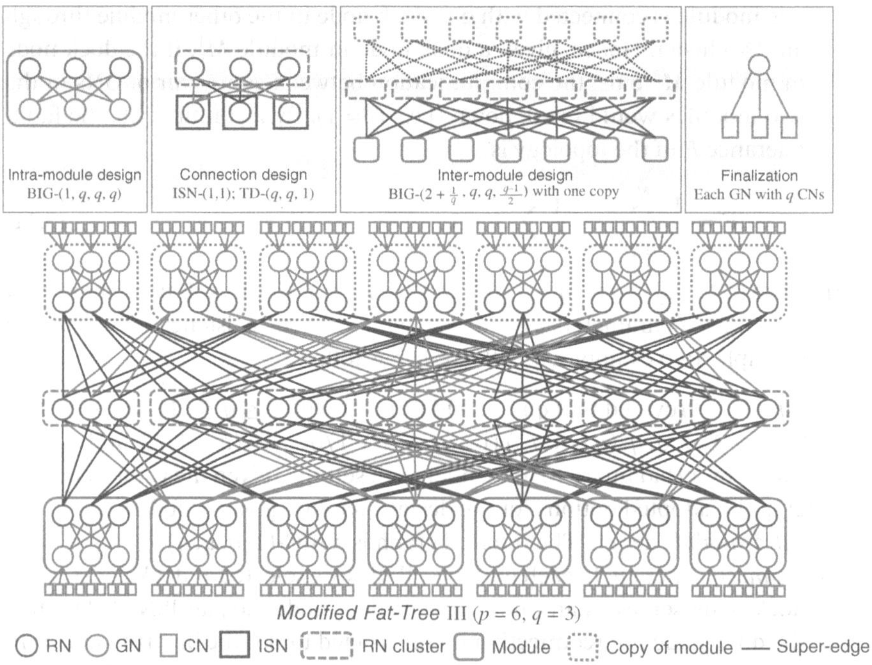

Figure 3.10 The complete design of a modified Fat-Tree III topology with $p = 6$ and $q = 3$.

connected with at least $(q-1)/2$ length-2 paths over super-edges under the bipartite mode. By the property of 1-parallelized TD-$(q, q, 1)$ interconnection, each length-2 path of super-edges between two modules consists of q edge-disjoint paths. Hence, the minimum number of edge-disjoint paths A_2 between any two modules is equal to $(q-1)/2 \cdot q = (q^2 - q)/2$. Meanwhile, the internal graph of each module is fixed as a BIG-$(1, q, q, q)$ that contains q GNs. For each GN, it is directly connected with q CNs. Hence, the attachment number B_2 of each module is $q \cdot q = q^2$. By Definition 2.8, the inter-module throughput T_2 of such a topology is equal to

$$T_2 = \frac{A_2}{B_2} = \frac{\frac{q^2-q}{2}}{q^2} = \frac{q-1}{2q}. \tag{3.13}$$

- *Fault-tolerance.* Let N denote the number of distinct pairs of modules. Note that each group of modules are connected with a common RN cluster via $2q$ super-edges using a 1-parallelized TD-$(2q, q, 1)$ interconnection. Therefore, for each pair of modules, any dock node in one module is connected with all dock nodes of one ISN (i.e., $t = 1$) in the other module. Since each ISN contains one dock node, for the i-th pair of modules, each dock node in one module is connected with one dock node in the other module through an RN cluster. Hence, given a dock node in module M_i^1, if the dock node in module M_i^2 fails, the communication between some pair of GNs in the two modules would break down (i.e., $f_i = 1$). By Definition 2.9, the fault-tolerance F of the topology is equal to

$$F = \frac{1}{N}\sum_{i=1}^{N} f_i = \frac{1}{N}\sum_{i=1}^{N} 1 = 1. \tag{3.14}$$

- *Example 2.* Provided that q is an even number with $2 \le q \le 212$, there exists a BIG-$(2 - 1/q, q, q, q/2)$ (i.e., $\theta_2 = 2 - 1/q$, $\lambda_2 = q/2$, and $k_2 = q$) induced by a Hadamard BIBD. Under such settings, given the value of parameter q, the adopted options for the design steps are given as follows.
  - *Intra-module design.* The internal graph of each module is fixed as a BIG-$(1, q, q, q)$.
  - *Connection design.* In each module, all RNs are selected as dock nodes. Each dock node forms an intermediate super-node, i.e., an ISN-$(1, 1)$. Then the dock node interconnection is constructed under the bipartite mode. To be specific, each composite-edge among q ISNs is designed as a TD-$(q, q, 1)$ interconnection through an RN cluster.
  - *Inter-module design.* The skeleton graph is given as a BIG-$(2 - 1/q, q, q, q/2)$ with one copy. For each copy of modules, they are connected with a common RN cluster via q super-edges. Such q super-edges are constructed using a 1-parallelized TD-$(q, q, 1)$ interconnection.

- *Finalization.* Each GN is directly connected with q CNs.

We call the resulting topology based on these designs a *modified Fat-Tree IV*. An instance of such a topology with $p = 4$ and $q = 2$ is presented in Figure 3.11. The performances of the topology are analyzed as follows.

- *Scalability.* Given the value of parameter q, the skeleton graph is fixed as a BIG-$(2 - 1/q, q, q, q/2)$ with one copy that contains $2 \cdot (2 - 1/q) \cdot q = 4q - 2$ modules. For each module, its internal graph is given by a BIG-$(1, q, q, q)$ that contains q GNs. Each GN is directly connected with q CNs. Hence, the scalability S of the topology (i.e., the total number of CNs contained in the topology) is equal to

$$S = (4q - 2) \cdot q \cdot q = 4q^3 - 2q^2. \tag{3.15}$$

- *Inter-module throughput.* Recall that in the topology, the skeleton graph is designed as a BIG-$(2 - 1/q, q, q, q/2)$ with one copy. Each pair of modules are connected with at least $q/2$ length-2 paths over super-edges under the bipartite mode. By the property of 1-parallelized TD-$(q, q, 1)$ interconnection, each length-2 path of super-edges between two modules consists of q edge-disjoint paths. Hence, the minimum number of edge-disjoint paths A_2 between any two modules is equal to $q/2 \cdot q = q^2/2$. Meanwhile, the internal graph of each module is fixed as a BIG-$(1, q, q, q)$ that contains q GNs. Each GN is directly connected with q CNs. Hence, the attachment number B_2 of each module is $q \cdot q = q^2$. By Definition 2.8, the inter-module throughput T_2 of such a topology is equal to

$$T_2 = \frac{A_2}{B_2} = \frac{\frac{q^2}{2}}{q^2} = \frac{1}{2}. \tag{3.16}$$

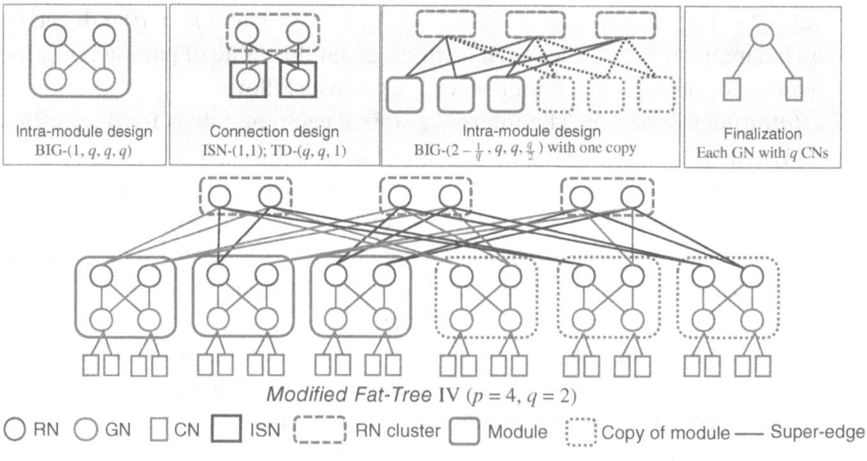

Figure 3.11 The complete design of a modified Fat-Tree IV topology with $p = 4$ and $q = 2$.

- *Fault-tolerance.* Let N denote the total number of distinct pairs of modules in the topology. Each group of modules are connected with a common RN cluster via $2q$ super-edges using a 1-parallelized TD-$(2q, q, 1)$ interconnection. Then for each pair of modules, any dock node in one module is connected with all dock nodes of one ISN (i.e., $t = 1$) in the other module. Since each ISN contains one dock node, for the i-th pair of modules, each dock node in one module is connected with one dock node in the other module through an RN cluster. Hence, given a dock node in module M_i^1, if the dock node in module M_i^2 fails, the communication between some pair of GNs in the two modules would break down (i.e., $f_i = 1$). By Definition 2.9, the fault-tolerance F of the topology is equal to

$$F = \frac{1}{N}\sum_{i=1}^{N} f_i = \frac{1}{N}\sum_{i=1}^{N} 1 = 1. \tag{3.17}$$

Remark 3.2 Some other feasible settings of parameters θ_2 and λ_2 may also lead to topologies with a high scalability and a high throughput. Further analysis to explore such topology designs can be an interesting direction for future work.

Case 2: $t = q$.

In this case, equation 3.1 can be equivalently rewritten as

$$\lambda_2 = \frac{k_2 - 1}{\theta_2 k_2 - 1}, \tag{3.18}$$

where parameters λ_2 and θ_2 are both positive integers. Note that if $\theta_2 > 1$, we have $\lambda_2 < 1$. Therefore, the only feasible settings are $\theta_2 = 1$ and $\lambda_2 = 1$. Accordingly, the skeleton graph is given by a BIG-$(1, 1, k_2, 1)$ with $2/k_2 - 1$ copies for which k_2 should be a factor of 2. By Definition 1.23, in the BIG-$(1, 1, k_2, 1)$, we always have $k_2 \geq 2$. Therefore, the only feasible setting is $k_2 = 2$. In other words, the skeleton graph adopts a BIG-(1, 1, 2, 1) with no copies. Under such settings, given the value of parameter q, the adopted options for the design steps are given as follows.

- *Intra-module design.* The internal graph of each module is fixed as a BIG-$(1, q, q, q)$.
- *Connection design.* In each module, all RNs are selected as dock nodes. Each dock node forms an intermediate super-node, i.e., an ISN-(1, 1). Then the dock node interconnection is constructed under the bipartite mode. To be specific, each composite-edge is designed as a TD-$(2, q, 1)$ interconnection between two ISNs through an RN cluster.
- *Inter-module design.* The skeleton graph is given as a BIG-(1, 1, 2, 1) with no copies. For any two modules, they are connected with a common RN cluster via two super-edges. Such two super-edges are constructed using a q-parallelized TD-$(2, q, 1)$ interconnection.
- *Finalization.* Each GN is directly connected with q CNs.

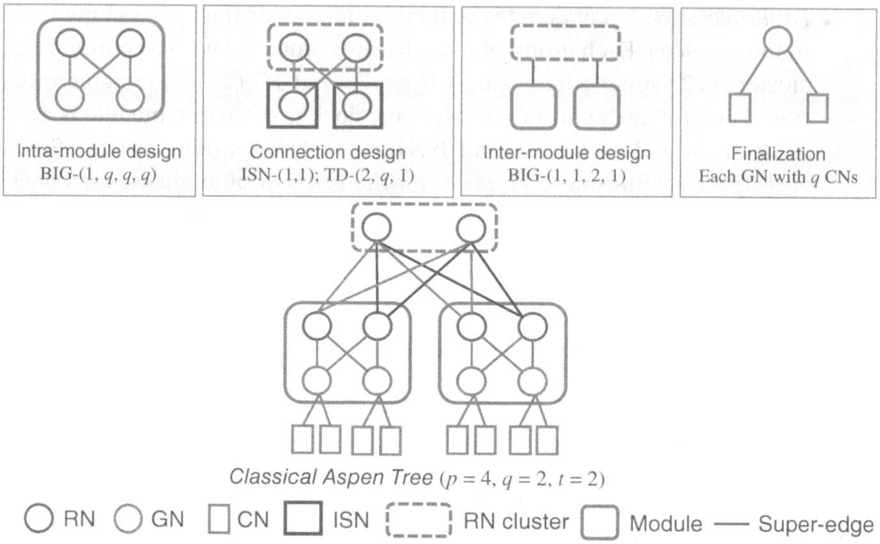

Figure 3.12 The complete design of an Aspen Tree topology with $p = 4$ and $t = q = 2$.

Note that the resulting topology based on these designs is a kind of Aspen Trees. An instance of such a topology with $p = 4$ and $q = t = 2$ is presented in Figure 3.12. The performances of the topology are analyzed as follows.

- *Scalability.* Given the value of parameter q, the skeleton graph is fixed as a BIG-(1, 1, 2, 1) with no copies that contains $2 \cdot 1 \cdot q/q = 2$ modules. For each module, its internal graph is fixed as a BIG-$(1, q, q, q)$ that contains q GNs. Each GN is directly connected with q CNs. Hence, the scalability S of the topology (i.e., the total number of CNs contained in the topology) is equal to

$$S = 2 \cdot q \cdot q = 2q^2. \tag{3.19}$$

- *Inter-module throughput.* Recall that the skeleton graph is given by a BIG-(1, 1, 2, 1) with no copies. For each pair of modules, they are connected with one length-2 path over super-edges under the bipartite mode. By the property of q-parallelized TD-$(2, q, 1)$ interconnection, each length-2 path of super-edges between two modules consists of $q \cdot q = q^2$ edge-disjoint paths. Hence, the minimum number of edge-disjoint paths A_2 between any two modules is q^2. Meanwhile, the internal graph of each module is a BIG-$(1, q, q, q)$ that contains q GNs. Each GN is directly connected with q CNs. Hence, the attachment number B_2 of each module is equal to $q \cdot q = q^2$. By Definition 2.8, the inter-module throughput T_2 of the topology is

$$T_2 = \frac{A_2}{B_2} = \frac{q^2}{q^2} = 1. \tag{3.20}$$

- *Fault-tolerance.* Let N denote the number of distinct pairs of modules. Note that each group of modules are connected with a common RN cluster via two super-edges using a q-parallelized $(2, q, 1)$ interconnection. Therefore, for each ordered pair of modules, any dock node in one module is connected with all dock nodes of q ISNs in the other module. Since each ISN contains one dock node, for the i-th pair of modules M_i^1 and M_i^2, each dock node in module M_i^1 is connected with q dock nodes in module M_i^2 through an RN cluster. Hence, given a dock node in module M_i^1, only when all of the q dock nodes in module M_i^2 fail simultaneously, the communication between some pair of GNs in the two modules would break down (i.e., $f_i = q$). Therefore, the fault-tolerance F of the topology is equal to

$$F = \frac{1}{N}\sum_{i=1}^{N} f_i = \frac{1}{N}\sum_{i=1}^{N} q = q. \tag{3.21}$$

Case 3: $1 < t < q$ **with** t **as a factor of** q.

In this case, the induced topologies with performance trade-offs between scalability and throughput are given in Figure 3.13. The detailed analysis is as follows.

First, recall that $\theta_2, \lambda_2 \geq 1$. Then we consider the following three sub-cases in which parameter t is supposed to be a factor of q.

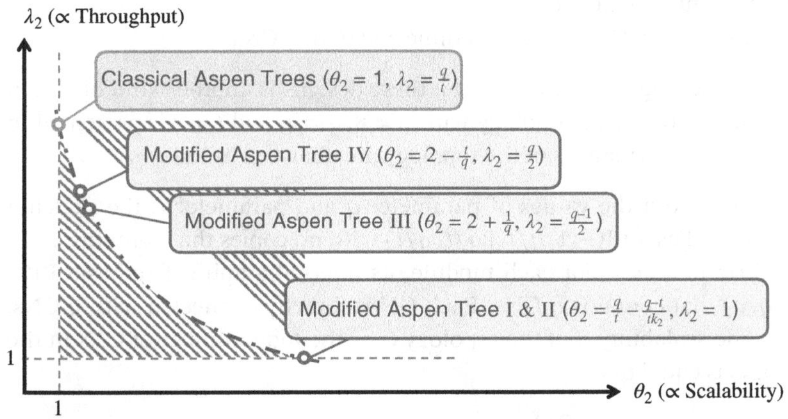

Figure 3.13 The trade-off between scalability and throughput in terms of parameters θ_2 and λ_2.

- Sub-case 1: $\theta_2 = 1$ and $\lambda_2 = q/t$, in which a higher throughput is achieved at the cost of a lower scalability.
- Sub-case 2: $\theta_2 = q/t - (q-t)/tk_2$ and $\lambda_2 = 1$, in which a higher scalability is achieved at the cost of a lower throughput.
- Sub-case 3: $1 < \theta_2 < q/t - (q-t)/tk_2$ and $1 < \lambda_2 < q/t$.

Further discussions are given as follows.

Sub-case 1: $\theta_2 = 1$ and $\lambda_2 = q/t$. In this sub-case, the skeleton graph is given by a BIG-$(1, q/t, k_2, q/t)$ with $2q/tk_2 - 1$ copies. Such a graph is isomorphic to a BIG-$(1, q/t, 2q/t, q/t)$ with no copies for any value of parameter k_2. Hence, the skeleton graph can be equivalently regarded as a BIG-$(1, q/t, 2q/t, q/t)$. Under such settings, given the value of parameter q, the adopted options for the design steps are summarized as follows.

- *Intra-module design*. The internal graph of each module is fixed as a BIG-$(1, q, q, q)$.
- *Connection design*. In each module, all RNs are selected as dock nodes. Each dock node forms an intermediate super-node, i.e., an ISN-$(1, 1)$. Then the dock node interconnection is constructed under the bipartite mode. To be specific, each composite-edge is designed as a TD-$(2q/t, q, 1)$ interconnection among $2q/t$ ISNs through an RN cluster.
- *Inter-module design*. The skeleton graph is given as a BIG-$(1, q/t, 2q/t, q/t)$. For any $2q/t$ modules, they are connected with a common RN cluster via $2q/t$ super-edges. Such $2q/t$ super-edges are constructed using a t-parallelized TD-$(2q/t, q, 1)$ interconnection.
- *Finalization*. Each GN is directly connected with q CNs.

Note that the resulting topology based on these designs is another kind of Aspen Trees. An instance of such a topology with $p = 8$, $q = 4$, and $t = 2$ is presented in Figure 3.14. The performances of the topology are analyzed as follows.

- *Scalability*. Given the values of parameter q and parameter t, the skeleton graph is fixed as a BIG-$(1, q/t, 2q/t, q/t)$ with no copies that contains $2 \cdot 1 \cdot q/t = 2q/t$ modules. For each module, its internal graph is fixed as a BIG-$(1, q, q, q)$ that contains q GNs. Each GN is directly connected with q CNs. Hence, the scalability S of the topology (i.e., the total number of CNs in the topology) is equal to

$$S = 2q/t \cdot q \cdot q = \frac{2q^3}{t}. \tag{3.22}$$

- *Inter-module throughput*. Given a BIG-$(1, q/t, 2q/t, q/t)$ as the skeleton graph, for each pair of modules, they are connected with at least q/t length-2 paths of super-edges under the bipartite mode. By the property of

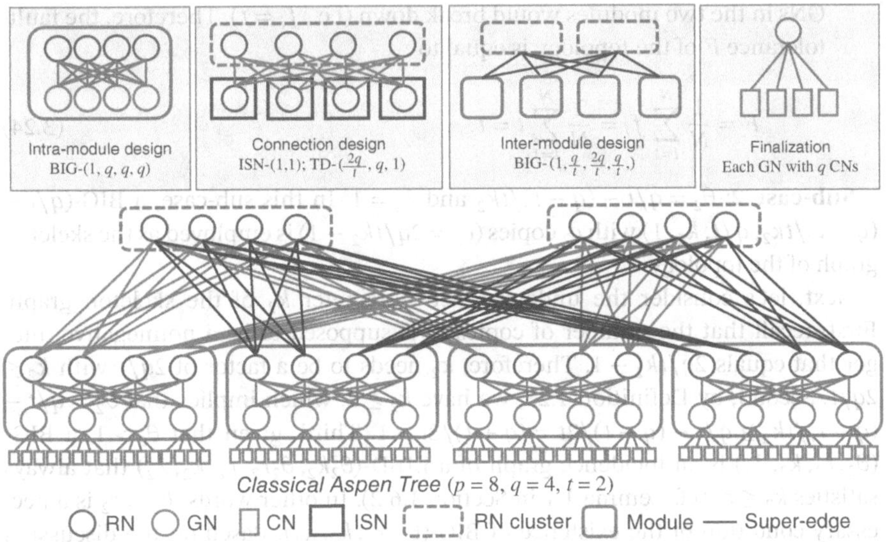

Figure 3.14 The complete design of an Aspen Tree topology with $p = 8$, $q = 4$, and $t = 2$.

t-parallelized TD-$(2q/t, q, 1)$ interconnection, each length-2 path of super-edges between two modules consists of tq edge-disjoint paths. Hence, the minimum number of edge-disjoint paths A_2 between any two modules is equal to $q/t \cdot tq = q^2$. Meanwhile, the internal graph of each module is fixed as a BIG-$(1, q, q, q)$ that contains q GNs. For each GN, it is directly connected with q CNs. Hence, the attachment number B_2 of each module is equal to $q \cdot q = q^2$. By Definition 2.8, the inter-module throughput T_2 of the topology is equal to

$$T_2 = \frac{A_2}{B_2} = \frac{q^2}{q^2} = 1. \tag{3.23}$$

- *Fault-tolerance.* Let N denote the number of distinct pairs of modules. Note that each group of modules are connected with a common RN cluster via q/t super-edges which form a t-parallelized $(q/t, q, 1)$ interconnection. Therefore, for each ordered pair of modules, any dock node in one module is connected with all dock nodes of t ISN in the other module. Since each ISN contains one dock node, for the i-th pair of modules M_i^1 and M_i^2, each dock node in module M_i^1 is connected with t dock nodes in module M_i^2 through an RN cluster. Hence, given a dock node in module M_i^1, only when t dock nodes in module M_i^2 fail simultaneously, the communication between some pair of

GNs in the two modules would break down (i.e., $f_i = t$). Therefore, the fault-tolerance F of the topology is equal to

$$F = \frac{1}{N}\sum_{i=1}^{N} f_i = \frac{1}{N}\sum_{i=1}^{N} t = t. \tag{3.24}$$

Sub-case 2: $\theta_2 = q/t - (q-t)/tk_2$ and $\lambda_2 = 1$. In this sub-case, a BIG-$(q/t - (q-t)/tk_2, q/t, k_2, 1)$ with c_2 copies ($c_2 = 2q/tk_2 - 1$) is employed as the skeleton graph of the topology.

Next, let's consider the undetermined parameter k_2 of the skeleton graph. First, recall that the number of copies c_2 is supposed to be a nonnegative integer that equals $2q/tk_2 - 1$. Therefore, k_2 needs to be a factor of $2q/t$ with $k_2 \leq 2q/t$. Second, by Definition 1.23, we have $k_2 \geq 2$, which implies that $\theta_2 = q/t - (q-t)/tk_2 \geq q/t - (q-t)/2t = (q+t)/2t > 1$. Third, given that $\theta_2 > 1$, a BIG-$(\theta_2, r_2, k_2, \lambda_2)$ is an incidence graph of a BIBD-$(\theta_2 k_2, \theta_2 r_2, r_2, k_2, \lambda_2)$ that always satisfies $k_2 \leq r_2$ (cf. Lemma 1.1 in Section 1.6.2). In other words, $k_2 \leq r_2$ is a necessary condition of the existence of BIG-$(\theta_2, r_2, k_2, \lambda_2)$. Based on the discussion in Section 2.4.3.1, such a necessary condition also holds for BIGs with copies. Hence, given a BIG-$(q/t - (q-t)/tk_2, q/t, k_2, 1)$ with c_2 copies, we have $k_2 \leq q/t$. In summary, the value of parameter k_2 is supposed to be a factor of $2q/t$ with $2 \leq k_2 \leq q/t$.

Based on these settings, we focus on two extreme conditions where $k_2 = 2$ and $k_2 = q/t$, respectively.[5]

- *Condition 1:* $k_2 = 2$. Under this condition, the skeleton graph is given as a BIG-$((q+t)/2t, q/t, 2, 1)$ with $q/t - 1$ copies, which exists if q/t is a positive even number. Under such settings, given the values of parameters q and t, the adopted options for the four design steps are given as follows, respectively.
  - *Intra-module design.* The internal graph of each module is fixed as a BIG-$(1, q, q, q)$.
  - *Connection design.* In each module, all RNs are selected as dock nodes. Each dock node forms an intermediate super-node, i.e., an ISN-(1, 1). Then the dock node interconnection is constructed under the bipartite mode. To be specific, each composite-edge is designed as a TD-$(2, q, 1)$ interconnection between two ISNs through an RN cluster.
  - *Inter-module design.* The skeleton graph is given as a BIG-$((q+t)/2t, q/t, 2, 1)((q+t)/2t, q/t, 2, 1)$ with $q/t - 1$ copies. For any two modules, they are connected with a common RN cluster via two

5 For cases with $2 < k_2 < q/t$, further analysis is required on the value selections of parameters q and t to guarantee the existence of BIGs-$(q/t - (q-t)/tk_2, q/t, k_2, 1)$. Such analysis can be left as interesting future work.

super-edges. Such two super-edges are constructed using a t-parallelized TD-(2, q, 1) interconnection.

- *Finalization.* Each GN is directly connected with q CNs.

We call the resulting topology based on these designs a *modified Aspen Tree I*. An instance of such a topology with $p = 8$, $q = 4$, and $t = 2$ is presented in Figure 3.15. The performances of the topology are analyzed as follows.

- *Scalability.* Given the value of parameter q, the skeleton graph is fixed as a BIG-($(q + t)/2t, q/t, 2, 1$) with $q/t - 1$ copies that contains $2 \cdot (q + t)/2t \cdot q/t = (q^2 + tq)/t^2$ modules. For each module, its internal graph is fixed as a BIG-(1, q, q, q) that contains q GNs. Each GN is directly connected with q CNs. Hence, the scalability S of the topology (i.e., the total number of CNs contained in the topology) is equal to

$$S = (q^2 + tq)/t^2 \cdot q \cdot q = \frac{q^4 + tq^3}{t^2}. \tag{3.25}$$

- *Inter-module throughput.* Recall that the skeleton graph is fixed as a BIG-($(q + t)/2t, q/t, 2, 1$) with $q/t - 1$ copies. For each pair of modules, they are connected with at least one length-2 path of super-edges under the bipartite mode. By the property of t-parallelized TD-(2, q, 1) interconnection, each length-2 path of super-edges between two modules consists of tq

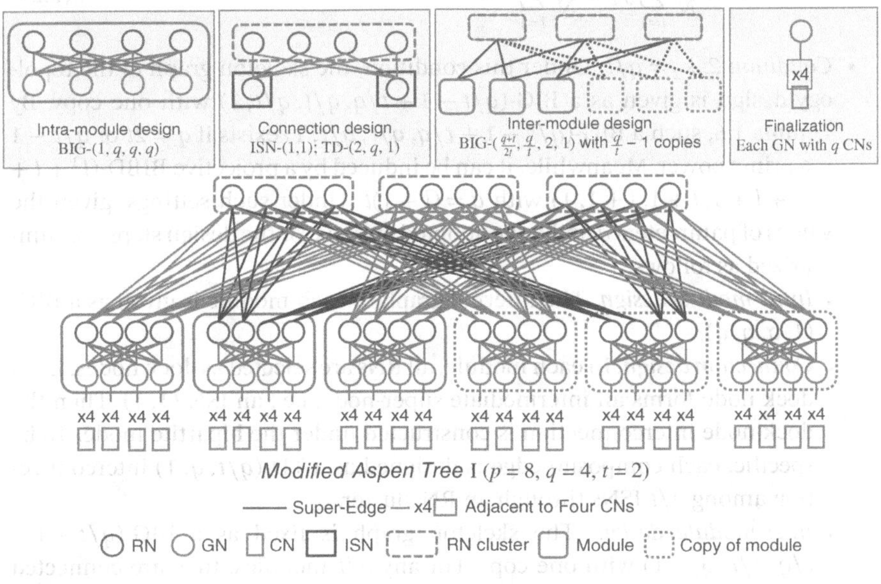

Figure 3.15 The complete design of a modified Aspen Tree I topology with $p = 8$, $q = 4$, and $t = 2$.

edge-disjoint paths. Hence, the minimum number of edge-disjoint paths A_2 between any two modules is $1 \cdot tq = tq$. Meanwhile, the internal graph of each module is given as a BIG-$(1, q, q, q)$ that contains q GNs. For each GN, it is directly connected with q CNs. Hence, the attachment number B_2 of each module is $q \cdot q = q^2$. By Definition 2.8, the inter-module throughput T_2 of the topology is equal to

$$T_2 = \frac{A_2}{B_2} = \frac{tq}{q^2} = \frac{t}{q}. \tag{3.26}$$

- *Fault-tolerance.* Let N denote the number of distinct pairs of modules. Note that each group of modules are connected with a common RN cluster via q/t super-edges using a t-parallelized $(q/t, q, 1)$ interconnection. Therefore, for each ordered pair of modules, any dock node in one module is connected with all dock nodes of t ISNs in the other module. Since each ISN contains one dock node, for the i-th pair of modules M_i^1 and M_i^2, each dock node in module M_i^1 is connected with t dock nodes in module M_i^2 through an RN cluster. Hence, given a dock node in module M_i^1, only when all of the t dock nodes in module M_i^2 fail simultaneously, the communication between some pair of GNs in the two modules would break down (i.e., $f_i = t$). By Definition 2.9, the fault-tolerance F of the topology is given by

$$F = \frac{1}{N}\sum_{i=1}^{N} f_i = \frac{1}{N}\sum_{i=1}^{N} t = t. \tag{3.27}$$

- *Condition 2:* $k_2 = q/t$. Under this condition, the skeleton graph of the topology design is given as a BIG-$(q/t - 1 + t/q, q/t, q/t, 1)$ with one copy. By Lemma 1.6, such a BIG-$(q/t - 1 + t/q, q/t, q/t, 1)$ exists if $q = 2t$ or $q/t - 1$ is a prime power. Meanwhile, it can be induced by a projective BIBD-$(l^2 + l + 1, l^2 + l + 1, l + 1, l + 1, 1)$ with $q = (l + 1)t$. Under such settings, given the values of parameters q and t, the adopted options for the design steps are summarized as follows.
  - *Intra-module design.* The internal graph of each module is given as a BIG-$(1, q, q, q)$.
  - *Connection design.* In each module, all RNs are selected as dock nodes. Each dock node forms an intermediate super-node, i.e., an ISN-$(1, 1)$. Then the dock node interconnection is constructed under the bipartite mode. To be specific, each composite-edge is designed as a TD-$(q/t, q, 1)$ interconnection among q/t ISNs through an RN cluster.
  - *Inter-module design.* The skeleton graph is fixed as a BIG-$(q/t - 1 + t/q, q/t, q/t, 1)$ with one copy. For any q/t modules, they are connected with a common RN cluster via q/t super-edges. Such q/t super-edges are constructed using a t-parallelized TD-$(q/t, q, 1)$ interconnection.

- *Finalization.* Each GN is directly connected with q CNs.

We call the resulting topology based on these designs a *modified Aspen Tree II*. An instance of such a topology with $p = 8$, $q = 4$, and $t = 2$ is presented in Figure 3.16. The performances of the topology are analyzed as follows.

- *Scalability.* Given the value of parameter q, the skeleton graph is fixed as a BIG-$(q/t - 1 + t/q, q/t, q/t, 1)$ with one copy that contains $2 \cdot (q/t - 1 + t/q) \cdot q/t = (2q^2 - 2tq + 2t^2)/t^2$ modules. For each module, its internal graph is given as a BIG-$(1, q, q, q)$ that contains q GNs. Each GN is directly connected with q CNs. Hence, the scalability S of the topology (i.e., the total number of CNs in the topology) is equal to

$$S = (2q^2 - 2tq + 2t^2)/t^2 \cdot q \cdot q = \frac{2q^4 - 2tq^3 + 2t^2q^2}{t^2}. \tag{3.28}$$

- *Inter-module throughput.* Recall that the skeleton graph is given as a BIG-$(q/t - 1 + t/q, q/t, q/t, 1)$ with one copy. For each pair of modules, they are connected with at least one length-2 paths of super-edges under the bipartite mode. By the property of t-parallelized TD-$(q/t, q, 1)$ interconnection, each length-2 path of super-edges between two modules consists of tq edge-disjoint paths. Hence, the minimum number of edge-disjoint paths A_2 between any two modules is $1 \cdot tq = tq$. Meanwhile, the internal graph

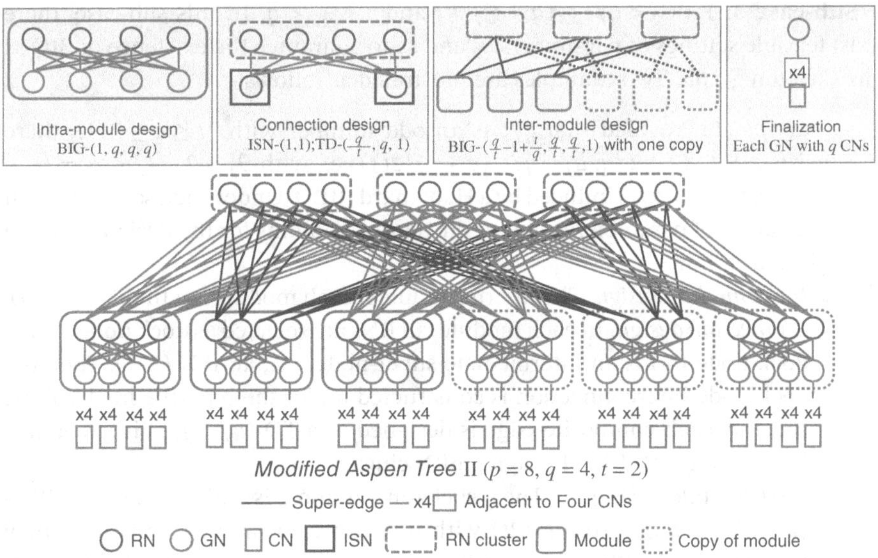

Figure 3.16 The complete design of a modified Aspen Tree II topology with $p = 8$, $q = 4$, and $t = 2$.

of each module is a BIG-$(1, q, q, q)$ that contains q GNs. For each GN, it is directly connected with q CNs. Hence, the attachment number B_2 of each module is $q \cdot q = q^2$. By Definition 2.8, the inter-module throughput T_2 of the topology is equal to

$$T_2 = \frac{A_2}{B_2} = \frac{tq}{q^2} = \frac{t}{q}. \tag{3.29}$$

- *Fault-tolerance.* Let N denote the number of distinct pairs of modules. Note that each group of modules are connected with a common RN cluster via q/t super-edges using a t-parallelized $(q/t, q, 1)$ interconnection. Therefore, for each ordered pair of modules, any dock node in one module is connected with all dock nodes of t ISNs in the other module. Since each ISN contains one dock node, for the i-th pair of modules M_i^1 and M_i^2, each dock node in module M_i^1 is connected with t dock nodes in module M_i^2 through an RN cluster. Hence, given a dock node in module M_i^1, only when all of the t dock nodes in module M_i^2 fail simultaneously, the communication between some pair of GNs in the two modules would break down (i.e., $f_i = t$). By Definition 2.9, it follows that the fault-tolerance F of this topology is given by

$$F = \frac{1}{N}\sum_{i=1}^{N} f_i = \frac{1}{N}\sum_{i=1}^{N} t = t. \tag{3.30}$$

Sub-case 3: $1 < \theta_2 < q/t - (q-t)/tk_2$ and $1 < \lambda_2 < q$. In this sub-case, there exist feasible settings of parameters θ_2 and λ_2 to guarantee the existence of BIG as the skeleton graph. Two examples are illustrated as follows.

- *Example 1.* Provided that q/t is an odd number with $3t \le q \le 211t$, there exists a BIG-$(2 + t/q, q/t, q/t, (q-t)/2t)$ (i.e., with $\theta_2 = 2 + t/q$, $\lambda_2 = (q - t)/2t$, and $k_2 = q/t$) induced by a Hadamard BIBD. Under such settings, given the values of parameters q and t, the adopted options for the design steps are given as follows.
  - *Intra-module design.* The internal graph of each module is a BIG-$(1, q, q, q)$.
  - *Connection design.* In each module, all RNs are selected as dock nodes. Each dock node forms an intermediate super-node, i.e., an ISN-$(1, 1)$. Then the dock node interconnection is constructed under the bipartite mode. To be specific, each composite-edge is designed as a TD-$(q/t, q, 1)$ interconnection among q/t ISNs through an RN cluster.
  - *Inter-module design.* The skeleton graph is given as a BIG-$(2 + t/q, q/t, q/t, (q-t)/2t)$ with one copy. For any q/t modules, they are connected with a common RN cluster via q/t super-edges. Such super-edges are constructed using a t-parallelized TD-$(q/t, q, 1)$ interconnection.
  - *Finalization.* Each GN is directly connected with q CNs.

We call the resulting topology based on these designs a *modified Aspen Tree III*. An instance of such a topology with $p = 12$, $q = 6$, and $t = 2$ is presented in Figure 3.17. The performances of the topology are analyzed as follows.

- *Scalability.* Given the value of parameter q, the skeleton graph is fixed as a BIG-$(2 + t/q, q/t, q/t, (q - t)/2t)$ with one copy that contains $2 \cdot (2 + t/q) \cdot q/t = 4q/t + 2$ modules. For each module, its internal graph is given as a BIG-$(1, q, q, q)$ that contains q GNs. Each GN is directly

Figure 3.17 The complete design of a modified Aspen Tree III topology with $p = 12$, $q = 6$, and $t = 2$.

connected with q CNs. Hence, the scalability S of the topology (i.e., the total number of CNs in the topology) is equal to

$$S = (4q/t + 2) \cdot q \cdot q = \frac{4q^3 + 2tq^2}{t}. \tag{3.31}$$

- *Inter-module throughput.* Recall that the skeleton graph is given as a BIG-$(2 + t/q, q/t, q/t, (q - t)/2t)$ with one copy. For each pair of modules, they are connected with at least $(q - t)/2t$ length-2 paths of super-edges under the bipartite mode. By the property of t-parallelized TD-$(q/t, q, 1)$ interconnection, each length-2 path of super-edges between two modules consists of tq edge-disjoint paths. Hence, the minimum number of edge-disjoint paths A_2 between any two modules is equal to $(q - t)/2t \cdot tq = (q^2 - tq)/2$. Meanwhile, the internal graph of each module is fixed as a BIG-$(1, q, q, q)$ that contains q GNs. Each GN is directly connected with q CNs. Hence, the attachment number B_2 of each module is $q \cdot q = q^2$. By Definition 2.8, the inter-module throughput T_2 of the topology is given by

$$T_2 = \frac{A_2}{B_2} = \frac{\frac{q^2-tq}{2}}{q^2} = \frac{q - t}{2q}. \tag{3.32}$$

- *Fault-tolerance.* Let N denote the number of distinct pairs of modules. Each group of modules are connected with a common RN cluster via q/t super-edges using a t-parallelized $(q/t, q, 1)$ interconnection. Then for each ordered pair of modules, any dock node in one module is connected with all dock nodes of t ISNs in the other module. Since each ISN contains one dock node, for the i-th pair of modules M_i^1 and M_i^2, each dock node in module M_i^1 is connected with t dock nodes in module M_i^2 through an RN cluster. Hence, given a dock node in module M_i^1, only when all of the t dock nodes in module M_i^2 fail simultaneously, the communication between two GNs in the two modules would break down (i.e., $f_i = t$). By Definition 2.9, the fault-tolerance F of the topology is equal to

$$F = \frac{1}{N}\sum_{i=1}^{N} f_i = \frac{1}{N}\sum_{i=1}^{N} t = t. \tag{3.33}$$

- *Example 2.* Provided that q/t is an even number with $2t \leq q \leq 212t$, there exists a BIG-$(2 - t/q, q/t, q/t, q/2t)$ (i.e., with $\theta_2 = 2 - t/q$, $\lambda_2 = q/2$, and $k_2 = q/2t$) induced by a Hadamard BIBD. Under such settings, given the values of parameters q and t, the adopted options for the design steps are given as follows.
  - *Intra-module design.* The internal graph of each module is fixed as a BIG-$(1, q, q, q)$.

- *Connection design.* In each module, all RNs are selected as dock nodes. Each dock node forms an intermediate super-node, i.e., an ISN-(1, 1). Then the dock node interconnection is constructed under the bipartite mode. To be specific, each composite-edge is designed as a TD-($q/t, q$, 1) interconnection among q/t ISNs through an RN cluster.
- *Inter-module design.* The skeleton graph is designed as a BIG-($2 - t/q, q/t, q/t, q/2t$) with one copy. For any q/t modules, they are connected with a common RN cluster via q/t super-edges. Such q/t super-edges are constructed using a t-parallelized TD-($q/t, q$, 1) interconnection.
- *Finalization.* Each GN is directly connected with q CNs.

We call the resulting topology based on these designs a *modified Aspen Tree IV*. An instance of such a topology with $p = 8, q = 4$, and $t = 2$ is presented in Figure 3.18. The performances of the topology are analyzed as follows.

- *Scalability.* Given the value of parameter q, the skeleton graph is fixed as a BIG-($2 - t/q, q/t, q/t, q/2t$) with one copy that contains $2 \cdot (2 - t/q) \cdot q/t = 4q/t - 2$ modules. For each module, its internal graph is given as a BIG-($1, q, q, q$) that contains q GNs. Each GN is directly connected with

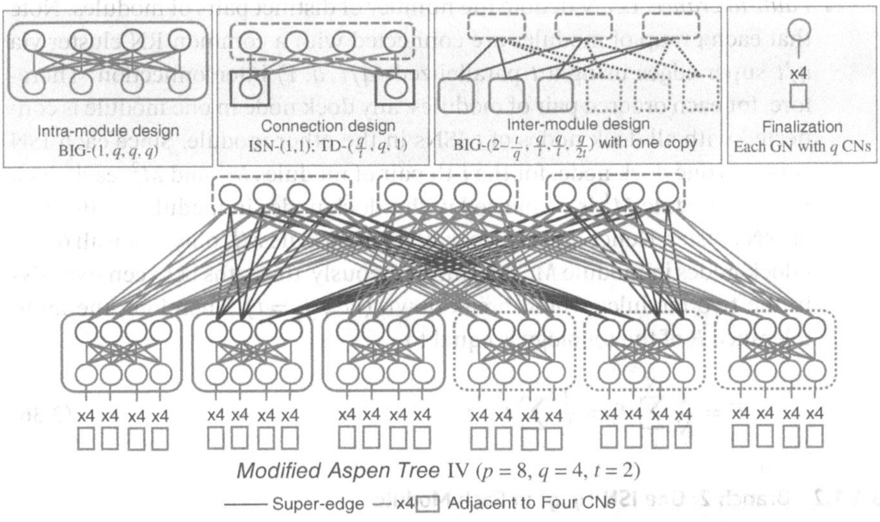

Figure 3.18 The complete design of a modified Aspen Tree IV topology with $p = 8$, $q = 4$, and $t = 2$.

q CNs. Hence, the scalability S of the topology (i.e., the total number of CNs in the topology) is equal to

$$S = (4q/t - 2) \cdot q \cdot q = \frac{4q^3 - 2tq^2}{t}. \tag{3.34}$$

- *Inter-module throughput.* Recall that the skeleton graph is fixed as a BIG-$(2 - t/q, q/t, q/t, q/2t)$ with one copy. For each pair of modules, they are connected with at least $q/2t$ length-2 paths of super-edges under the bipartite mode. By the property of t-parallelized TD-$(q/t, q, 1)$ interconnection, each length-2 path of super-edges between two modules consists of tq edge-disjoint paths. Hence, the minimum number of edge-disjoint paths A_2 between any two modules is equal to $q/2t \cdot tq = q^2/2$. Meanwhile, the internal graph of each module is fixed as a BIG-$(1, q, q, q)$ that contains q GNs. For each GN, it is directly connected with q CNs. Hence, the attachment number B_2 of each module is $q \cdot q = q^2$. By Definition 2.8, the inter-module throughput T_2 of the topology is given by

$$T_2 = \frac{A_2}{B_2} = \frac{\frac{q^2}{2}}{q^2} = \frac{1}{2}. \tag{3.35}$$

- *Fault-tolerance.* Let N denote the number of distinct pairs of modules. Note that each group of modules are connected with a common RN cluster via q/t super-edges using a t-parallelized $(q/t, q, 1)$ interconnection. Therefore, for each ordered pair of modules, any dock node in one module is connected with all dock nodes of t ISNs in the other module. Since each ISN contains one dock node, for the i-th pair of modules M_i^1 and M_i^2, each dock node in module M_i^1 is connected with t dock nodes in module M_i^2 through an RN cluster. Hence, given a dock node in module M_i^1, only when all of the t dock nodes in module M_i^2 fail simultaneously, the paths between two GNs in the two modules would break down (i.e., $f_i = t$). Therefore, the fault-tolerance F of the topology is equal to

$$F = \frac{1}{N}\sum_{i=1}^{N} f_i = \frac{1}{N}\sum_{i=1}^{N} t = t. \tag{3.36}$$

3.1.1.2 Branch 2: One ISN-(q, q) in Each Module

Parameter settings of connection design. Along with this branch, for connection design, dock node selection is fixed by using one ISN-(q, q) in each module $(x = q)$. For dock node interconnection, the parameter settings of the corresponding TD-(n, μ, q) interconnection are discussed as follows.

- *Parameter μ.* Under a TD-(n, μ, q) interconnection, each dock node is supposed to be adjacent to μq RNs in each RN cluster. Accordingly, the value of

μq is subject to the outer-degree of each dock node (i.e., $\mu q \leq q$). Therefore, the only feasible setting is $\mu = 1$ and thus a TD-$(n, 1, q)$ interconnection is employed as the composite edge.

- *Parameter n*. In general, for any integer $n > 1$, there exists a TD-$(n, 1, q)$ interconnection in which n ISNs are interconnected through an RN cluster. Each ISN contains q dock nodes and the RN cluster contains only one RN. For each RN in the RN cluster, it is adjacent to n dock nodes in distinct ISNs. Since the degree of each RN is p, we have $n \leq p$.

Following these settings, we show a TD-$(n, 1, q)$ interconnection with $n = 2$ and $q = 2$ in Figure 3.19.

Parameter settings of inter-module design. For inter-module design, the skeleton graph is given as a BIG-$(\theta_2, r_2, k_2, \lambda_2)$ with c_2 copies since each pair of modules are connected under the bipartite mode. In such a graph, element nodes and block nodes are instantiated as modules and RN clusters, respectively. By duplicating operation, modules are partitioned into $c_2 + 1$ groups such that each group of modules and all RN clusters form a BIG-$(\theta_2, r_2, k_2, \lambda_2)$. An example of a BIG-(1, 1, 2, 1) with one copy (i.e., $c_2 = 1$) is given in Figure 3.20a.

For each group of modules, an RN cluster is directly connected with k_2 modules in such a group via k_2 super-edges. Such super-edges form a t-parallelized TD-$(n, 1, q)$ interconnection where $n = k_2$. In Figure 3.20b, we illustrate the super-edges of the skeleton graph (see Figure 3.20a) as a 2-parallelized TD-(2, 1, 2) interconnection (i.e., with $t = 2$, $q = 2$, and $n = k_2 = 2$).

Under a t-parallelized TD-$(n, 1, q)$ interconnection, each super-edge between an RN cluster and a module consists of tq^2 edges. Each of such edges is incident with an RN in the RN cluster and a dock node in the module. In other words, each super-edge occupies tq^2 degrees of RNs in the RN cluster and tq^2 outer-degrees of dock nodes in the corresponding module. Accordingly, we have the following remark.

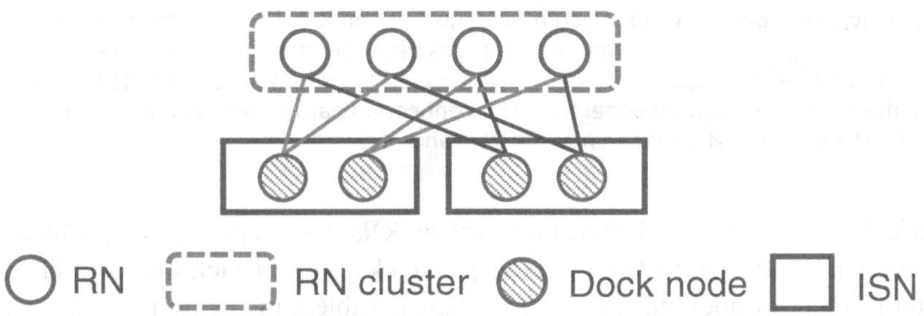

Figure 3.19 An example of a TD-$(n, 1, q)$ interconnection with $n = 2$ and $q = 2$ under which two ISNs-(q, q) are connected through a single RN cluster.

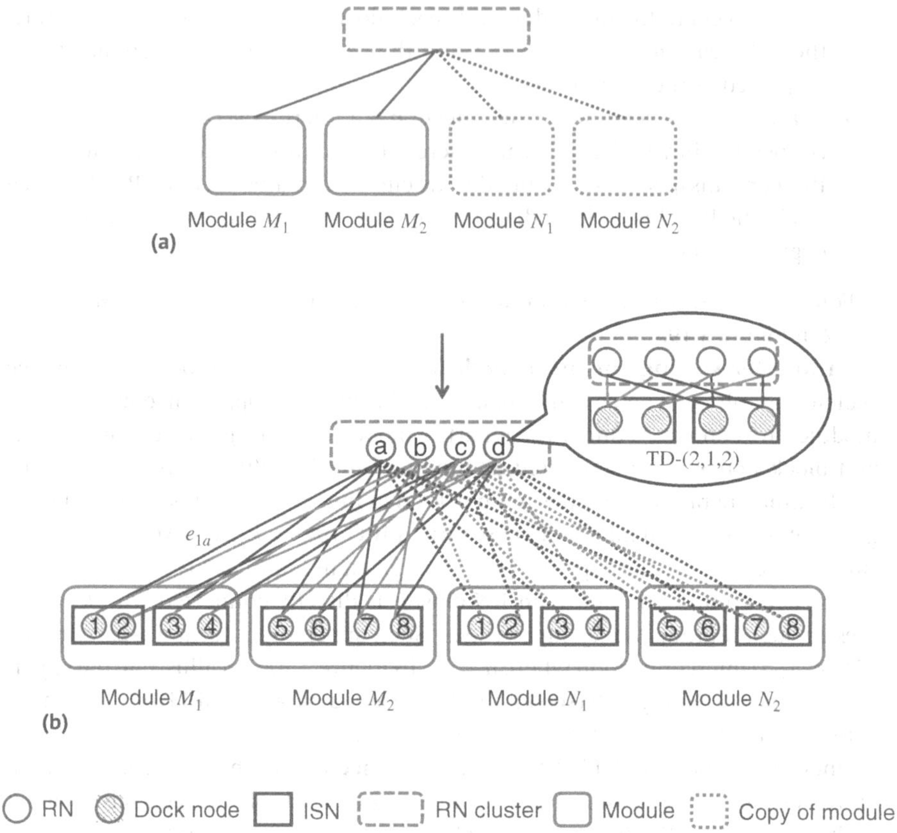

Figure 3.20 An illustration of the skeleton graph and super-edges instantiation. (a) A BIG-(1, 1, 2, 1) with one copy as the skeleton graph. Since c_2 is equal to 1, such a skeleton graph contains two groups of modules. One group consists of modules M_1 and M_2. The other group contains modules N_1 and N_2. (b) The two super-edges between an RN cluster and a group of modules are instantiated as a 2-parallelized TD-(2, 1, 2) interconnection. To be specific, the super-edges between module M_1 and an RN cluster contains edges $\{e_{1a}, e_{1b}, e_{2c}, e_{2d}, e_{3a}, e_{3b}, e_{4c}, e_{4d}\}$, while super-edges between module M_2 and an RN cluster consists of edges $\{e_{5a}, e_{5c}, e_{6b}, e_{6d}, e_{7a}, e_{7c}, e_{8b}, e_{8d}\}$. Such two super-edges form a 2-parallelized TD-(2, 1, 2) interconnection. The connection pattern between modules N_1 and N_2 is the same as that between modules M_1 and M_2.

Remark 3.3 In each RN cluster, there are q^2 RNs with degree $p = 2q$. Meanwhile, there are q RNs with outer-degree q as dock nodes in each module. Since the total number of dock nodes' outer-degrees is subject to the number of RNs in each RN cluster (i.e., $tq^2 \leq q^2$), the only feasible value for parameter t is one. Through such dock nodes, each module is connected with $q^2/tq^2 = 1$ RN cluster. In the meantime, each RN cluster is connected with $2q^3/tq^2 = 2q/t = 2q$

modules. By the properties of BIG-$(\theta_2, r_2, k_2, \lambda_2)$ with copies, we have $r_2 = 1$ and $k_2(c_2 + 1) = 2q$. As a result, the adopted skeleton graph (i.e., a BIG-$(\theta_2, r_2, k_2, \lambda_2)$ with c_2 copies) can be equivalently regarded as a BIG-$(\theta_2, 1, k_2, \lambda_2)$ with $2q/k_2 - 1$ copies.

By the aforementioned remark and Lemma 1.6, given that $r_2 = 1$, we have

$$\lambda_2 = \frac{r_2(k_2 - 1)}{\theta_2 k_2 - 1} = \frac{k_2 - 1}{\theta_2 k_2 - 1}, \tag{3.37}$$

where parameters λ_2 and θ_2 are both positive integers. Note that if $\theta_2 > 1$, we have $\lambda_2 < 1$. Therefore, the only feasible setting is $\theta_2 = 1$ and hence $\lambda_2 = 1$. In other words, the skeleton graph is given as a BIG-$(1, 1, k_2, 1)$ with $2q/k_2 - 1$ copies, where $k_2 = n > 2$ and the value of k_2 is supposed to be a factor of $2q$.

Based on the prior analysis, we consider two cases where $n = k_2 = 2$ and where $n = k_2 > 2$, respectively.

Case 1: $n = k_2 = 2$.

In this case, the skeleton graph is given as a BIG-$(1, 1, 2, 1)$ with $q - 1$ copies. Under such settings, given the value of parameter q, the adopted options of the design steps are given as follows.

- *Intra-module design*: The internal graph is a BIG-$(1, q, q, q)$.
- *Connection design*: In each module, all RNs are selected as dock nodes. Each group of q RNs serve as an intermediate super-node, i.e., an ISN-(q, q). Then the dock node interconnection is constructed under the bipartite mode. The composite-edge between any two ISNs is designed as a TD-$(2, 1, q)$ interconnection through an RN cluster.
- *Inter-module design*: The skeleton graph adopts a BIG-$(1, 1, 2, 1)$ with $q -$ 1 copies. The p modules are split into q groups, each of which contains two modules. For any two modules in the same group, they are connected with a common RN cluster via two super-edges. Such super-edges form a 1-parallelized TD-$(2, 1, q)$ interconnection between the two modules.
- *Finalization*: Each GN is directly connected with q CNs.

In fact, the resulting topology based on these designs is an AB Fat-Tree topology with p-port switches. In Figure 3.21, we show an instance of such a topology with $p = 4$, $q = 2$, and $n = k_2 = 2$. The performances of the topology are analyzed as follows.

- *Scalability.* Given the value of parameter q, the skeleton graph is given as a BIG-$(1, 1, 2, 1)$ with $q - 1$ copies that contains $2q$ modules. For each module, its internal graph is fixed as a BIG-$(1, q, q, q)$ that contains q GNs. Each GN

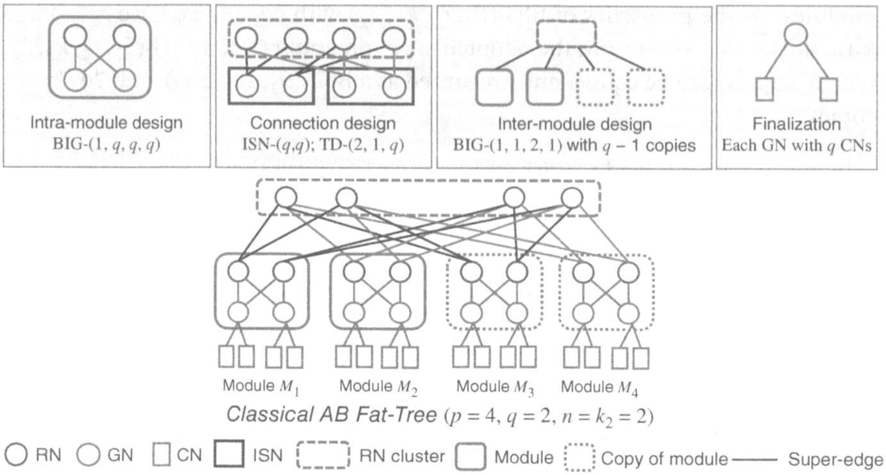

Figure 3.21 The complete design of an AB Fat-Tree topology with $p = 4$, $q = 2$, and $n = k_2 = 2$.

is directly connected with q CNs. Hence, the scalability S of the topology (i.e., the total number of CNs in the topology) is equal to

$$S = 2q \cdot q \cdot q = 2q^3. \tag{3.38}$$

- *Inter-module throughput.* Recall that the skeleton graph is given as a BIG-(1, 1, 2, 1) with $q - 1$ copies. For each pair of modules, they are connected with at least one super-edge under the bipartite mode. By the property of 1-parallelized TD-(2, 1, q) interconnection, each super-edge between two modules consists of q^2 edge-disjoint path. Hence, the minimum number of edge-disjoint paths A_2 between any two modules is equal to q^2. Meanwhile, the internal graph of each module is fixed as a BIG-(1, q, q, q) that contains q GNs. For each GN, it is directly connected with q CNs. Hence, the attachment number B_2 of each module is $q \cdot q = q^2$. By Definition 2.8, the inter-module throughput T_2 of the topology is equal to

$$T_2 = \frac{A_2}{B_2} = \frac{q^2}{q^2} = 1. \tag{3.39}$$

- *Fault-tolerance.* By duplicating operation, the topology contains a total number of $2q$ modules that are partitioned into q groups. Each group consists of two modules and corresponds to a copy of modules. Accordingly, there are $N = 2q(2q - 1)$ ordered pairs of modules that can be classified into two kinds depending on their connection patterns. The following are the detailed discussions and analysis.

- Case 1: In this case, two modules are the copies of a common module and have the same connection pattern with the RN cluster. Recall that in an AB Fat-Tree topology, there are q groups of modules. Hence, there are $\binom{q}{2}$ group pairs. For each group pair, there are four ordered pairs of modules with such a connection pattern. Hence, in this case, there are $4 \cdot \binom{q}{2} = 2q(q-1)$ ordered pairs of modules. Specifically, for the i-th ordered pair of modules $M_i^{(1)}$ and $M_i^{(2)}$, each dock node in $M_i^{(1)}$ is connected with one and only one dock node in $M_i^{(2)}$ through an RN cluster. Hence, given a dock node (denoted by symbol a) in module $M_i^{(1)}$, the failure of the dock node in module $M_i^{(2)}$ would disconnect the communications between two GNs in distinct modules through dock node a (i.e., $f_i = 1$).
- Case 2: In this case, two modules are not the copies of a common module and have distinct connection patterns with the RN cluster. Recall that there are $2q(q-1)$ ordered pairs of modules in case 1. Hence, the number of ordered module pairs in this case is equal to $N - 2q(q-1) = 2q(2q-1) - 2q(q-1) = 2q^2$. Specifically, for the i-th ordered pair of modules $M_i^{(1)}$ and $M_i^{(2)}$, each dock node in $M_i^{(1)}$ is connected with all q dock nodes in $M_i^{(2)}$ through an RN cluster. Hence, given a dock node (denoted by symbol b) in module $M_i^{(1)}$, only when the q dock nodes in module $M_i^{(2)}$ fail simultaneously, the communications between two GNs in distinct modules through dock node b in module $M_i^{(1)}$ would break down (i.e., $f_i = q$).

Based on this analysis and by Definition 2.9, the fault-tolerance F of the topology is equal to

$$\begin{aligned} F &= \frac{1}{N}\sum_{i=1}^{N} f_i \\ &= \frac{1}{2q(2q-1)}\left[2q(q-1)\cdot 1 + 2q^2 \cdot q\right] \\ &= \frac{q^2+q-1}{2q-1}. \end{aligned} \tag{3.40}$$

Case 2: $n = k_2 > 2$.

In this case, by Lemma 2.1, provided that q is a prime power, then there exists a TD-$(n, 1, q)$ if and only if $2 \le n \le q+1$. In particular, given $q = a^j$ as a prime power with a as a prime number and j as a positive integer, the feasible setting of n is $n = a^i$ if $0 < i \le j$ or $q = 2a^i$ if $0 < i < j$. Under such settings, the adopted options for the design steps are given as follows.

- *Intra-module design*: The internal graph of each module is fixed as a BIG-$(1, q, q, q)$.

- *Connection design*: In each module, all RNs are selected as dock nodes. Each group of q RNs form an intermediate super-node, i.e., an ISN-(q, q). Then the dock node interconnection is constructed under the bipartite mode. The composite-edge among q ISNs is designed as a TD-$(q, 1, q)$ interconnection through an RN cluster.
- *Inter-module design*: The skeleton graph is given as a BIG-$(1, 1, q, 1)$ with one copy. The p modules are split into q groups, each containing two modules. For any two modules in the same group, they are connected with a common RN cluster via q super-edges. Such q super-edges are constructed using a 1-parallelized TD-$(q, 1, q)$ interconnection.
- *Finalization*: Each GN is directly connected with q CNs.

We call the resulting topology based on these designs a *modified AB Fat-Tree*. In Figure 3.22, we provide an example of the topology with $p = 6$, $q = 3$, and $n = k_2 = 3$ for illustration. The performances of the topology are as follows.

- *Scalability*. Given the value of parameter q, the skeleton graph is fixed as a BIG-$(1, 1, q, 1)$ with one copy that contains $2q$ modules. For each module, its internal graph is given as a BIG-$(1, q, q, q)$ that contains q GNs. Each GN is directly connected with q CNs. Hence, the scalability S of the topology (i.e., the total number of CNs in the topology) is equal to

$$S = 2q \cdot q \cdot q = 2q^3. \tag{3.41}$$

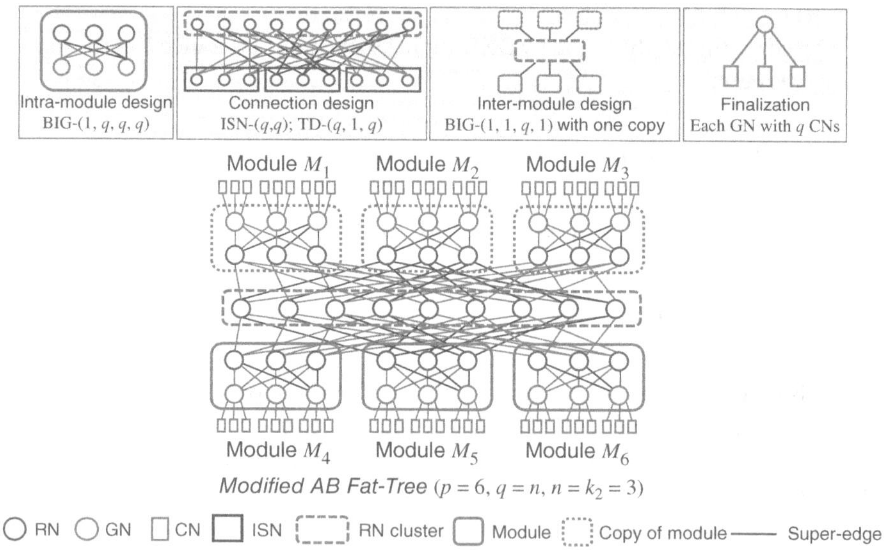

Figure 3.22 The complete design of a modified AB Fat-Tree topology with $p = 6$, $q = 3$, and $n = k_2 = 3$.

- *Inter-module throughput.* Recall that the skeleton graph is given as a BIG-$(1, 1, q, 1)$ with one copy. For each pair of modules, they are connected with at least one super-edge under the bipartite mode. By the property of 1-parallelized TD-$(q, 1, q)$ interconnection, each super-edge between two modules consists of q^2 edge-disjoint path. Hence, the minimum number of edge-disjoint paths A_2 between any two modules is equal to q^2. Meanwhile, the internal graph of each module is fixed as a BIG-$(1, q, q, q)$ that contains q GNs. For each GN, it is directly connected with q CNs. Hence, the attachment number B_2 of each module is equal to $q \cdot q = q^2$. By Definition 2.8, the inter-module throughput T_2 of the topology is equal to

$$T_2 = \frac{A_2}{B_2} = \frac{q^2}{q^2} = 1. \tag{3.42}$$

- *Fault-tolerance.* By duplicating operation, the topology contains a total number of $2q$ modules that are partitioned into $c_2 + 1$ groups. Each group consists of k_2 modules and corresponds to a copy of modules. Accordingly, there are $N = 2q(2q - 1)$ ordered pairs of modules. In particular, there are two kinds of connection pattern among all module pairs. The following are the descriptions and analysis of fault-tolerance under such two cases.
  - Case 1: In this case, two modules are the copies of a common module and have the same connection pattern with the RN cluster. Recall that in a modified AB Fat-Tree topology, there are $c_2 + 1$ groups of modules. Hence, there are $\binom{c_2+1}{2}$ group pairs. For each group pair, there are $2k_2$ ordered pairs of modules with such a connection pattern. Hence, in this case, there are $2k_2 \cdot \binom{c_2+1}{2}$ ordered pairs of modules. Specifically, for the i-th ordered pair of modules $M_i^{(1)}$ and $M_i^{(2)}$, each dock node in $M_i^{(1)}$ is connected with one and only one dock node in $M_i^{(2)}$ through an RN cluster. Hence, given a dock node (denoted by symbol a) in module $M_i^{(1)}$, the failure of *one* dock node in module $M_i^{(2)}$ would disconnect the communications between two GNs in distinct modules through dock node a (i.e., $f_i = 1$).
  - Case 2: In this case, two modules are not the copies of a common module and have distinct connection patterns with the RN cluster. Recall that there are $2q(q - 1)$ ordered pairs of modules in case 1. Hence, the number of ordered module pairs in this case is equal to $N - 2k_2 \binom{c_2+1}{2} = 2q(2q - 1) - 2k_2 \binom{c_2+1}{2}$. Specifically, for the i-th ordered pair of modules $M_i^{(1)}$ and $M_i^{(2)}$, each dock node in $M_i^{(1)}$ is connected with all q dock nodes in $M_i^{(2)}$ through an RN cluster. Hence, given a dock node (denoted by symbol b) in module $M_i^{(1)}$, only when the q dock nodes in module $M_i^{(2)}$ fail simultaneously, the communications between two GNs in distinct modules through dock node b in module $M_i^{(1)}$ would break down (i.e., $f_i = q$).

As a result, the fault-tolerance F of this topology is given by

$$\begin{aligned}
F &= \frac{1}{N}\sum_{i=1}^{N} f_i \\
&= \frac{1}{2q(2q-1)}\left\{2k_2 \cdot \binom{c_2+1}{2} \cdot 1 + \left[2q(2q-1) - 2k_2 \cdot \binom{c_2+1}{2}\right] \cdot q\right\} \\
&= q + \frac{2k_2 \cdot \binom{c_2+1}{2}}{2q(2q-1)} \cdot (1-q) \\
&= q + \frac{k_2 \cdot (c_2+1)c_2}{2q(2q-1)} \cdot (1-q) \\
&= 1 + (q-1) \cdot \frac{2q}{2q-1} \cdot \frac{k_2-1}{k_2}.
\end{aligned} \tag{3.43}$$

3.1.1.3 Performance Comparisons Among Tree-like Topologies

In the evolution of tree-like topology family, with different parameter settings for connection design and inter-module design, a number of tree-like topology designs are derived and parameterized in terms of a common set of parameters. Such topologies include existing well-known designs (i.e., Fat-Tree [Al-Fares et al. 2008], AB Fat-Tree [Liu et al. 2013], and Aspen Tree [Walraed-Sullivan et al. 2013]) and their modifications. These results allow us not only to understand and modify existing tree-like topologies in a systematic fashion but also to compare their performances in a unified manner.

In the rest of this subsection, we first demonstrate the parameter settings and discuss their impacts on the performances of such topologies. Next, we justify the design of existing tree-like topologies and compare them with their modifications. For ease of comparison, we summarize the performance comparisons of Fat-Tree (and its modifications), Aspen Tree (and its modifications), and AB Fat-Tree (and its modifications) in Tables 3.1–3.3, respectively.

The parameter settings and their impacts on performances. Based on the results presented in Tables 3.1–3.3, the parameter settings and their impacts on performances are illustrated as follows, respectively.

- **Parameter** x ($1 \leq x \leq q$). Given the value of t, the larger the value of x, the higher the fault-tolerance. Particularly, when $x > 1$, some constraints are required on the port number of switches in the topologies to guarantee the existence of TD-(n, μ, x) interconnections. Specifically, Tables 3.1 and 3.3 show that Fat-Tree and its modifications have the same value of t as AB Fat-Tree and its modifications (i.e., $t = 1$). Given this setting, we see the value of parameter x equals 1 under Fat-Tree and its modifications, while its value becomes q under AB Fat-Tree (for $q > 1$) and its modifications (given q as

Table 3.1 A summary of performance metrics of tree-like topologies with $x = 1$ and $t = 1$.

$x = 1, t = 1$	Fat-Tree	Modified Fat-Tree I	Modified Fat-Tree II	Modified Fat-Tree III	Modified Fat-Tree IV
θ_2	1	$\frac{q+1}{2}$	$q-1+\frac{1}{q}$	$2+\frac{1}{q}$	$2-\frac{1}{q}$
λ_2	q	1	1	$\frac{q-1}{2}$	$\frac{q}{2}$
Scalability	$2q^3$	q^4+q^3	$2q^4-2q^3+2q^2$	$4q^3+2q^2$	$4q^3-2q^2$
Latency	6	6	6	6	6
Intra-module throughput	1	1	1	1	1
Inter-module throughput	1	$\frac{1}{q}$	$\frac{1}{q}$	$\frac{1}{2}-\frac{1}{2q}$	$\frac{1}{2}$
Fault-tolerance	1	1	1	1	1
Existence condition	$q > 1$	$q > 1$	$q = 2$ or $q - 1$ is a prime power	q: odd number $3 \le q \le 211$	q: even number $2 \le q \le 212$

Table 3.2 A summary of performance metrics of tree-like topologies with $x = 1$ and $1 < t < q$.

$x = 1, 1 < t < q$ t is a factor of q	Aspen Tree*	Modified Aspen Tree I	Modified Aspen Tree II	Modified Aspen Tree III	Modified Aspen Tree IV
θ_2	1	$\frac{q+t}{2t}$	$\frac{q}{t}-1+\frac{t}{q}$	$2+\frac{t}{q}$	$2-\frac{t}{q}$
λ_2	$\frac{q}{t}$	1	1	$\frac{q-t}{2t}$	$\frac{q}{2t}$
Scalability	$\frac{2q^3}{t}$	$\frac{q^4+tq^3}{t}$	$\frac{2q^4-2tq^3+2t^2q^2}{t^2}$	$\frac{4q^3+2tq^2}{t}$	$\frac{4q^3-2tq^2}{t}$
Latency	6	6	6	6	6
Intra-module throughput	1	1	1	1	1
Inter-module throughput	1	$\frac{1}{q}$	$\frac{1}{q}$	$\frac{1}{2}-\frac{t}{2q}$	$\frac{1}{2}$
Fault-tolerance	t	t	t	t	t
Existence condition	$q > 1$	$q > 1$	$q = 2t$ or $\frac{q}{t} - 1$ is a prime power	q: odd number $3 \le q \le 211$	q: even number $2 \le q \le 212$

* The performance metrics of Aspen Tree also apply to the case with $t = q$.

Table 3.3 A summary of performance metrics of tree-like topologies with $x = q$ and $t = 1$.

$x = q, t = 1$ $\theta_2 = 1, \lambda_2 = 1$	AB Fat-Tree $k_2 = 2$	Modified AB Fat-Tree $2 < k_2 \leq q$ and k_2 is a factor of $2q$
Scalability	$2q^3$	$2q^3$
Latency	6	6
Intra-module throughput	1	1
Inter-module throughput	1	1
Fault-tolerance	$1 + (q-1) \cdot \frac{q}{2q-1}$	$1 + (q-1) \cdot \frac{2q}{2q-1} \cdot \frac{k_2 - 1}{k_2}$
Existence condition	$q > 1$	q is a prime power

a prime number). Such results verify that AB Fat-Tree and its modifications outperform Fat-Tree and its modifications in terms of higher fault-tolerances.

- **Parameter** t ($1 \leq t \leq q$ and t is a factor of q). Given the value of x, the larger the value of t, the higher the fault-tolerance and the lower the scalability. Specifically, Tables 3.1 and 3.2 show that Fat-Tree and its modifications have the same value of x as the Aspen Tree and its modifications (i.e., $x = 1$). Given this setting, we see that the fault-tolerance of Fat-Tree and its modifications equals one, whereas Aspen Tree and its modifications all have a fault-tolerance of t with t as a factor of q and $t > 1$. In other words, Aspen Tree and its modifications outperform Fat-Tree and its modifications in terms of a higher fault-tolerance. Nonetheless, we also see that the scalabilities of Aspen Tree and its modifications are generally lower than Fat-Tree and its modifications. For example, for Fat-Tree, the scalability equals $2q^3$ while for Aspen Tree, the scalability equals $2q^3/t$ that is smaller than $2q^3$ as $t > 1$. Besides, the scalability of Modified Fat-Tree III equals $4q^3 + 2q^2$ while for Modified Fat-Tree III, its scalability is $(4q^3 + 2tq^2)/t = 4q^3/t + 2q^2$ which is smaller than $4q^3 + 2q^2$ as $t > 1$. Such results verify that a larger value of t conduces to a higher fault-tolerance at the cost of a lower scalability.
- **Parameters** θ_2 **and** λ_2 ($\theta_2 \geq 1, \lambda_2 \geq 1$). By equation 3.1, the value of parameter θ_2 is inversely related to the value of parameter λ_2. Given fixed values of x and t, the larger value of θ_2 (resp. the smaller value of λ_2), the higher scalability and the lower inter-module throughput (see Figures 3.6 and 3.13). Specifically, as shown in Table 3.1, Fat-Tree and its modifications have the same values of x and t (i.e., $x = 1$ and $t = 1$) but differ in the values of θ_2 and

λ_2. Among such topologies, a topology with a larger value of θ_2 (resp. a larger value of λ_2) has a higher scalability (resp. a higher inter-module throughput). Similar conclusions can be drawn from Table 3.2 for Aspen Tree and its modifications. Such results verify that a larger value of θ_2 conduces to a higher scalability, while a larger value of λ_2 leads to a higher inter-module throughput.

Justification of existing tree-like topology designs and comparisons with their modifications. Based on the preceding discussions, the parameter selections of existing tree-like topology designs can be justified. Meanwhile, the differences between such topologies and their modifications can also be demonstrated. The details are presented as follows.

- *Fat-Tree and its modifications.* First, we justify the parameter selections of the classical Fat-Tree topology. Specifically, under the design of Fat-Tree:
  - The minimum value of t (i.e., $t = 1$) is achieved with a higher scalability at the cost of a lower fault-tolerance.
  - The minimum value of θ_2 (i.e., $\theta_2 = 1$) is achieved with a higher inter-module throughput at the cost of a lower scalability.
  - The values of x and θ_2 are set as $x = 1$ and $\theta_2 = 1$ so that the composite-edge and skeleton graph always exist for any $q > 1$. In other words, Fat-Tree has no extra restrictions on the port number of switches.

  Next, we discuss the differences between the classical Fat-Tree topology and its modifications based on results in Table 3.1.

  - First, the modified Fat-Trees have the same values of x and t as the classical Fat-Tree (i.e., $x = t = 1$), thus their fault-tolerance performances remain unchanged.
  - Second, compared with the classical Fat-Tree ($\theta_2 = 1$ and $\lambda_2 = q$), the modified Fat-Trees have larger values of θ_2 and smaller values of λ_2. For example, for Modified Fat-Tree I with $\theta_2 = (q + 1)/2$ and $\lambda_2 = 1$, we have $\theta_2 > 1$ as $q > 1$ and $\lambda_2 < q$. According to our previous discussion, such settings of Fat-Tree modifications lead to higher scalabilities and lower inter-module throughputs than the classical Fat-Tree.
  - Third, provided that $\theta_2 > 1$ for modified Fat-Trees, in some particular cases (e.g., when $q - 1$ is not a prime power for Modified Fat-Tree II and when q is an even number for Modified Fat-Tree III), there may not exist any BIG-$(\theta_2, r_2, k_2, \lambda_2)$ as the skeleton graph. In other words, the design of modified Fat-Trees may have extra restrictions on the port number of switches (refer to the existence conditions in Table 3.1).

- *Aspen Tree and its modifications.* Next, we justify the parameter selections of the classical Aspen Tree topology. Specifically, in the design of Aspen Tree:
  - The value of t is selected with $1 < t \leq q$. Compared to Fat-Tree with $t = 1$, such a setting implies a higher scalability at the cost of a lower fault-tolerance.
  - The minimum value of θ_2 (i.e., $\theta_2 = 1$) is achieved with a higher inter-module throughput at the cost of a lower scalability.
  - The values of x and θ_2 are set as $x = 1$ and $\theta_2 = 1$ so that the composite-edge and skeleton graph always exist for any value of q. In other words, Aspen Tree also has no extra restriction on the port number of switches.

  Next, we discuss the differences between the classical Aspen Tree topology and its modifications based on the results in Table 3.2.

  - First, the modified Aspen Trees have the same values of x and t as the Aspen Tree with $x = 1$ and $1 < t < q$, thus their fault-tolerance performances remain unchanged.
  - Second, compared to the classical Aspen Tree with $\theta_2 = 1$ and $\lambda_2 = q/t$, the modified Aspen Trees have larger values of θ_2 and smaller values of λ_2. Based on our previous discussion, with such parameter settings, modified Aspen Trees generally have higher scalabilities and lower inter-module throughputs than the classical Aspen Tree.
  - Third, provided that $\theta_2 > 1$ for modified Aspen Trees, in some particular cases (e.g., when $\frac{q}{t} - 1$ is not a prime power for Modified Fat-Tree II and when q is an even number for Modified Fat-Tree III), there may not exist any BIG-($\theta_2, r_2, k_2, \lambda_2$) as the skeleton graph. In other words, the design of modified Aspen Trees may have extra restrictions on the port number of switches (refer to the existence conditions in Table 3.2).
- *AB Fat-Tree and its modification.* Next, we justify the parameter selections of the classical AB Fat-Tree topology. Specifically, under the design of AB Fat-Tree:
  - The minimum value of t (i.e., $t = 1$) is achieved with a higher scalability at the cost of a lower fault-tolerance.
  - Compared to Fat-Tree and Aspen Tree (both with $x = 1$), a larger value of x (i.e., $x = q$) is selected under the classical AB Fat-Tree with a higher fault-tolerance. Meanwhile, to guarantee that a TD-(k_2, 1, q) interconnection as the composite-edge exists for any $q > 1$, the value of k_2 is set equal to two. In other words, the design of the classical AB Fat-Tree topology has no extra restriction on the port number of switches.

  Next, we discuss the differences between classical AB Fat-Tree topology and its modification based on the results in Table 3.3.

- First, Modified AB Fat-Tree has the same values of θ_2 and λ_2 as the classical AB Fat-Tree with $\theta_2 = 1$ and $\lambda_2 = q$, thus their scalability and inter-module throughput performances remain unchanged.
- Second, Modified AB Fat-Tree has the same values of x and t as the classical AB Fat-Tree (i.e., $x = q$ and $t = 1$). However, Modified AB Fat-Tree achieves an even higher fault-tolerance than the classical AB Fat-Tree. Note that the only difference in their parameter settings lies in the value selection of parameter k_2. From Table 3.3, we see that Modified AB Fat-Tree has a larger value of k_2. This implies that the value of k_2 may have an impact on the fault-tolerance of such topologies, which will be further analyzed in the next chapter.
- Third, Modified AB Fat-Tree has a larger value of k_2 than AB Fat-Tree with $k_2 = 2$. Nonetheless, when $k_2 > 2$, a TD-$(k_2, 1, q)$ interconnection as the composite-edge may not exist for some values of q. In other words, Modified AB Fat-Tree has extra restrictions on the port number of switches.

Remark 3.4 In the prior discussions, we analyzed tree-like topologies by demonstrating their parameter settings, justifying their parameter selections, and comparing existing classical designs with their modifications in a qualitative manner. However, some intriguing questions remain unanswered. For example, as larger values of t and x both conduce to higher fault-tolerances, which parameter has a greater impact? Besides, given that the classical AB Fat-Tree and Modified AB Fat-Tree have the same values of x and t, how can Modified AB Fat-Tree achieve a higher fault-tolerance? In addition, now that larger values of λ_2 conduce to higher inter-module throughputs, why do Fat-Tree with $\lambda_2 = q$ and the classical AB Fat-Tree with $\lambda_2 = 1$ both achieve the same inter-module throughput? To answer such questions, a *quantitative* characterization of the trade-offs among scalability, throughput, and fault-tolerance is required, which will be elaborated in Chapter 4.

3.1.2 Evolution of HyperX-like Topologies

For HyperX-like topologies, the port number of each switch is denoted by $p = p_1 + p_2 + p_3$, where p_1, p_2, and p_3 are all positive integers. In the evolution of such topologies, the intra-module design and part of connection design remain unchanged. Typically, the design procedure of HyperX-like topologies can be described as follows.

- Step 1: *Intra-module design.* A complete graph (CG) with $p_1 + 1$ nodes is employed as the internal graph for each module, in which each node is instantiated as a GN. Notably, the inner-degree of each GN is equal to p_1. An example with the internal graph fixed as a CG with four nodes (i.e., $p_1 = 3$) is presented in Figure 3.23.

- Step 2: *Connection design.*
  - *Dock node selection.* In each module, all of the $p_1 + 1$ GNs are selected as dock nodes. Such dock nodes are partitioned into $(p_1 + 1)/x$ ISNs-(x, w, l) (x should be a factor of $p_1 + 1$). Each dock node is adjacent to all GNs with its outer-degree as p_2. By Theorem 2.1, the parameter settings for ISNs-(x, w, l) satisfy $w = x$ and $l = 1$, aka ISNs-(x, x). Figure 3.24 shows the dock node selection in the case where $x = 1$.
  - *Dock node interconnection.* Dock nodes in distinct modules are all interconnected under the plain mode. Specifically, two ISNs-(x, w, l) are interconnected by a composite-edge that is designed as an all-to-all interconnection.
- Step 3: *Inter-module design.*
  - *Skeleton graph design.* Based on the connection design, the skeleton graph is designed as a CG with all nodes instantiated as modules. Let m denote the number of nodes (modules) in the skeleton graph. By the definition of CG, the degree of each node (module) is $m - 1$. Figure 3.25a shows such a skeleton graph with three modules (i.e., $m = 3$).
  - *Super-edge instantiation.* In the skeleton graph, each pair of modules are interconnected by a super-edge that is instantiated as a t-parallelized all-to-all interconnection. Under such a design, $2t$ ISNs-(x, x) in two distinct modules interconnected (i.e., t ISNs-(x, x) in each module). Each dock node in the ISNs of one module is directly connected with all the x dock nodes in one ISN-(x, x) of the other module. Therefore, each super-edge occupies tx^2 outer-degrees of tx dock nodes in each module. Meanwhile, each module is incident to $m - 1$ super-edges and each module contains $p_1 + 1$ GNs with outer-degree p_2. As a result, we have $(m - 1) \cdot tx^2 = (p_1 + 1)p_2$ or equivalently,

$$m = \frac{(p_1 + 1)p_2}{tx^2} + 1. \tag{3.44}$$

    Since m is a positive integer, tx^2 should be a factor of $(p_1 + 1)p_2$. In Figure 3.25b, we show a t-parallelized all-to-all interconnection between two modules with $t = 1$.
- Step 4: *Finalization.* Each GN in the modules is directly connected with p_3 CNs.

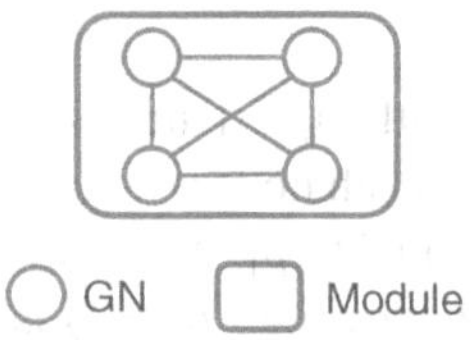

Figure 3.23 An example with the internal graph as a CG with four GNs ($p_1 = 3$).

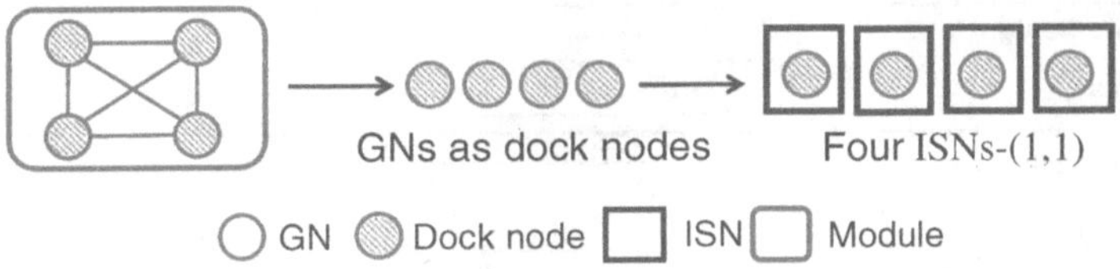

Figure 3.24 An example of dock node selection. In the given module, the four GNs in the module are all selected as dock nodes. Such dock nodes are grouped as four ISNs-(1, 1) ($x = 1$).

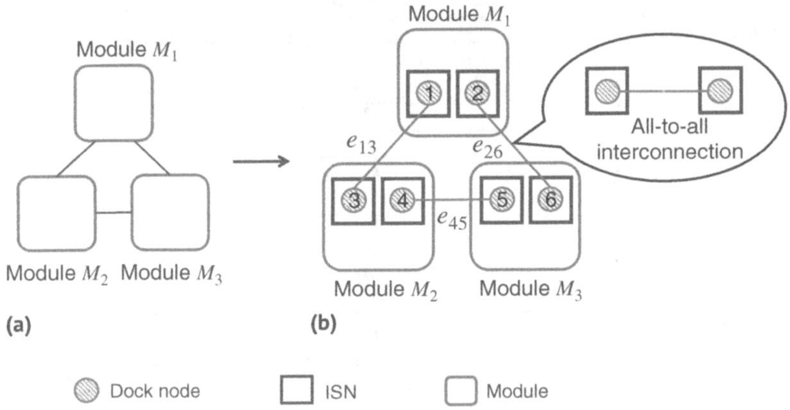

Figure 3.25 An illustration of skeleton graph design and super-edge instantiation. (a) A CG with three modules as the skeleton graph. (b) The super-edge between each pair of modules is instantiated as a 1-parallelized all-to-all interconnection (i.e., $t = 1$). To be specific, the super-edge between module M_1 and module M_2 consists of one edge e_{13}. The super-edge between module M_1 and module M_3 consists of one edge e_{26}. The super-edge between module M_2 and module M_3 consists of one edge e_{45}. Each module contains two GNs (i.e., $p_1 + 1 = 2$) with outer-degree 1 (i.e., $p_2 = 1$). The two GNs in each module are partitioned into two ISNs. Each ISN contains one dock node (i.e., $x = 1$). By equation (3.44), we have $m = 3$ (i.e., the number of modules in the skeleton graph equals three).

The design procedure implies that different HyperX-like topologies can be derived and modified by adjusting the parameter settings in the second and the third design steps.[6] A roadmap of the evolution of HyperX-like topologies is presented in Figure 3.26. To provide a guideline for parameter selections, we illustrate impacts of the parameters on the performance metrics as follows.

First, recall that the intra-module design and part of design options in connection design are fixed for HyperX-like topologies. Accordingly, the intra-module throughput and the value range of latency remain unchanged throughout the evolution of the topology family. This is demonstrated by the following theorem (see the proof in Section A.3.2).

6 A few examples can be found in Sections 2.5.5 and 2.5.6.

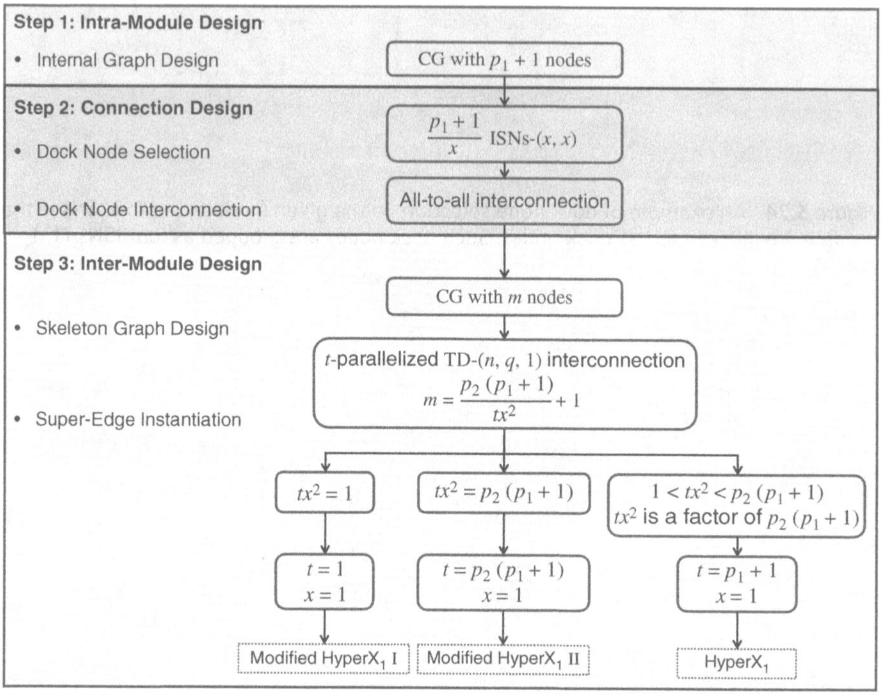

Figure 3.26 A roadmap of the evolution of HyperX-like topologies. The finalization step is omitted since in HyperX-like topologies, each GN in the modules is directly connected with p_3 CNs (servers).

Theorem 3.2 *Following the design procedure with one iteration, HyperX-like topologies have a constant intra-module throughput as* $\frac{p_1}{p_3}$. *Meanwhile, HyperX-like topologies have a latency equal* 3, 4, *or* 5. *Specifically, the latency is equal to three if and only if their parameter settings satisfy* $m = 1 + \frac{p_2}{p_1+1}$. *In the case where* $m \neq 1 + \frac{p_2}{p_1+1}$, *the latency is four if* $tx \geq p_1 + 1$.

When it comes to other performance metrics, i.e., the scalability, the inter-module throughput, and the fault-tolerance, our discussions are given as follows.

- *Scalability*. Since parameter m denotes the total number of modules in the topology, given the intra-module design, the larger the value of m, the more the modules (hence more CNs) and the higher the scalability.
- *Inter-module throughput*. Each pair of modules are interconnected by a super-edge using a t-parallelized all-to-all interconnection. Such a super-edge comprises tx^2 edge-disjoint paths between dock nodes in the two modules. Hence, a larger value of tx^2 leads to a higher inter-module throughput between modules.

- *Fault-tolerance.* Given a source GN in one module and a target GN in another module, consider their communications over the paths with the minimum length. Recall that in HyperX-like topologies, the interconnections between dock nodes are conducted under the plain mode as t-parallelized all-to-all interconnection. As a result, their communications would traverse through the direct connections between dock nodes in the two modules. Given a dock node in the same module as the source GN, by the property of t-parallelized all-to-all interconnection, such a dock node is adjacent to x dock nodes in the same module as the target GN. Under such settings, the simultaneous failures of x dock nodes in the same module as the target GN would disconnect the communications through the paths with the minimum length. By Definition 2.9, it follows that a larger value of x leads to a higher fault-tolerance.

Remark 3.5 The aforementioned discussions have two implications.

- First, given the constant values of m and tx^2, the scalability and the inter-module throughput are fixed, while the fault-tolerance can be improved by increasing the value of parameter x (and decreasing the value of parameter t since tx^2 is a constant). However, to ensure tx^2 as a factor of $(p_1 + 1)p_2$, the constraint that $x > 1$ would impose extra restrictions on the value selections of p_1 and p_2 with $p_1 + p_2 < p$.
- Second, there exists a trade-off between scalability and inter-module throughput. Specifically, recall that a larger value of m leads to a higher scalability, while a larger value of tx^2 leads to a higher inter-module throughput. Then by equation (3.44), given the values of parameters p_1 and p_2, there exists a trade-off between scalability and inter-module throughput, as illustrated in Figure 3.27.

For Hyper-like topologies, the intra-module design and part of design options in connection design are fixed. Nonetheless, parameter settings in the remaining steps still leave a few degrees of freedom for topology design. Particularly, by equation (3.44), for the inter-module design of such topologies, tx^2 should be a factor of $(p_1 + 1)p_2$. Then given fixed values of parameters p_1, p_2, and p_3, different value selections of tx^2 would lead to three branches of evolution for HyperX-like topologies.

- Branch 1: $tx^2 = 1$ and $m = (p_1 + 1)p_2 + 1$, in which a higher scalability is achieved at the cost of a lower inter-module throughput.
- Branch 2: $tx^2 = (p_1 + 1)p_2$ and $m = 2$, in which a higher inter-module throughput is achieved at the cost of a lower scalability.
- Branch 3: $1 < tx^2 < (p_1 + 1)p_2$ and $2 < m < (p_1 + 1)p_2 + 1$.

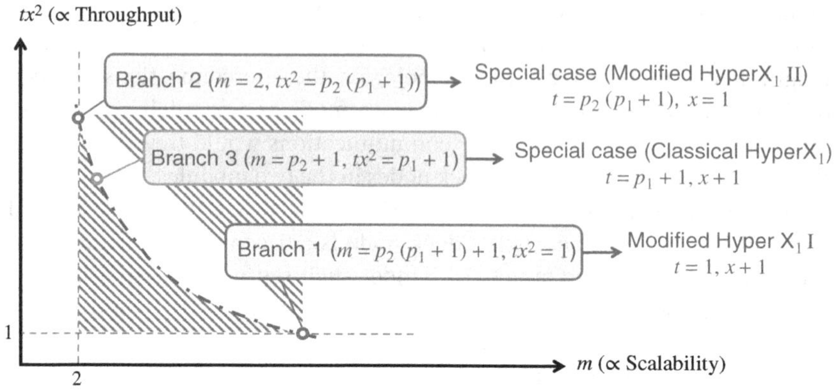

Figure 3.27 The trade-off between scalability and throughput in terms of parameters tx^2 and m.

The following are the detailed discussions.

3.1.2.1 Branch 1: $tx^2 = 1$ and $m = (p_1 + 1)p_2 + 1$.

Along with this branch, the only feasible parameter settings are $t = 1$ and $x = 1$ since $tx^2 = 1$ and they are both positive integers. Under such settings, given fixed values of parameters p_1, p_2, and p_3, the adopted options for the design steps are summarized as follows.

- *Intra-module design*. The internal graph is a CG with $p_1 + 1$ GNs.
- *Connection design*. In each module, all GNs are selected as dock nodes. Each dock node forms an intermediate super-node (i.e., an ISN-(1, 1)). The dock node interconnection is constructed under the plain mode. To be specific, each composite-edge between two ISNs is designed as an all-to-all interconnection.
- *Inter-module design*. The skeleton graph is given as a CG with $(p_1 + 1)p_2 + 1$ modules. Each pair of modules are directly connected through a super-edge. Such a super-edge is constructed by a 1-parallelized all-to-all interconnection.
- *Finalization*. Each GN is directly connected with p_3 CNs.

We call the resulting topology based on these designs a *modified HyperX$_1$ I*. An instance of such a topology with $p_1 = 1$, $p_2 = 2$, and $p_3 = 1$ is presented in Figure 3.28. The performance metrics of the topology are analyzed as follows.[7]

---

7 By Theorem 3.2, for Hyper-like topology with one iteration, given the values of parameters p_1 and p_3, the intra-module throughput T_1 is always equal to p_1/p_3. Unless otherwise stated, it is omitted in the subsequent analysis for Hyper-like topologies in the rest of this subsection.

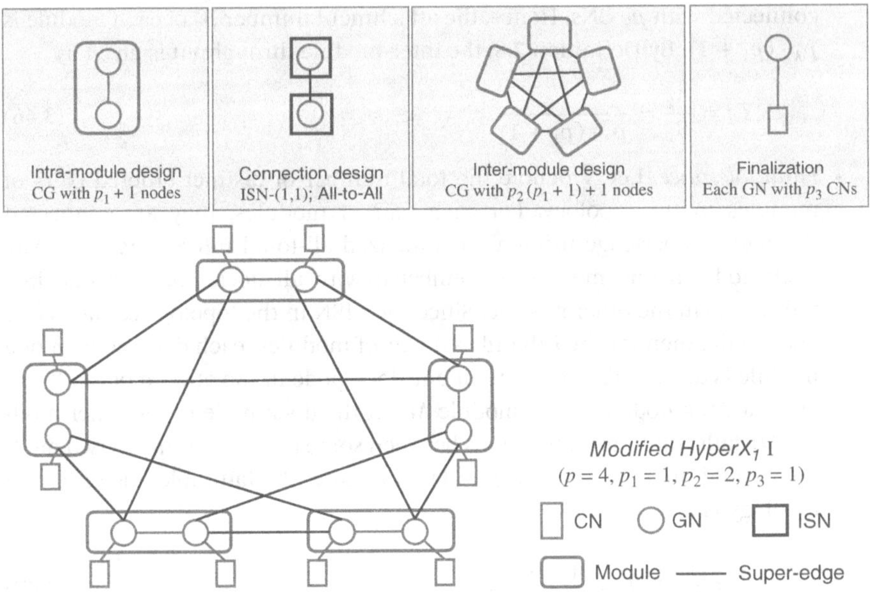

Figure 3.28 The complete design of a modified HyperX$_1$ I topology with $p_1 = 1$, $p_2 = 2$, and $p_3 = 1$.

- *Scalability.* For the topology, the skeleton graph is fixed as a CG with $(p_1 + 1)p_2 + 1$ modules. For each module, its internal graph is fixed as a CG with $p_1 + 1$ GNs. Each GN is directly connected with p_3 CNs. Hence, the scalability S of the topology (i.e., the total number of attached CNs in the topology) equals

$$S = [(p_1 + 1)p_2 + 1] \cdot (p_1 + 1) \cdot p_3. \tag{3.45}$$

- *Latency.* For the modified HyperX$_1$ I topology, we have $m = (p_1 + 1)p_2 + 1$. Since $p_1 > 0$ (i.e., $p_1 + 1 > 1$), then we have $m > 1 + \dfrac{p_2}{p_1 + 1}$. Meanwhile, given that $x = 1$, we have $tx = tx^2 = 1 < p_1 + 1$. By Theorem 3.2, the latency of the topology equals four or five in this case.
- *Inter-module throughput.* Recall that the skeleton graph is designed as a CG where each pair of modules are connected through a super-edge that is designed as a 1-parallelized all-to-all interconnection. By the property of such interconnection, each super-edge between two modules consists of tx^2 edge-disjoint paths. Hence, the minimum number of edge-disjoint paths A_2 between any two modules is $tx^2 = 1$. Meanwhile, the internal graph of each module is fixed as a CG that contains $p_1 + 1$ GNs. For each GN, it is directly

connected with p_3 CNs. Hence, the attachment number B_2 of each module is $p_3 \cdot (p_1 + 1)$. By Definition 2.8, the inter-module throughput is given by

$$T_2 = \frac{A_2}{B_2} = \frac{1}{p_3 \cdot (p_1 + 1)}. \tag{3.46}$$

- *Fault-tolerance.* Let N denote the total number of distinct ordered pairs of modules in the topology. For each pair of modules, they are connected through a super-edge using a 1-parallelized all-to-all interconnection. Any dock node in one module is connected with all dock nodes of one ISN (i.e., $t = 1$) in the other module. Since each ISN in the topology contains one dock node, then for the i-th ordered pair of modules, each dock node in one module is directly connected with one dock node in the other module. Then given a dock node in some module M_i^1, if the dock node in the other module M_i^2 fails, the communications between some pair of GNs in the two modules would break down (i.e., $f_i = 1$). Therefore, the fault-tolerance F of the topology is

$$F = \frac{1}{N}\sum_{i=1}^{N} f_i = \frac{1}{N}\sum_{i=1}^{N} 1 = 1. \tag{3.47}$$

3.1.2.2 Branch 2: $tx^2 = (p_1 + 1)p_2$ and $m = 2$.

Along with this branch, we consider a special case in which $t = (p_1 + 1)p_2$ and $x = 1$.[8] In this case, given the fixed values of parameters p_1, p_2, and p_3, the adopted options for the design steps are given as follows.

- *Intra-module design.* The internal graph is designed as a CG with $p_1 + 1$ GNs.
- *Connection design.* In each module, all GNs are selected as dock nodes. Each dock node forms an intermediate super-node (i.e., an ISN-(1, 1)). The dock node interconnection is constructed under the plain mode. To be specific, each composite-edge between two ISNs is designed as an all-to-all interconnection.
- *Inter-module design.* The skeleton graph is designed as a CG with two modules. Such two modules are directly connected through a super-edge. This super-edge is constructed by a $p_2(p_1 + 1)$-parallelized all-to-all interconnection.
- *Finalization.* Each GN is directly connected with p_3 CNs.

We call the resulting topology based on these designs a *modified HyperX$_1$ II* topology. In Figure 3.29, we show an instance of such a topology with $p_1 = 1$, $p_2 = 2$, and $p_3 = 1$. The performances of the topology are analyzed as follows.

8 Some other feasible settings of parameters t and x may lead to topologies with higher fault-tolerances. Further analysis to explore such designs is left to interested readers for future work.

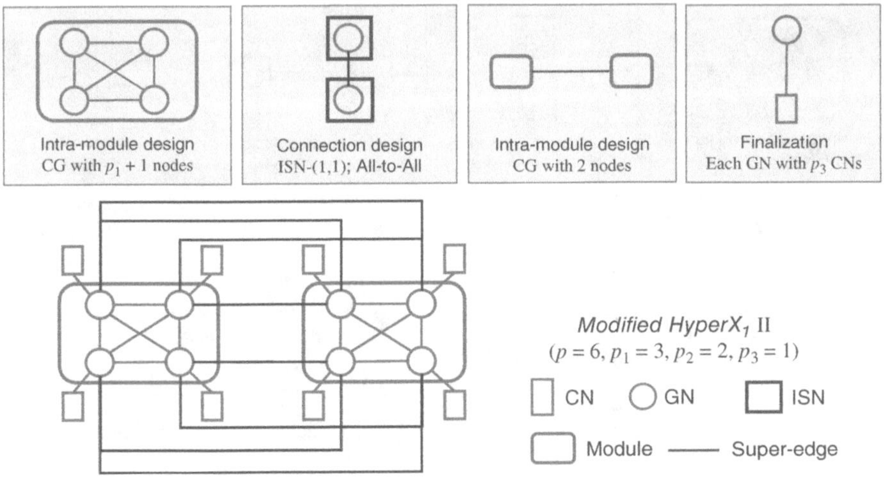

Figure 3.29 The complete design of a modified HyperX$_1$ II topology with $p_1 = 3$, $p_2 = 2$, and $p_3 = 1$. The latency of such a topology is four.

- *Scalability.* Recall that the skeleton graph is designed as a CG with two modules. For each module, its internal graph is fixed as a CG with $p_1 + 1$ GNs. Each GN is directly connected with p_3 CNs. Hence, the scalability S of the topology (i.e., the total number of attached CNs in the topology) is given by

$$S = 2(p_1 + 1) \cdot p_3. \tag{3.48}$$

- *Latency.* For the modified HyperX$_1$ II topology, we have $m = 2$. Particularly, we consider two sub-cases.
  - If $p_2 \neq p_1 + 1$ (see the example in Figure 3.29), then $m > 1 + \frac{p_2}{p_1 + 1}$. Since $x = 1$ and $p_2 \geq 1$, we have $tx = tx^2 = p_2(p_1 + 1) \geq p_1 + 1$. In this sub-case, by Theorem 3.2, the latency of the topology equals four.
  - If $p_2 = p_1 + 1$, then $m = 1 + \frac{p_2}{p_1 + 1} = 2$ (see the example in Figure 3.30). In this sub-case, by Theorem 3.2, the latency of the modified HyperX$_1$ II topology equals three.
- *Inter-module throughput.* Recall that the skeleton graph is designed as a CG where each pair of modules are connected through a super-edge. By the property of 1-parallelized all-to-all interconnection, each super-edge between two modules consists of tx^2 edge-disjoint paths. Hence, the minimum number of edge-disjoint paths A_2 between any two modules is $tx^2 = p_2(p_1 + 1)$. Meanwhile, the internal graph of each module is fixed as a CG that contains $p_1 + 1$ GNs. Each GN is directly connected with p_3 CNs. Hence, the attachment

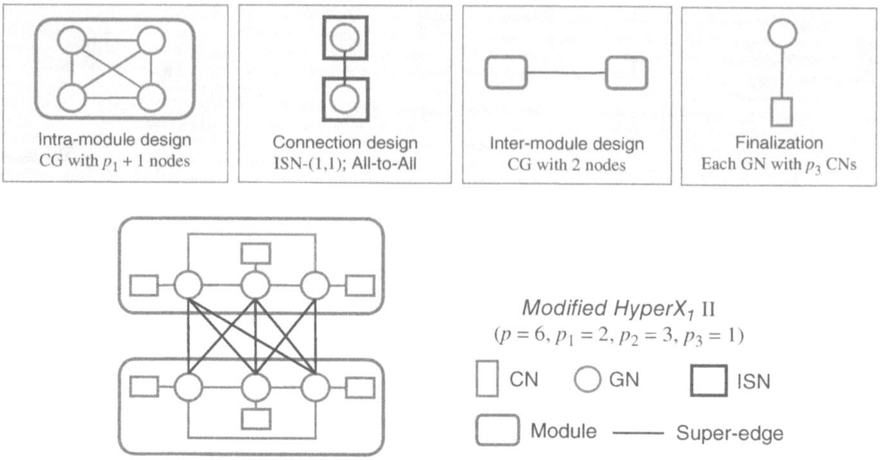

Figure 3.30 An example of the modified HyperX$_1$ II topology with $m = 2$, $p_1 = 2$, and $p_2 = 3$. The latency of such a topology equals three.

number B_2 of each module is $p_3 \cdot (p_1 + 1)$. By Definition 2.8, the inter-module throughput equals

$$T_2 = \frac{A_2}{B_2} = \frac{p_2}{p_3}. \tag{3.49}$$

- *Fault-tolerance.* Let N denote the total number of ordered pairs of modules in the topology. Each pair of modules are connected through a super-edge using a 1-parallelized all-to-all interconnection. Accordingly, any dock node in one module is connected with all dock nodes of one ISN (i.e., $t = 1$) in the other module. Since each ISN contains one dock node, for the i-th ordered pair of modules, each dock node in one module is directly connected with one dock node in the other module. Therefore, given a dock node in some module M_i^1, if the dock node in the other module M_i^2 fails, the communication between some pair of GNs in the two modules would break down (i.e., $f_i = 1$). As a result, the fault-tolerance F of the topology is equal to

$$F = \frac{1}{N}\sum_{i=1}^{N} f_i = \frac{1}{N}\sum_{i=1}^{N} 1 = 1. \tag{3.50}$$

3.1.2.3 Branch 3: $1 < tx^2 < (p_1 + 1)p_2$ and $2 < m < (p_1 + 1)p_2 + 1$.

Along with this branch, there exist a number of feasible settings for parameters m, t, and x. For example, we can set $t = p_1 + 1$ and $x = 1$. Under such settings, given the fixed values of parameters p_1, p_2, and p_3, the adopted options for the design steps are given as follows.

- *Intra-module design*. The internal graph is designed as a CG with $p_1 + 1$ GNs.
- *Connection design*. In each module, all GNs are selected as dock nodes. Each dock node forms an intermediate super-node (i.e., an ISN-(1, 1)). The dock node interconnection is constructed under the plain mode. To be specific, each composite-edge between two ISNs is designed as an all-to-all interconnection.
- *Inter-module design*. The skeleton graph is designed as a CG with $p_2 + 1$ modules. Each pair of modules are directly connected through a super-edge. Each super-edge is constructed by a $(p_1 + 1)$-parallelized all-to-all interconnection.
- *Finalization*. Each GN is directly connected with p_3 CNs.

In fact, the resulting topology based on these designs is a classical HyperX$_1$ topology. For ease of understanding, we show an instance of such a topology with $p_1 = 1$, $p_2 = 2$, and $p_3 = 1$ in Figure 3.31. The performance metrics of such a topology are analyzed as follows.

- *Scalability*. For the topology, the skeleton graph is a CG with $p_2 + 1$ modules. For each module, its internal graph is fixed as a CG with $p_1 + 1$ GNs. Each

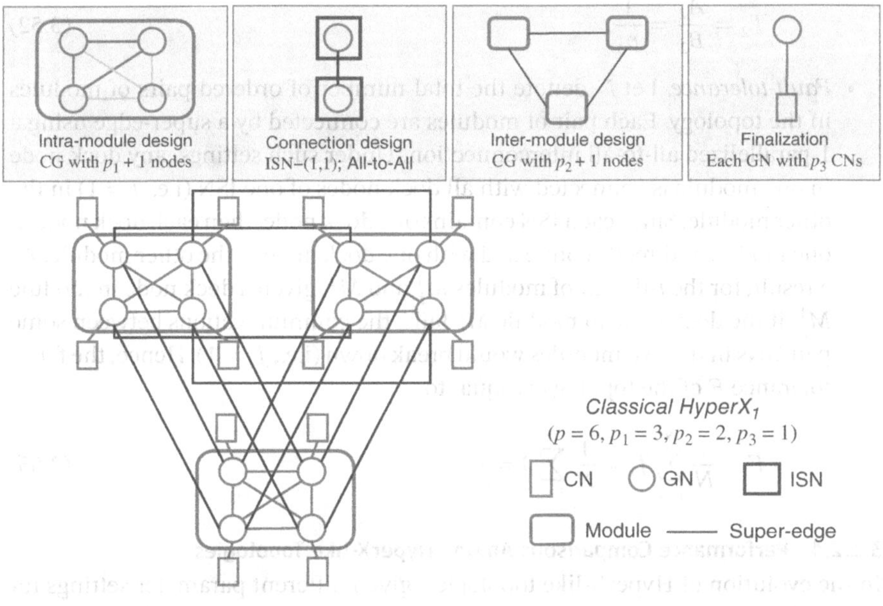

Figure 3.31 The complete design of a classical HyperX$_1$ topology with $p_1 = 3$, $p_2 = 2$, and $p_3 = 1$.

GN is directly connected with p_3 CNs. Hence, the scalability S of the topology (i.e., the total number of attached CNs in the topology) is equal to

$$S = (p_2 + 1) \cdot (p_1 + 1) \cdot p_3. \tag{3.51}$$

- *Latency.* For the topology, its parameter settings satisfy $m = p_2 + 1$. Since $p_1 > 0$ (i.e., $p_1 + 1 > 1$), we have $m = p_2 + 1 > \frac{p_2}{p_1 + 1} + 1$. Meanwhile, since $t = p_1 + 1$ and $x = 1$, we have $tx = p_1 + 1$. By Theorem 3.2, the latency of HyperX_1 topologies is four.
- *Inter-module throughput.* In the topology, each pair of modules are connected by a super-edge that is constructed as a 1-parallelized all-to-all interconnection. By the property of such interconnection, each super-edge between two modules comprises tx^2 edge-disjoint paths. Hence, the minimum number of edge-disjoint paths A_2 between any two modules is $tx^2 = p_1 + 1$. Meanwhile, the internal graph of each module is fixed as a CG with $p_1 + 1$ GNs. Each GN is directly connected with p_3 CNs. Hence, the attachment number B_2 of each module is $p_3 \cdot (p_1 + 1)$. By Definition 2.8, the inter-module throughput is equal to

$$T_2 = \frac{A_2}{B_2} = \frac{1}{p_3}. \tag{3.52}$$

- *Fault-tolerance.* Let N denote the total number of ordered pairs of modules in the topology. Each pair of modules are connected by a super-edge using a 1-parallelized all-to-all interconnection. Under such settings, any dock node in one module is connected with all dock nodes of one ISN (i.e., $t = 1$) in the other module. Since each ISN contains one dock node, then each dock node in one module is directly connected with one dock node in the other module. As a result, for the i-th pair of modules M_i^1 and M_i^2, given a dock node in module M_i^1, if the dock node in module M_i^2 fails, the communications between some pair GNs in the two modules would break down (i.e., $f_i = 1$). Hence, the fault-tolerance F of the topology is equal to

$$F = \frac{1}{N} \sum_{i=1}^{N} f_i = \frac{1}{N} \sum_{i=1}^{N} 1 = 1. \tag{3.53}$$

3.1.2.4 Performance Comparisons Among HyperX-like Topologies

In the evolution of HyperX-like topologies, given different parameter settings for connection design and inter-module design, a number of HyperX-like topologies are derived and analyzed in terms of a common set of parameters (i.e., parameters x, m, and t). Such topologies include the classical HyperX_1 topology (Ahn et al. 2009) and its modifications. The analysis and discussions allow us not only to understand and modify existing HyperX-like topology designs in a systematic fashion but also to compare their performances in a unified manner.

In the rest of this subsection, we first demonstrate the parameter settings and discuss their impacts on the performance metrics of such topologies. Then we exploit such settings to justify the design of the classical HyperX_1 topology and compare it with its modifications. For ease of reference, we summarize the performances of the classical HyperX_1 topology and its modifications in Table 3.4.

The parameter settings and their impacts on performance. Based on the results in Table 3.4, the parameter settings and their impacts on performance metrics are illustrated as follows.

- **Parameter** m ($2 \leq m \leq p_2(p_1+1)+1$). Table 3.4 shows that given the value of x, the larger the value of m, the higher the scalability. Specifically, the classical HyperX_1 topology has a value of m that is less than the modified HyperX_1 I topology (as $p_2+1 < p_2(p_1+1)+1$) and greater than the modified HyperX_1 II topology (as $p_2+1 > 2$). Accordingly, the classical HyperX_1 topology achieves a scalability that is higher than the modified HyperX_1 II and lower than the modified HyperX_1 I. Such results imply that the scalability of the classical HyperX_1 can be improved by adopting the design of the modified HyperX_1 I topology.
- **Parameter** tx^2 (as a factor of $p_2(p_1+1)$ with $t \geq 1$ and $x \geq 1$). Table 3.4 shows that given the value of x, the value of parameter t is positively proportional to the inter-module throughput of the topologies. Specifically, the classical HyperX_1 topology has a value of t that is greater than the modified HyperX_1 I (as $p_1+1 > 1$) and less than the modified HyperX_1 II (as $p_1+1 < p_2(p_1+1)$). Accordingly, the classical HyperX_1 topology achieves an inter-module throughput that is higher than the modified HyperX_1 I

Table 3.4 A summary of performance metrics of HyperX-like topologies.

$x=1$	HyperX_1	Modified HyperX_1 I	Modified HyperX_1 II
m	p_2+1	$p_2(p_1+1)+1$	2
t	p_1+1	1	$p_2(p_1+1)$
Scalability	$(p_2+1)(p_1+1)p_3$	$[p_2(p_1+1)+1](p_1+1)p_3$	$2(p_1+1)p_3$
Latency	4	4 or 5	3 or 4
Intra-module throughput	$\frac{p_1}{p_3}$	$\frac{p_1}{p_3}$	$\frac{p_1}{p_3}$
Inter-module throughput	$\frac{1}{p_3}$	$\frac{1}{p_3(p_1+1)}$	$\frac{p_2}{p_3}$
Fault-tolerance	1	1	1
Existence conditions	$p_1, p_2, p_3 > 1$	$p_1, p_2, p_3 > 1$	$p_1, p_2, p_3 > 1$

$\left(\text{as } \frac{1}{p_3} > \frac{1}{p_3(p_1+1)}\right)$ and lower than the modified HyperX$_1$ II $\left(\text{as } \frac{1}{p_3} < \frac{p_2}{p_3}\right)$. Such results imply that the inter-module throughput of the classical HyperX$_1$ topology can be improved by adopting the design of the modified HyperX$_1$ II topology.

Justification of the design of the classical HyperX$_1$ topology and comparisons with its modifications. Based on the results in Table 3.4, the parameter settings of the classical HyperX$_1$ topology can be justified and its differences from the modifications can be demonstrated. Specifically, under the design of the classical HyperX$_1$ topology:

- The parameter settings $x = 1$ and $t = p_1 + 1$ are adopted to achieve a balance between scalability and inter-module throughput.
- The value of tx^2 is set as $p_1 + 1$ (i.e., always a factor of $(p_1 + 1)p_2$). Hence, the design of the classical HyperX$_1$ topology has no extra restrictions on the port number of switches.

Next, we discuss the differences between the classical HyperX$_1$ topology and its modifications based on results in Table 3.4.

- First, the modified HyperX$_1$ topologies have the same value of x as the classical HyperX$_1$ topology (i.e., $x = 1$). Hence, their fault-tolerance performances remain unchanged.
- Second, for the classical HyperX$_1$ topology and its modifications, the value of tx^2 is always a factor of $p_2(p_1 + 1)$ for any values of p_1, p_2, and p_3. Therefore, the designs of such topologies have no extra restrictions on the port number of switches (refer to the existence conditions in Table 3.4).
- Third, compared with the classical HyperX$_1$ with $m = p_2 + 1$ and $tx^2 = p_1 + 1$, the modified HyperX$_1$ I topology has a larger value of m (i.e., $m = p_2(p_1 + 1) + 1$) and a smaller value of tx^2 (i.e., $tx^2 = 1$), while the modified HyperX$_1$ II topology has a smaller value of m (i.e., $m = 2$) and a larger value of tx^2 (i.e., $tx^2 = p_2(p_1 + 1)$). Based on our previous discussion, such settings of the modified HyperX$_1$ I (resp. the modified HyperX$_1$ II) lead to a higher scalability (resp. a lower scalability) and a lower inter-module throughput (resp. a higher scalability) than the classical HyperX$_1$ topology.

3.2 Evolution of Server-centric Topologies

In this section, our focus is shifted onto the evolution of server-centric topologies. In a typical server-centric topology, servers are equipped with multiple network interface cards (NICs) and employed to provide routing function.[9]

9 In practice, server-centric topologies (e.g., the classical DCell topology [Guo et al. 2008] and BCube topology [Guo et al. 2009]) are usually constructed by interconnecting several

In the rest of this section, for ease of analysis, we consider the evolution of server-centric topologies under the unified design procedure with *one* iteration. By default, all switches are assumed to have the same number of p ports ($p \geq 2$) and modeled as RNs with degree p. Meanwhile, since all servers are responsible for routing function, they are modeled as HNs and each HN is assumed to have a degree of two. In other words, there are no CNs in such topologies.

Based on the described settings, the design procedure for server-centric topologies is summarized as follows.

- Step 1: *Intra-module design.* A BIG-($\theta_1, r_1, k_1, \lambda_1$) with c_1 copies is employed as the internal graph for each module. In such a graph, the element nodes and the block nodes are modeled as HNs and RNs, respectively. By Definition 1.23, the inner-degrees of each HN and each RN equal r_1 and $k_1(c_1 + 1)$, respectively.
- Step 2: *Connection design.*
  - *Dock node selection.* In each module, all HNs are selected as dock nodes and grouped as ISNs-(x, w, l). Recall that the degree of each HN equals two, while the degree of each RN is p. Then for each HN, its outer-degree and inner-degree are both set to be one since they are supposed to be positive. For each RN in modules, its inner-degree and outer-degree are set as p and zero, respectively, as RNs are not selected as dock nodes.
  - *Dock node interconnection.* In this sub-step, different choices of interconnection modes lead to two families of topology designs. When dock nodes in distinct modules are connected under the plain mode (i.e., given two ISNs-(x, w, l) in different modules, they are interconnected through a composite-edge as an all-to-all interconnection), then the topology design would evolve toward the level-1 DCell topology and its variants (collectively referred to as *DCell-like* topologies). When dock nodes in distinct modules are connected under the bipartite mode (i.e., given a number of n ISNs-(x, w, l), they are interconnected by a composite-edge as a TD-(n, μ, x) interconnection), then the topology design would evolve toward the level-1 BCube topology and its variants (collectively referred to as *BCube-like* topologies).
- Step 3: *Inter-module design.*
  - *Skeleton graph design.* With this connection design, a complete graph (CG) with m modules is employed as the skeleton graph for DCell-like topologies, while the skeleton graph for BCube-like topologies is a BIG-($\theta_2, r_2, k_2, \lambda_2$) with c_2 copies.

---

homogeneous containers, each containing a number of servers and switches. Therefore, such topologies are also known as *container-based* topologies.

  - *Super-edge instantiation*. The super-edge instantiations are also different between DCell-like topologies and BCube-like topologies. In DCell-like topologies, the skeleton graph is a CG in which each super-edge is introduced under the plain mode as a t-parallelized all-to-all interconnection between two modules. In BCube-like topologies, the skeleton graph is a BIG-$(\theta_2, r_2, k_2, \lambda_2)$ with c_2 copies in which each super-edge is introduced under the bipartite mode between a module and an RN cluster. For each group of n super-edges that are incident to a common RN cluster, they are instantiated using a t-parallelized TD-(n, μ, x) interconnection.
- Step 4: *Finalization*. This step is omitted since such a topology contains no CNs.

Based on this design procedure, the intra-module design and the dock node selection in connection design are generally fixed for server-centric topologies. Their corresponding parameter settings are specified as follows, respectively.

Parameter settings of intra-module design. Regarding intra-module design, a BIG-$(\theta_1, r_1, k_1, \lambda_1)$ with c_1 copies is fixed as the internal graph for each module. Therein, the inner-degrees of each HN (element node) and each RN (block node) are set to be one and p, respectively. This implies that $r_1 = 1$ and $k_1(c_1 + 1) = p$. By the property of BIG (cf. Lemma 1.6), it is established that

$$\lambda_1 = \frac{r_1(k_1 - 1)}{\theta_1 k_1 - 1} = \frac{k_1 - 1}{\theta_1 k_1 - 1},$$

where $\lambda_1, \theta_1 \geq 1$. In this case, the only feasible settings are $\theta_1 = 1$ and $\lambda_1 = 1$; i.e., the corresponding internal graph is a BIG-$(1, 1, k_1, 1)$ with c_1 copies. Since the inner-degree of each RN is p, the BIG-$(1, 1, k_1, 1)$ with c_1 copies is isomorphic to a BIG-$(1, 1, p, 1)$ with no copies for any values of k_1 and c_1 (see the example in Figure 3.32 with $p = 4$ for illustration). In the following analysis, without loss of generality, we set $k_1 = p$ and $c_1 = 0$.

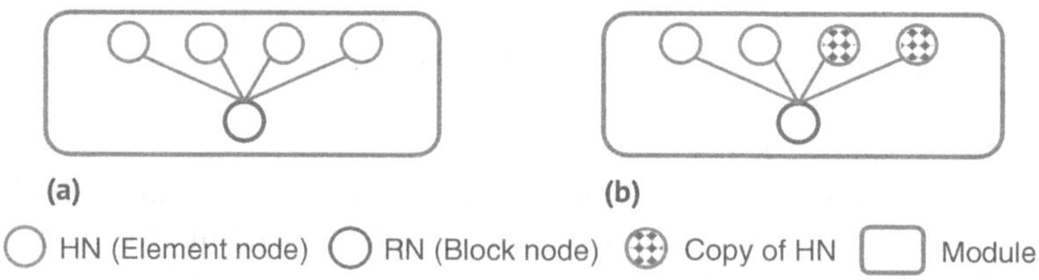

Figure 3.32 Instances of BIG-$(1, 1, k_1, 1)$ with c_1 copies such that $k_1(c_1 + 1) = p = 4$. The sub-figures show two cases with distinct values of k_1 and c_1. Nonetheless, they are both isomorphic to a BIG-$(1, 1, p, 1)$ with no copies ($p = 4$). (a) A BIG-$(1, 1, k_1, 1)$ with no copies, where $k_1 = 4$ and $c_1 = 0$. (b) A BIG-$(1, 1, k_1, 1)$ with c_1 copies, where $k_1 = 2$ and $c_1 = 2$.

When it comes to the dock node interconnection in connection design, different design options and parameter settings can lead to two families of topology designs, i.e., the DCell-like topologies and the BCube-like topologies. Specifically,

- when the plain mode is employed for dock node interconnection, the design procedure would induce DCell-like topologies;
- when the bipartite mode is employed for dock node interconnection, the design procedure would induce BCube-like topologies.

The evolutions of such two families of topology designs are presented in Figure 3.33 and demonstrated in Sections 3.2.1 and 3.2.2, respectively.

Parameter settings of dock node selection in connection design. Note that the internal graph of each module is a BIG-(1, 1, p, 1) that contains p HNs (all of

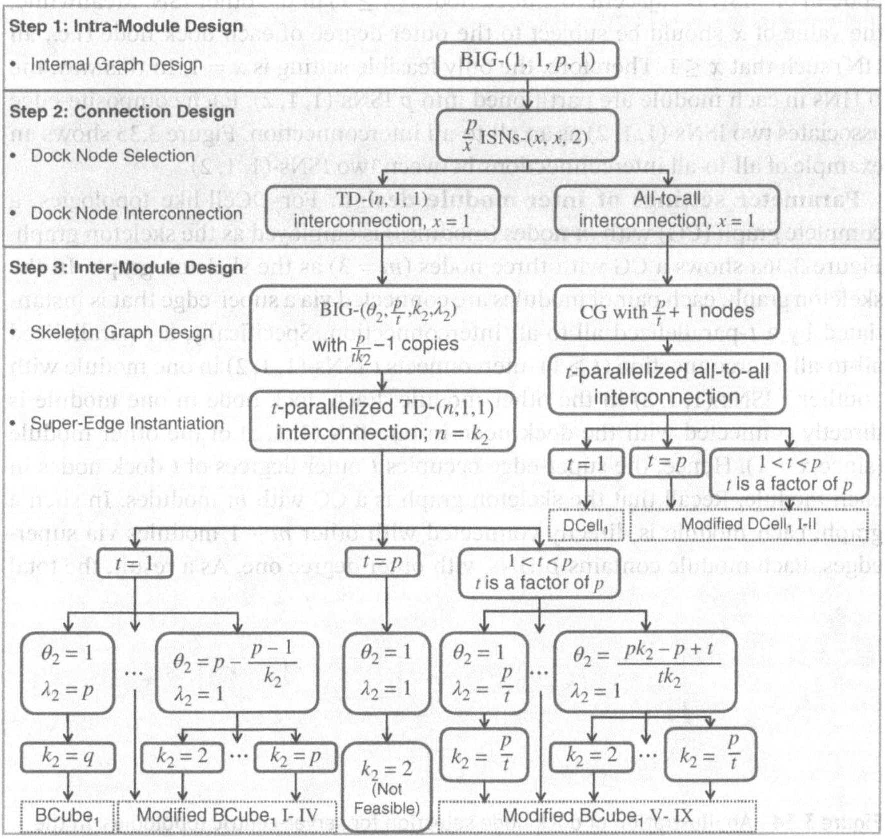

Figure 3.33 A roadmap of the evolution of server-centric topologies. The finalization step is omitted since there are no CNs in such topologies.

them are element nodes) as dock nodes. Such p HNs are grouped into p/x ISNs-(x, w, l) where integer x is a factor of p. Since the dock nodes are selected from the element nodes of the BIG, by Theorem 2.3, the parameters of ISNs-(x, w, l) satisfy $x = w$ and $l = 2$. In other words, given any HN and an ISN-(x, w, l) in the same module, all x dock nodes (HNs) in the ISN-(x, w, l) are within a distance of two from the HN. Figure 3.34 shows the dock node selection when $x = 2$, provided that the internal graph is a BIG-(1, 1, 4, 1).

3.2.1 Evolution of DCell-like Topologies

Parameter settings of dock node interconnection in connection design. In a DCell-like topology, an all-to-all interconnection is employed to interconnect two ISNs-$(x, x, 2)$ in distinct modules. Under an all-to-all interconnection, each dock node in one ISN is adjacent to x dock nodes ($x \geq 1$) in the other ISN. Meanwhile, the value of x should be subject to the outer-degree of each dock node (i.e., an HN) such that $x \leq 1$. Therefore, the only feasible setting is $x = 1$. In this way, the p HNs in each module are partitioned into p ISNs-(1, 1, 2). Each composite-edge associates two ISNs-(1, 1, 2) as an all-to-all interconnection. Figure 3.35 shows an example of all-to-all interconnections between two ISNs-(1, 1, 2).

Parameter settings of inter-module design. For DCell-like topologies, a complete graph (CG) with m nodes (modules) is employed as the skeleton graph. Figure 3.36a shows a CG with three nodes ($m = 3$) as the skeleton graph. In the skeleton graph, each pair of modules are connected via a super-edge that is instantiated by a t-parallelized all-to-all interconnection. Specifically, a t-parallelized all-to-all interconnection ($t \geq 1$) interconnects t ISNs-(1, 1, 2) in one module with another t ISNs-(1, 1, 2) in the other module. Each dock node in one module is directly connected with the dock node in one ISN-(1, 1, 2) of the other module (since $x = 1$). Hence, the super-edge occupies t outer-degrees of t dock nodes in each module. Recall that the skeleton graph is a CG with m modules. In such a graph, each module is directly connected with other $m - 1$ modules via super-edges. Each module contains p HNs with outer-degree one. As a result, the total

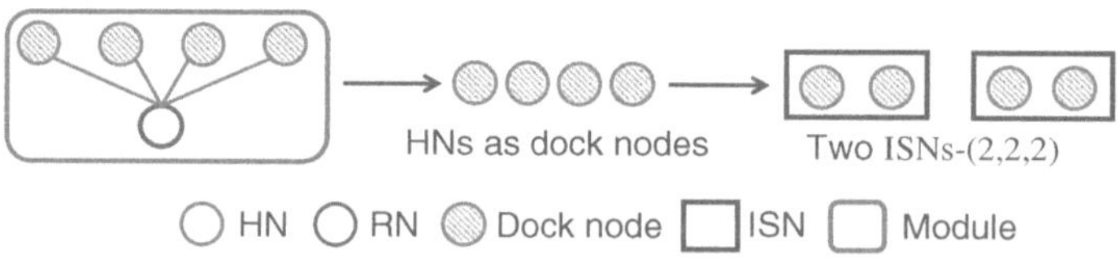

Figure 3.34 An illustration of dock node selection for server-centric topologies. In the given module, all HNs in the module are selected as dock nodes and grouped into two ISNs-(2, 2, 2) (i.e., $x = 2$).

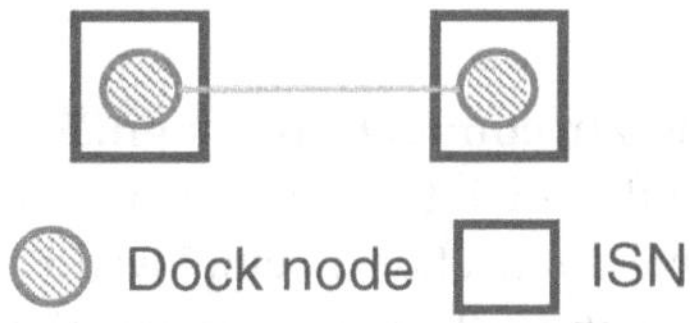

Figure 3.35 An all-to-all interconnection between two ISNs-(1, 1, 2).

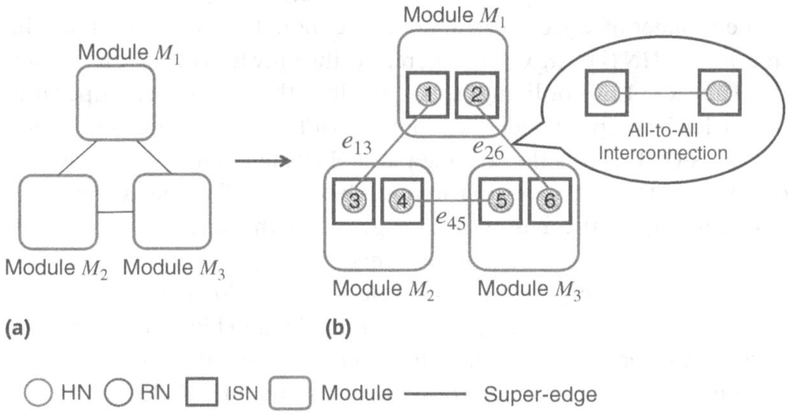

Figure 3.36 An illustration of skeleton graph design and super-edge instantiation. (a) A CG with three nodes (modules) as the skeleton graph. (b) The super-edge between each pair of modules is instantiated as a 1-parallelized all-to-all interconnection. Specifically, the super-edge between modules M_1 and M_2 comprises edge e_{13}, the super-edge between modules M_1 and M_3 comprises edge e_{26}, and the super-edge between modules M_2 and M_3 comprises edge e_{45}.

outer-degree of HNs in each module is equal to the total outer-degree occupied by the $m - 1$ t-parallelized interconnections; i.e., $p = (m - 1)t$, or equivalently,

$$m = \frac{p}{t} + 1. \tag{3.54}$$

where the number of modules m is an integer and thus t should be a factor of p.

In Figure 3.36b, we show how super-edges in the skeleton graph (see Figure 3.36a) are constructed using t-parallelized all-to-all interconnections with $t = 1$. Note that there are three modules, each of which contains two HNs (i.e., $m = 3$ and $p = 2$). By equation (3.54), we have $t = p/(m - 1) = 1$.

These discussions imply that based on equation (3.54), a number of DCell-like topologies can be derived and modified by assigning different values to parameters m and t. As the method for analyzing the guideline for parameter selections mirrors that of the evolution of switch-centric topologies, it is omitted here.

3.2.2 Evolution of BCube-like Topologies

Parameter settings of dock node interconnection in connection design. For BCube-like topologies, dock nodes in distinct modules are interconnected under the bipartite mode. Specifically, a TD-(n, μ, x) interconnection is employed to interconnect ISNs-(x, x, 2) in distinct modules through an RN cluster, where the values of parameters n, μ, and x are to be determined. For each TD-(n, μ, x) interconnection, each dock node is adjacent to μx dock nodes ($\mu, x \geq 1$) in the RN cluster. Hence, the number of adjacent dock nodes μx should be no greater than the outer-degree of each HN (i.e., $\mu, x \leq 1$). Therefore, the only feasible parameter settings are $\mu = 1$ and $x = 1$. Accordingly, in each module, the p HNs are grouped into p ISNs-(1, 1, 2). Each composite-edge is designed as a TD-(n, 1, 1) interconnection to connect n ISNs-(1, 1, 2) together through an RN cluster. Figure 3.37 shows an example of TD-(n, 1, 1) interconnection among n ISNs-(1, 1, 2) in the case of $n = 3$.

Parameter settings of inter-module design. Since the dock nodes are interconnected under the bipartite mode, for inter-module design, a BIG-($\theta_2, r_2, k_2, \lambda_2$) with c_2 copies ($c_2 \geq 0$) is employed as the skeleton graph. In such a graph, element nodes and block nodes are instantiated as modules and block nodes, respectively. Particularly, by employing duplicating operation, modules are partitioned into $c_2 + 1$ groups, such that each group of modules and all RN clusters form a BIG-($\theta_2, r_2, k_2, \lambda_2$). An example with the skeleton graph as a BIG-(1, 1, 2, 1) with one copy (i.e., $c_2 = 1$) is presented in Figure 3.38a.

For each RN cluster employed in the inter-module design, it is directly connected with k_2 modules via a number of k_2 super-edges. Such super-edges are constructed using a t-parallelized TD-(n, 1, 1) interconnection with $n = k_2$. In Figure 3.38b, we show an example in which super-edges in a skeleton graph (see Figure 3.38a) are constructed using a 2-parallelized TD-(2, 1, 1) interconnection (i.e., with the parameter settings $t = 2$ and $n = k_2 = 2$).

Under a t-parallelized TD-(n, 1, 1) interconnection, each super-edge between an RN cluster and a module comprises t edges. Each edge is incident to an RN in the RN cluster and a dock node in the module. In other words, each super-edge

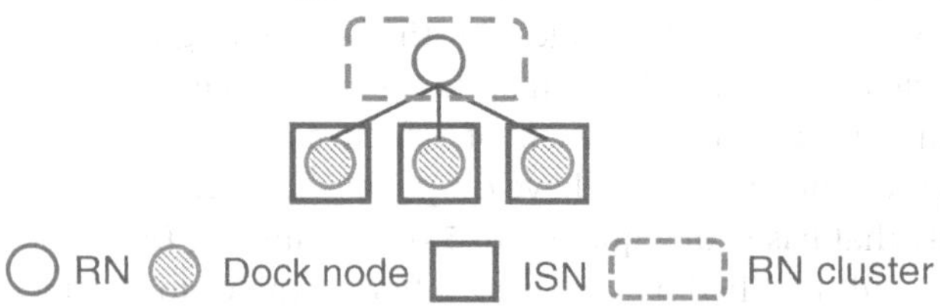

Figure 3.37 A TD-(n, 1, 1) interconnection with $n = 3$. Under this interconnection, three ISNs-(1, 1, 2) are connected through a single RN cluster with one RN.

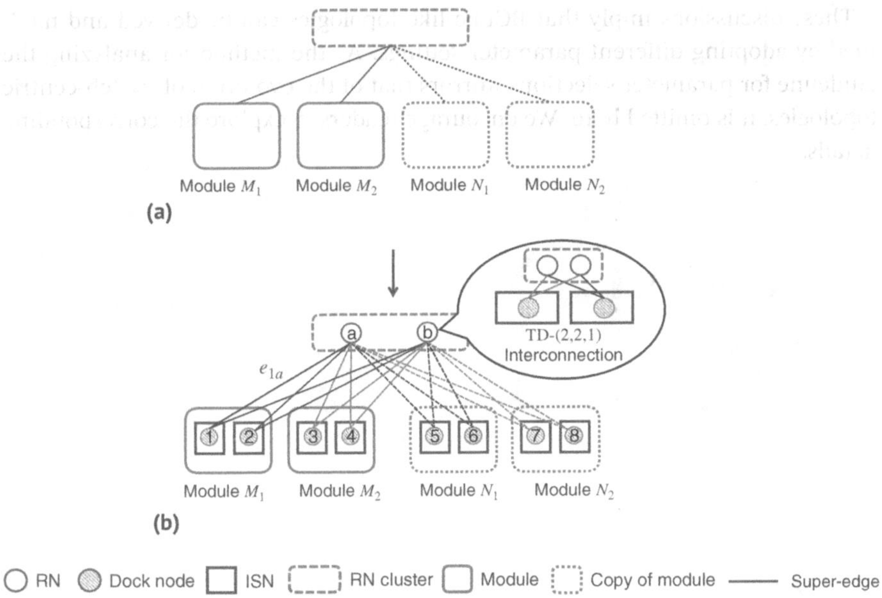

Figure 3.38 An illustration of skeleton graph design and super-edge instantiation. (a) A BIG-(1, 1, 2, 1) with one copy (i.e., $c_2 = 1$) as the skeleton graph. Such a graph contains two groups of modules. One group comprises modules M_1 and M_2. The other group comprises modules N_1 and N_2. (b) The two super-edges between an RN cluster and each group of modules are instantiated as a 2-parallelized TD-(2, 2, 1) interconnection. To be specific, the super-edge between module M_1 and an RN cluster consists of four edges $\{e_{1a}, e_{1b}, e_{2a}, e_{2b}\}$, while the super-edge between module M_2 and the RN cluster consists of another four edges $\{e_{3a}, e_{3b}, e_{4a}, e_{4b}\}$. These two super-edges jointly form a 2-parallelized TD-(2, 2, 1) interconnection. The connection pattern with the RN cluster for module N_1 (resp. module N_2) is the same as that for module M_1 (resp. module M_2).

occupies t degrees of RNs in the RN cluster and t outer-degrees of dock nodes in their corresponding module. Accordingly, we have the following remark.

Remark 3.6 Each RN cluster is directly connected with every p/t modules through an RN with degree p. Meanwhile, each module contains a number of p HNs with outer-degree one as dock nodes. Through such dock nodes, the module is directly connected with p/t RN clusters. By the properties of BIG with copies, we have $r_2 = p/t$ and $k_2(c_2 + 1) = p/t$. As a result, the graph can be equivalently written as a BIG-($\theta_2, p/t, k_2, \lambda_2$) with $p/tk_2 - 1$ copies where parameter t is a factor of p.

These discussions imply that BCube-like topologies can be derived and modified by adopting different parameter settings. As the method for analyzing the guideline for parameter selections mirrors that of the evolution of switch-centric topologies, it is omitted here. We encourage readers to explore the corresponding details.

4

Systematic Framework for Generic Topology Design

4.1 Overview

In Chapter 3, we demonstrate how systematic topology design can be conducted under the framework, provided that options in some design steps are *subject to* existing design. By following the steps in the design procedure, we study how parameters of the design options impact the performance metrics and what trade-offs are achieved under distinct parameter settings in a *stepwise* fashion. In the course of analysis, we not only uncover the positions of existing topologies in the performance trade-off spectrum, but also derive a number of feasible modifications for such topologies with various performances.

Yet, the study on those restricted cases merely serves as an appetizer of *generic* topology design under the framework. For now, the analysis in Chapter 3 only sheds light on a corner in the topology design space with *qualitative* understanding of how performance trade-offs are regulated individually by parameters in the design options. A *thorough* exploration of the topology design space and a *quantitative* comprehension of performance trade-offs achieved by the combinations of design options under distinct parameter settings are still missing.

In this chapter, we broaden the horizon by considering *generic* topology design for large-scale computer networks in a *clean-slate* manner. Following the design procedure with *one* iteration, such a systematic design not only considers all feasible combinations of design options proposed in Chapter 2, but also performs a comprehensive quantitative analysis to provide guidelines for topology design. In particular, we identify a total number of 24 *feasible* combinations of design options (with undetermined parameters) for steps in the design procedure with one iteration (see Section 4.2.1). Each combination is referred to as a *design pattern* that can be recognized in the topology design space with a unique naming (see Section 4.2.2). Next, we discuss the parameters and their fundamental relationships under the design patterns in Section 4.3. Following the discussion, the performance metrics are analyzed under different design patterns in terms of their parameters (see Section 4.4). Settling all the above work readily paves the way for our investigation of generic topology design. In Section 4.5, we exploit the derived quantitative results to provide guidelines for design pattern selection. To this end, in Section 4.5.1, we compare the impacts of different options on the performance metrics. With such a comparison, we suggest which option should be selected to achieve distinct design goals. Based on the comparison results, in Section 4.5.2, we further demonstrate how such guidelines apply to the design pattern selection of existing topologies. Next, in Section 4.6, we treat design patterns as iteration-level templates for systematic topology design and provide guidelines for determining the parameter values under design patterns. Specifically, we introduce a parameter determination scheme that employ the quantitative analytical results to generate topologies whose parameters can satisfy designers' performance requirements. To demonstrate the scheme, we not only reveal the implicit performance requirements behind existing topologies to reverse-engineer them, but also set other performance requirements to forward-engineer a number of new topologies.

Meanwhile, we make the following five assumptions for ease of analysis.

- *Assumption 1:* All nodes of the same type have the same node degrees, inner-degrees, and outer-degrees.
- *Assumption 2:* All intermediate super-nodes (ISNs) contain the same number of dock nodes with the same type.
- *Assumption 3:* All modules in the topology have the same internal graph design.
- *Assumption 4:* Each module does not contain GNs and HNs at the same time.
- *Assumption 5:* All RN clusters contain the same number of RNs.

We encourage readers to explore more topology designs by enriching the design space under more general settings: e.g., by introducing other viable design tools, relaxing the assumptions being made, and considering topology design with more than one iteration.

4.2 Patterns of Topology Design

From the lens of the design procedure, topology design can be essentially viewed as a sequential procedure of design tool selection for the design steps. Each *feasible* combination of such tools can be employed as a basic template on which concrete topologies would be generated under different parameter settings. For ease of description, we refer to each combination as a *design pattern* in the rest of this monograph.

To conduct a comprehensive investigation of systematic topology design, we first explore *all* design patterns whose derived topologies compose the design space (see Section 4.2.1). Then we characterize them with specific naming in Section 4.2.2 so that each pattern can be conveniently recognized and uniquely described in subsequent sections.

4.2.1 Exploration of Design Patterns

Recall that the design procedure with one iteration typically contains *four* steps: the intra-module design step, the connection design step, the inter-module design step, and the finalization step. For each step, a few design tools have been introduced as options in Chapter 2 and are summarized as follows:

- *Intra-module design.* Regarding intra-module design, there are two key decisions to make. One is regarding what kind of internal graph should be employed for intra-module design. The other is regarding which type of nodes should appear in the internal graph. For such decisions, the following are their feasible options.
  - A complete graph (CG) in which nodes are all GNs or all HNs.
  - A complete bipartite graph (CBG) in which nodes are all GNs or all HNs.
  - A block incidence graph (BIG) with copies, where block nodes are RNs and element nodes are all GNs or all HNs.
  - A transversal design graph (TDG) with copies, where block nodes are RNs and element nodes are all GNs or all HNs.
- *Connection design.* Connection design includes two sub-steps, i.e., dock node selection and dock node interconnection. For each sub-step, their feasible options are presented as follows:
  - *Dock node selection.* The set of feasible options in this sub-step is constrained by the option selected for intra-module design. Specifically:
    - When the internal graph is a CG or a CBG, any type of nodes (GNs or HNs) can be selected as dock nodes. Depending on the internal graph type, we denote the option in this case as "CG(any)" or "CBG(any)."
    - When the internal graph is a BIG or a TDG, there are two options for dock node selection. One is to employ element nodes (GNs or HNs) as dock nodes, while the other is to employ block nodes (RNs) as dock nodes. Depending on the internal graph type, we denote the first option as "BIG(element)" or "TDG(element)" accordingly. Likewise, for the second option, we denote it as "BIG(block)" or "TDG(block)."
  - *Dock node interconnection.* Regardless of which options are selected for intra-module design and dock node selection, there are two options for dock node interconnection. *One* is to interconnect dock nodes under the plain mode by employing all-to-all interconnection as the composite edge. *The other* is to interconnect them under the bipartite mode by employing TD interconnection as the composite edge.

- *Inter-module design.* The set of feasible options in this step is constrained by the option selected for dock node interconnection. Specifically:
  - When dock nodes are interconnected under the plain mode, the only option is to design the skeleton graph as a CG with each super-edge fixed as a parallelized all-to-all interconnection.
  - When dock nodes are interconnected under the bipartite mode, the only option is to design the skeleton graph as a BIG with copies. Meanwhile, super-edges in the skeleton graph are constructed as parallelized TD interconnections.
- *Finalization.* This step involves no design tools as it only considers the number of CNs to be attached to each GN (or HN) in the topology.

By the preceding discussions, we identify a total number of 24 feasible combinations of options for the design steps (i.e., 24 design patterns). Depending on which type of nodes are contained in the internal graph (i.e., GNs or HNs), they are summarized in Tables 4.1 and 4.2, respectively.

Remark 4.1 Note that each design pattern can be viewed as a *template* for a particular class of concrete topologies. In other words, a design pattern can be cast to a concrete topology under fixed parameter settings of its design options.

Table 4.1 A summary of design patterns with GNs in the internal graph.

Numbering of design pattern	Intra-module design	Dock node selection	Dock node interconnection	Inter-module design
1	CG	Any nodes as dock nodes	Plain mode	CG
2			Bipartite mode	BIG
3	CBG	Any nodes as dock nodes	Plain mode	CG
4			Bipartite mode	BIG
5	BIG	Element nodes as dock nodes	Plain mode	CG
6			Bipartite mode	BIG
7		Block nodes as dock nodes	Plain mode	CG
8			Bipartite mode	BIG
9	TDG	Element nodes as dock nodes	Plain mode	CG
10			Bipartite mode	BIG
11		Block nodes as dock nodes	Plain mode	CG
12			Bipartite mode	BIG

Table 4.2 A summary of design patterns with HNs in the internal graph.

Numbering of design pattern	Intra-module design	Dock node selection	Dock node interconnection	Inter-module design
13	CG	Any nodes as dock nodes	Plain mode	CG
14			Bipartite mode	BIG
15	CBG	Any nodes as dock nodes	Plain mode	CG
16			Bipartite mode	BIG
17	BIG	Element nodes as dock nodes	Plain mode	CG
18			Bipartite mode	BIG
19		Block nodes as dock nodes	Plain mode	CG
20			Bipartite mode	BIG
21	TDG	Element nodes as dock nodes	Plain mode	CG
22			Bipartite mode	BIG
23		Block nodes as dock nodes	Plain mode	CG
24			Bipartite mode	BIG

4.2.2 Naming of Design Patterns

In this subsection, we provide a unified naming scheme for the design patterns, in order to facilitate the references to them in our subsequent analysis and discussions. Recall that for some design (sub-)step, the set of its feasible options is dependent on the option selected for the previous (sub-)step and independent of options for the other (sub-)steps. Specifically, the options for dock node selection are dependent on the chosen type of internal graph, while the feasible options for skeleton graph design are dependent on the selected mode for dock node interconnection. Meanwhile, options for such steps are all independent of the type of nodes in the internal graph.

Following the previously mentioned idea, we adopt a *triple* $< d_1, d_2, d_3 >$ to characterize each design pattern. Each term in the triple describes the design choice made for one or few design (sub-)steps with dependency between their feasible options. Design choices in different terms are independent of each other. The detailed descriptions of the three terms are given as follows:

- The first term d_1 describes the selected type for nodes in the internal graph. Recall that modules either contain several GNs or several HNs. Therefore, the first term is written as symbol G if the node type is GN or symbol H otherwise.

- The second term d_2 describes the selected options for internal graph design and dock node selection. A total number of six options are available for choice, denoted by CG(all), CBG(all), BIG(element), BIG(block), TDG(element), and TDG(block), respectively.
- The third term d_3 describes the selected options for dock node interconnection and inter-module design. Two options are available for choice, denoted by symbol P for dock node interconnection under the plain mode and B for that under the bipartite mode. Note that when the interconnection mode is selected, the inter-module design is also determined.

Using the naming scheme in Figure 4.1, we visualize the design space based on a three-dimensional (3D) reference frame. Each dimension corresponds to one term in the triples of the design patterns. Every design pattern delimits a two-dimensional (2D) surface that contains a class of concrete topologies (denoted by points).

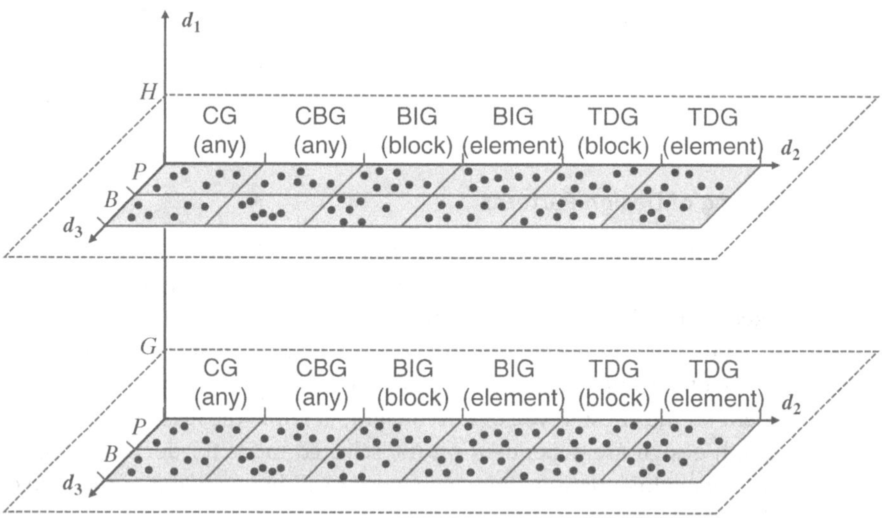

Figure 4.1 A visual illustration of the topology design space and design patterns. The three dimensions of the reference frame correspond to the three terms in the naming of design patterns. Each design pattern delimits a 2D surface. A point in the surface denotes a concrete topology that can be generated under the current design pattern with particular parameter settings. Note that all design patterns that have all nodes as HNs in the internal graph compose the upper plane, while other patterns jointly form the lower plane.

4.3 Parameters of Design Patterns

In this section, we first investigate the design patterns' parameters (refer to Table 4.3 for an overview) and perform analysis on their relationships and constraints (see Section 4.3.1). For ease of reference, we then summarize such results for each design pattern, respectively, in Section 4.3.2.

4.3.1 Relationships and Constraints Among Parameters

Recall that under the naming scheme, each design pattern is characterized by a triple $< d_1, d_2, d_3 >$ that employs three terms to describe selected options for the design steps. Following such a scheme, in this subsection, we discuss the parameters of feasible options for each of the three terms, respectively.

Parameters of options for the first term d_1. Recall that this term describes the type of nodes in the internal graph of modules. Two options are available for this term,

Table 4.3 Parameters of options and their descriptions.

Parameter	Description
	Node degrees and their proportions
$p_R/p_G/p_H$	Degree of each RN/GN/HN
$\alpha_R/\alpha_G/\alpha_H$	Proportion of degrees for connections with other nodes in the same module for each RN/GN/HN in modules
$\beta_R/\beta_G/\beta_H$	Proportion of degrees for connections with other nodes in distinct modules or RN clusters for each RN/GN/HN in modules
γ_G/γ_H	Proportion of nodes for connections with CNs for each GN/HN
	Parameters for internal graph design
m_1	Number of nodes when the internal graph is a CG
m_1'	Number of nodes when the internal graph is a CBG
$\theta_1, r_1, k_1, \lambda_1, c_1$	Parameters of the skeleton graph when it is a BIG with copies
n_1, μ_1, v_1, c_1'	Parameters of the skeleton graph when it is a TDG with copies
	Parameters for super-edge design
x, w, l	Parameters of ISNs
n, μ	Parameters of TD-(n, μ, x) interconnection
t	Number of ISNs in each module contained in each super-edge
	Parameters for skeleton graph design
m_2	Number of modules when the skeleton graph is a CG
$\theta_2, r_2, k_2, \lambda_2, c_2$	Parameters of the skeleton graph when it is a BIG with copies

namely symbol G if GN is the node type in the internal graph and symbol H if HN is the node type. For them, a number of parameters are introduced as follows:

First, let symbols p_R (resp. p_G, p_H) denote the degree of each RN (resp. GN, HN) in modules. Then we define the following parameters to characterize the connections for different types of nodes.

1. *Parameters* α_R, α_G, and α_H. For each RN (resp. each GN, each HN) in modules, we use parameter α_R (resp. α_G, α_H) to denote the proportion of its degrees employed to connect with other nodes in the same module. The value of parameter α_R (resp. α_G, α_H) should be bounded within range $(0, 1]$[1] such that the value of $\alpha_R p_R$ (resp. $\alpha_G p_G$, $\alpha_H p_H$), i.e., the inner-degree of each RN (resp. GN, HN), is a positive integer.
2. *Parameters* β_R, β_G, and β_H. For each RN (resp. each GN, each HN) in modules, we use parameter β_R (resp. β_G, β_H) to denote the proportion of its degrees employed to connect with nodes in other modules or RN clusters. The value of parameter β_R (resp. β_G, β_H) should be bounded within range $[0, 1)$[2] such that the value of $\beta_R p_R$ (resp. $\beta_G p_G$, $\beta_H p_H$), i.e., the outer-degree of each RN (resp. GN, HN), is a non-negative integer.
3. *Parameters* γ_G and γ_H. For each GN (resp. HN), we use parameter γ_G (resp. γ_H) to denote the proportion of its degrees employed to connect with CNs. For the value of γ_G, it should be bounded within range $(0, 1)$ such that the value of $\gamma_G p_G$ is a positive integer. Note that the value is greater than zero because by definition, each GN must be connected with at least one CN. For the value of $\gamma_H p_H$, it should be bounded within range $[0, 1)$ such that the value of $\gamma_H p_H$ is a non-negative integer. Note that the value can be zero because by definition, each HN is allowed to be connected with no CNs. Meanwhile, the values of γ_G and γ_H are both less than one as each GN or HN has to be connected with non-CN nodes (e.g., other nodes in modules).

By the preceding definitions, these parameters should satisfy the following equations.

$$\alpha_R + \beta_R = 1, \tag{4.1}$$

$$\alpha_G + \beta_G + \gamma_G = 1, \tag{4.2}$$

$$\alpha_H + \beta_H + \gamma_H = 1. \tag{4.3}$$

---

1 Their values cannot be zero as they need to connect with other nodes in the same module.
2 Note that RNs (resp. GNs, HNs) in modules are selected as dock nodes if and only if $\beta_R > 0$ (resp. $\beta_G > 0$, $\beta_H > 0$). Besides, their values cannot be one as they need to connect with other nodes in the same module.

Based on the equations (4.1)–(4.3), in subsequent analysis, we focus on the parameters for the 12 design patterns with the first term d_1 as G, i.e., p_G, α_G, β_G, and γ_G. In this case, our analysis is performed based on equations (4.1) and (4.2) as no HNs would appear in the internal graph of modules. As for the other 12 design patterns with the first term d_1 as H, by replacing parameter p_G (resp. α_G, β_G, and γ_G) with parameter p_H (resp. α_H, β_H, and γ_H), similar analysis can be performed based on equations (4.1) and (4.3).

Parameters of options for the second term d_2. Recall that this term describes the selected type of internal graph and the way of dock node selection. Six options are available for this term. Their parameters are introduced and analyzed as follows:

- *CG(any).* Under this option, the internal graph is designed as a CG in which all nodes (GNs) are selected as dock nodes. In this case, there are seven parameters, including the three parameters $\alpha_G, \beta_G, \gamma_G$ for GNs, one parameter m_1 that denotes the total number of GNs in the internal graph, and the three parameters x, w, and l for ISNs-(x, w, l). The following are their relationships and constraints.
  - Since GNs are selected as dock nodes, the proportion of their degrees for connections with nodes in other modules, i.e., the value of β_G, should be positive. Meanwhile, by equation (4.1), we have $\alpha_G + \beta_G + \gamma_G = 1$ so that $0 < \alpha_G, \beta_G, \gamma_G < 1$.
  - By the property of CG, the inner-degree of each GN (i.e., $\alpha_G p_G$) is equal to the total number of GNs (i.e., m_1) minus one. In other words, we have $m_1 = \alpha_G p_G + 1$.
  - By Theorem 2.1, we have $x = w$ and $l = 1$ for ISNs-(x, w, l) in this case.
  - Since the $\alpha_G p_G + 1$ GNs are partitioned into a number of ISNs-(x, w, l), the value of x (the number of dock nodes in each ISN) should be a factor of $\alpha_G p_G + 1$; i.e., $x \mid \alpha_G p_G + 1$.
- *CBG(any).* Under this option, the internal graph is designed as a CBG in which all nodes (GNs) are selected as dock nodes. There are also seven parameters for this option, including the three parameters $\alpha_G, \beta_G, \gamma_G$ for GNs, one parameter m_1' that denotes the total number of GNs in the internal graph, and the three parameters x, w, and l for ISNs-(x, w, l). The following are their relationships and constraints.
  - Since GNs are selected as dock nodes, the proportion of their degrees for connections with nodes in other modules, i.e., the value of β_G, should be positive. Meanwhile, by equation (4.1), we have $\alpha_G + \beta_G + \gamma_G = 1$ so that $0 < \alpha_G, \beta_G, \gamma_G < 1$.
  - By the property of CBG, the number of nodes in each node partition is equal to the inner-degree of each node in the other node partition. Since each GN has the same inner-degree of $\alpha_G p_G$ in the graph, we have $m_1' = 2\alpha_G p_G$.

  - By Theorem 2.2, if $x = 1$, we have $w = 1$ and $l = 2$; otherwise, we have $w = \lfloor \frac{x}{2} \rfloor$ and $l = 1$.
  - Since the $2\alpha_G p_G$ GNs are partitioned into a number of ISNs-(x, w, l), the value of x (the number of dock nodes in each ISN) should be a factor of $2\alpha_G p_G$; i.e., $x \mid 2\alpha_G p_G$.
- *BIG(block)*. Under this option, the internal graph is designed as a BIG with copies in which only the block nodes (RNs) are selected as dock nodes. There are 13 parameters for this option, including the 3 parameters $\alpha_G, \beta_G, \gamma_G$ for GNs (element nodes), the 2 parameters α_R, β_R for RNs (block nodes), the 5 parameters $\theta_1, r_1, k_1, \lambda_1, c_1$ for the internal graph BIG-$(\theta_1, r_1, k_1, \lambda_1)$ with c_1 copies, and the 3 parameters x, w, and l for ISNs-(x, w, l). Their relationships and constraints are discussed as follows:
  - Given that only the block nodes (RNs) are selected as dock nodes, it follows that $\beta_G = 0$ and $\beta_R > 0$. Meanwhile, by equations (4.1) and (4.2), we have $\alpha_G + \gamma_G = 1$ and $\alpha_R + \beta_R = 1$ so that $0 < \alpha_G, \gamma_G, \alpha_R, \beta_R < 1$.
  - By Lemma 1.6, the parameters of the internal graph (a BIG) satisfy

$$\lambda_1 = \frac{r_1(k_1 - 1)}{\theta_1 k_1 - 1}. \tag{4.4}$$

  - By the property of BIG with copies, each GN has an inner-degree of r_1. Since the inner-degree of each GN also equals $\alpha_G p_G$, we have $r_1 = \alpha_G p_G$.
  - By the property of BIG with copies, each RN has an inner-degree of $k_1(c_1 + 1)$. Since the inner-degree of each RN also equals $\alpha_R p_R$, we have $k_1(c_1 + 1) = \alpha_R p_R$.
  - Since we have $k_1(c_1 + 1) = \alpha_R p_R$ and c_1 is an integer, the value of k_1 should be a factor of $\alpha_R p_R$ (i.e., $k_1 \mid \alpha_R p_R$).
  - By Theorem 2.4, we have $x \geq \theta_1 w$ and $l = 1$ for the ISNs-(x, w, l). To provide as many dock nodes as possible for each GN within distance one (i.e., the largest possible value of w), we set $x = \theta_1 w$ and $l = 1$, which requires the BIG-$(\theta_1, r_1, k_1, \lambda_1)$ to be resolvable.
  - Provided that $x = \theta_1 w$ and w should be an integer, the value of θ_1 needs to be a factor of x; i.e., $\theta_1 \mid x$.
  - Since the $\theta_1 r_1 = \theta_1 \alpha_G p_G$ block nodes are partitioned into a number of ISNs-(x, w, l), the value of x (the number of dock nodes in each ISN) should be a factor of $\theta_1 \alpha_G p_G$; i.e., $x \mid \theta_1 \alpha_G p_G$.
- *BIG(element)*. Under this option, the internal graph is designed as a BIG with copies in which only the element nodes (GNs) are selected as dock nodes. There are 13 parameters for this option, including the 3 parameters $\alpha_G, \beta_G, \gamma_G$ for GNs (element nodes), the 2 parameters α_R, β_R for RNs (block nodes), the 5 parameters $\theta_1, r_1, k_1, \lambda_1, c_1$ for the internal graph BIG-$(\theta_1, r_1, k_1, \lambda_1)$ with

c_1 copies, and the 3 parameters x, w, and l for ISNs-(x, w, l). Their relationships and constraints are discussed as follows:

- Since only the element nodes (GNs) are selected as dock nodes, it follows that $\beta_R = 0$ and $\beta_G > 0$. Meanwhile, by equations (4.1) and (4.2), we have $\alpha_R = 1 - \beta_R = 1$ and $\alpha_G + \beta_G + \gamma_G = 1$ with $0 < \alpha_G, \beta_G, \gamma_G < 1$.
- By Lemma 1.6, the parameters of the internal graph (a BIG) satisfy

$$\lambda_1 = \frac{r_1(k_1 - 1)}{\theta_1 k_1 - 1}. \tag{4.5}$$

- By the property of BIG with copies, each GN has an inner-degree of r_1. Since the inner-degree of each GN also equals $\alpha_G p_G$, we have $r_1 = \alpha_G p_G$.
- By the property of BIG with copies, each RN has an inner-degree of $k_1(c_1 + 1)$. Since the inner-degree of each RN also equals $\alpha_R p_R$, we have $k_1(c_1 + 1) = \alpha_R p_R$.
- Since we have $k_1(c_1 + 1) = \alpha_R p_R$ and c_1 is an integer, the value of k_1 should be a factor of $\alpha_R p_R$ (i.e., $k_1 \mid \alpha_R p_R$).
- By Theorem 2.3, we have $x = w$ and $l = 2$.
- There are $\theta_1 k_1(c_1 + 1) = \theta_1 \alpha_R p_R$ element nodes in the graph. They are partitioned into a number of ISNs-(x, w, l). Accordingly, the value of x (the number of dock nodes in each ISN) should be a factor of $\theta_1 \alpha_R p_R$; i.e., $x \mid \theta_1 \alpha_R p_R$.

- *TDG(block).* Under this option, the internal graph is designed as a TDG with copies in which only the block nodes (RNs) are selected as dock nodes. There are 12 parameters for this option, including the 3 parameters $\alpha_G, \beta_G, \gamma_G$ for GNs (element nodes), the 2 parameters α_R, β_R for RNs (block nodes), the 4 parameters n_1, μ_1, v_1, c'_1 for the TDG-(n_1, μ_1, v_1) with c'_1 copies, and the 3 parameters x, w, and l for ISNs-(x, w, l). The following are the relationships and constraints they are subject to.
  - Given that only the block nodes (RNs) are selected as dock nodes, it follows that $\beta_G = 0$ and $\beta_R > 0$. Meanwhile, by equations (4.1) and (4.2), we have $\alpha_G + \gamma_G = 1$ and $\alpha_R + \beta_R = 1$ with $0 < \alpha_G, \gamma_G, \alpha_R, \beta_R < 1$.
  - By the property of TDG with copies, each GN has an inner-degree of $\mu_1 v_1$. Since the inner-degree of each GN also equals $\alpha_G p_G$, we have $\mu_1 v_1 = \alpha_G p_G$.
  - Provided that $\mu_1 v_1 = \alpha_G p_G$ and the value of μ_1 is an integer, the value of v_1 needs to be a factor of $\alpha_G p_G$; i.e., $v_1 \mid \alpha_G p_G$.
  - By the property of TDG with copies, the inner-degree of each RN is equal to $n_1(c'_1 + 1)$. Since the inner-degree of each RN is denoted by $\alpha_R p_R$, we have $n_1(c'_1 + 1) = \alpha_R p_R$.
  - Since we have $n_1(c'_1 + 1) = \alpha_R p_R$ and c'_1 is an integer, the value of n_1 should be a factor of $\alpha_R p_R$ (i.e., $n_1 \mid \alpha_R p_R$).

- By Theorem 2.6, we have $l = 1$ and $x \geq v_1 w$. To provide as many dock nodes as possible for each GN within distance one (i.e., the largest possible value of w), we set $x = v_1 w$ and $l = 1$, which requires the TDG-(n_1, μ_1, v_1) to be resolvable.
- Provided that $x = v_1 w$ and w is an integer, the value of v_1 should be a factor of x; i.e., $v_1 \mid x$.
- There are $\mu_1 v_1^2$ (also equal to $\alpha_G p_G v_1$ as $\mu_1 v_1 = \alpha_G p_G$) block nodes in the internal graph. They are partitioned into a number of ISNs-(x, w, l). Accordingly, the value of x needs to be a factor of $\alpha_G p_G v_1$; i.e., $x \mid \alpha_G p_G v_1$.

• *TDG(element)*. Under this option, the internal graph is designed as a TDG with copies in which only the element nodes (GNs) are selected as dock nodes. There are 12 parameters for this option, including the 3 parameters $\alpha_G, \beta_G, \gamma_G$ for GNs (element nodes), the 2 parameters α_R, β_R for RNs (block nodes), the 4 parameters n_1, μ_1, v_1, c_1' for the TDG-(n_1, μ_1, v_1) with c_1' copies, and the 3 parameters x, w, and l for ISNs-(x, w, l). The following are their relationships and constraints.
  - Since only the element nodes (GNs) are selected as dock nodes, it follows that $\beta_R = 0$ and $\beta_G > 0$. By equations (4.1) and (4.2), we have $\alpha_R = 1 - \beta_R = 1$ and $\alpha_G + \beta_G + \gamma_G = 1$ with $0 < \alpha_G, \beta_G, \gamma_G < 1$.
  - By the property of TDG with copies, each GN has an inner-degree of $\mu_1 v_1$. Since the inner-degree of each GN also equals $\alpha_G p_G$, we have $\mu_1 v_1 = \alpha_G p_G$.
  - Given that $\mu_1 v_1 = \alpha_G p_G$ and μ_1 is an integer, the value of v_1 needs to be a factor of $\alpha_G p_G$; i.e., $v_1 \mid \alpha_G p_G$.
  - By the property of TDG with copies, each RN has an inner-degree of $n_1(c_1' + 1)$. Since the inner-degree of each RN also equals $\alpha_R p_R$, we have $n_1(c_1' + 1) = \alpha_R p_R$.
  - Provided that $n_1(c_1' + 1) = \alpha_R p_R$ and c_1 is an integer, the value of n_1 needs to be a factor of $\alpha_R p_R$; i.e., $n_1 \mid \alpha_R p_R$.
  - By Theorem 2.5, we have $l \geq 2$. Meanwhile, if $l = 2$, then we have $w \leq x - 1$. To provide as many dock nodes as possible for each GN within a short distance (i.e., the largest value of w and the smallest value of l), we set $w = x - 1$ and $l = 2$.
  - Note that each ISN-(x, w, l) is formed by selecting the same number of element nodes from each group of element nodes. Since there are n_1 groups of element nodes, the value of x needs to be a factor of n_1.
  - There are $n_1 v_1 (c_1' + 1)$ (also equal to $v_1 \alpha_R p_R$ as $n_1(c_1' + 1) = \alpha_R p_R$) element nodes in the internal graph. They are partitioned into a number of ISNs-(x, w, l). Accordingly, the value of x needs to be a factor of $v_1 \alpha_R p_R$; i.e., $x \mid v_1 \alpha_R p_R$.

Parameters of options for the third term d_3. Recall that this term describes the atomic mode (either the plain mode denoted by symbol P or the bipartite mode by symbol B) selected for dock node interconnection and the selected type of skeleton graph for inter-module design. Meanwhile, once the mode for dock node interconnection is determined, the skeleton graph design is also fixed. Hence, there are two options for this term. Their parameters are analyzed as follows:

- *Option P.* Under this option, dock nodes are interconnected under the plain mode using t-parallelized all-to-all interconnections for super-edge design. Meanwhile, the skeleton graph is designed as a CG. For this option, there are two parameters, including the parameter t for t-parallelized all-to-all interconnection and the parameter m_2 that denotes the number of modules in the skeleton graph. Their constraints and relationships with the aforementioned parameters are analyzed as follows:
  - Recall that under the plain mode, a composite-edge interconnects two ISNs-(x, w, l) as an all-to-all interconnection. Such an interconnection occupies x outer-degrees of each dock node in the two ISNs. Hence, the value of x should be a factor of the outer-degree of each dock node. More specific details in terms of different options of term d_2 are given as follows:
    - For design patterns with d_2 = CG(any) or CBG(any) or BIG (element) or TDG(element), each GN is selected as a dock node with an outer-degree of $\beta_G p_G$, which implies that $x \mid \beta_G p_G$.
    - For design patterns with d_2 = BIG(block) or TDG(block), RNs with outer-degree $\beta_R p_R$ are selected as dock nodes, which implies $x \mid \beta_R p_R$.
  - In the skeleton graph (i.e., a CG with m_2 modules), each module is incident to $m_2 - 1$ super-edges. Each super-edge between two modules is instantiated as a t-parallelized all-to-all interconnection. Such an interconnection involves tx dock nodes in each module and each dock node is directly connected to all x dock nodes of some ISN-(x, w, l) in the other module. Hence, each super-edge occupies tx^2 outer-degrees of dock nodes in each module. Since each module is incident to $m_2 - 1$ super-edges, the sum of outer-degrees of all dock nodes in each module is equal to $tx^2 \cdot (m_2 - 1)$. For design patterns with distinct options of term d_2, more detailed analysis is presented as follows:
    - For design patterns with d_2 = CG(any), m_1 GNs with outer-degree $\beta_G p_G$ are selected as dock nodes. Hence, the sum of outer-degrees of all dock nodes in each module is equal to $m_1 \beta_G p_G$, which implies $m_1 \beta_G p_G = tx^2 \cdot (m_2 - 1)$. Since parameter m_2 is an integer, the value of tx^2 should be a factor of $m_1 \beta_G p_G$; i.e., $tx^2 \mid m_1 \beta_G p_G$.
    - For design patterns with d_2 = CBG(any), m_1' GNs with outer-degree $\beta_G p_G$ are selected as dock nodes. Hence, the sum of outer-degrees of all dock

nodes in each module is equal to $m'_1\beta_G p_G$, which implies $m'_1\beta_G p_G = tx^2 \cdot (m_2 - 1)$. Since parameter m_2 is an integer, the value of tx^2 should be a factor of $m'_1\beta_G p_G$; i.e., $tx^2 \mid m'_1\beta_G p_G$.

- For design patterns with $d_2 =$ BIG(block), $\theta_1\alpha_G p_G$ RNs with outer-degree $\beta_R p_R$ are selected as dock nodes. Hence, the sum of outer-degrees of all dock nodes in each module is equal to $\theta_1\alpha_G p_G\beta_R p_R$, which implies $\theta_1\alpha_G p_G\beta_R p_R = tx^2 \cdot (m_2 - 1)$. Since parameter m_2 is an integer, the value of tx^2 should be a factor of $\theta_1\alpha_G p_G\beta_R p_R$; i.e., $tx^2 \mid \theta_1\alpha_G p_G\beta_R p_R$.
- For design patterns with $d_2 =$ BIG(element), $\theta_1\alpha_R p_R$ GNs with outer-degree $\beta_G p_G$ are selected as dock nodes. Hence, the sum of outer-degrees of all dock nodes in each module is equal to $\theta_1\alpha_R p_R\beta_G p_G$, which implies $\theta_1\alpha_R p_R\beta_G p_G = tx^2 \cdot (m_2 - 1)$. Since parameter m_2 is an integer, the value of tx^2 should be a factor of $\theta_1\alpha_R p_R\beta_G p_G$; i.e., $tx^2 \mid \theta_1\alpha_R p_R\beta_G p_G$.
- For design patterns with $d_2 =$ TDG(block), $\mu_1 v_1^2$ RNs with outer-degree $\beta_R p_R$ are selected as dock nodes. Hence, the sum of outer-degrees of all dock nodes in each module is equal to $\mu_1 v_1^2\beta_R p_R$, which implies $\mu_1 v_1^2\beta_R p_R = tx^2 \cdot (m_2 - 1)$. Since parameter m_2 is an integer, the value of tx^2 should be a factor of $\mu_1 v_1^2\beta_R p_R$; i.e., $tx^2 \mid \mu_1 v_1^2\beta_R p_R$.
- For design patterns with $d_2 =$ TDG(element), $n_1 v_1(c'_1 + 1)$ GNs with outer-degree $\beta_G p_G$ are selected as dock nodes. Hence, the sum of outer-degrees of all dock nodes in each module is equal to $n_1 v_1(c'_1 + 1)\beta_G p_G$, which implies $n_1 v_1(c'_1 + 1)\beta_G p_G = tx^2 \cdot (m_2 - 1)$. Since parameter m_2 is an integer, the value of tx^2 should be a factor of $n_1 v_1(c'_1 + 1)\beta_G p_G$; i.e., $tx^2 \mid n_1 v_1(c'_1 + 1)\beta_G p_G$.

▪ Note that each super-edge is a t-parallelized all-to-all interconnection. Since such a super-edge involves tx dock nodes in each module, the value of tx should be less than or equal to the number of dock nodes in each module. The following is the detailed analysis for different options of term d_2.
  - For design patterns with $d_2 =$ CG(any), m_1 GNs are selected as dock nodes. Hence, we have $tx \leq m_1$.
  - For design patterns with $d_2 =$ CBG(any), m'_1 GNs are selected as dock nodes. Hence, we have $tx \leq m'_1$.
  - For design patterns with $d_2 =$ BIG(block), $\theta_1\alpha_G p_G$ RNs are selected as dock nodes. Hence, we have $tx \leq \theta_1\alpha_G p_G$.
  - For design patterns with $d_2 =$ BIG(element), $\theta_1\alpha_R p_R$ GNs are selected as dock nodes. Hence, we have $tx \leq \theta_1\alpha_R p_R$.
  - For design patterns with $d_2 =$ TDG(block), $\mu_1 v_1^2$ RNs are selected as dock nodes. Hence, we have $tx \leq \mu_1 v_1^2$.
  - For design patterns with $d_2 =$ TDG(element), $n_1 v_1(c'_1 + 1)$ GNs are selected as dock nodes. Hence, we have $tx \leq n_1 v_1(c'_1 + 1)$.

- *Option B*. Under this option, dock nodes are interconnected under the bipartite mode using t-parallelized TD interconnections. Meanwhile, the skeleton graph is designed as a BIG with copies. For this option, there are eight parameters, including the three parameters t, n, μ for t-parallelized TD-(n, μ, x) interconnection and the five parameters $\theta_2, r_2, k_2, \lambda_2, c_2$ for the skeleton graph which is a BIG-$(\theta_2, r_2, k_2, \lambda_2)$ with c_2 copies. Their constraints and relationships with the aforementioned parameters are analyzed as follows:
  - A composite-edge interconnects n ISNs-(x, w, l) through an RN cluster as a TD-(n, μ, x) interconnection. Such an interconnection occupies μx outer-degrees of each dock node in the n ISNs. Hence, the value of μx should be a factor of the outer-degree of each dock node. For distinct design patterns, the following are the detailed discussions.
    - For those design patterns with $d_2 =$ CG(any) or CBG(any) or BIG(element) or TDG(element), GNs with outer-degree $\beta_G p_G$ are selected as dock nodes, which implies that $\mu x \mid \beta_G p_G$.
    - For those design patterns with $d_2 =$ BIG(block) or TDG(block), RNs with outer-degree $\beta_R p_R$ are selected as dock nodes. In this case, we have $\mu x \mid \beta_R p_R$.
  - The skeleton graph is a BIG-$(\theta_2, r_2, k_2, \lambda_2)$ with c_2 copies. By Lemma 1.6, the parameters of such a BIG satisfy

$$\lambda_2 = \frac{r_2(k_2 - 1)}{\theta_2 k_2 - 1}. \tag{4.6}$$

  - Under the t-parallelized TD-(n, μ, x) interconnection, each super-edge between an RN cluster and a module contains $t\mu x^2$ edges. Since each RN cluster contains μx^2 RNs, each super-edge occupies t outer-degrees of each RN in the RN cluster. Meanwhile, in the skeleton graph (a BIG-$(\theta_2, r_2, k_2, \lambda_2)$ with c_2 copies), each RN cluster is incident to $k_2(c_2 + 1)$ super-edges. Hence, for each RN in the RN cluster, a number of $t \cdot k_2(c_2 + 1)$ degrees are occupied in total. Given that the degree of each RN is p_R, we have $t \cdot k_2(c_2 + 1) = p_R$. This implies that tk_2 needs to be a factor of p_R since c_2 is an integer, i.e., $tk_2 \mid p_R$.
  - Under the t-parallelized TD-(n, μ, x) interconnection, each super-edge between an RN cluster and a module contains of $t\mu x^2$ edges. In the skeleton graph, each module is incident to r_2 super-edges that occupy a total number of $r_2 \cdot t\mu x^2$ outer-degrees of dock nodes in the module. Meanwhile, super-edge involves tx dock nodes in each module. This implies that the value of tx should be less than or equal to the number of dock nodes in each module. Under distinct design patterns, the detailed discussions are given as follows:
    - For design patterns with $d_2 =$ CG(any), m_1 GNs with outer-degree $\beta_G p_G$ are selected as dock nodes. Hence, the sum of outer-degrees of all dock nodes in each module is equal to $m_1 \beta_G p_G$, which implies

$m_1\beta_G p_G = r_2 \cdot t\mu x^2$. Since parameter r_2 is an integer, the value of $t\mu x^2$ should be a factor of $m_1\beta_G p_G$; i.e., $t\mu x^2 \mid m_1\beta_G p_G$.

- For design patterns with $d_1 = G$ and $d_2 =$ CBG(any), m_1' GNs with outer-degree $\beta_G p_G$ are selected as dock nodes. Hence, the sum of outer-degrees of all dock nodes in each module is equal to $m_1'\beta_G p_G$, which implies $m_1'\beta_G p_G = r_2 \cdot t\mu x^2$. Since parameter r_2 is an integer, the value of $t\mu x^2$ should be a factor of $m_1'\beta_G p_G$; i.e., $t\mu x^2 \mid m_1'\beta_G p_G$.
- For design patterns with $d_1 = G$ and $d_2 =$ BIG(block), $\theta_1\alpha_G p_G$ RNs with outer-degree $\beta_R p_R$ are selected as dock nodes. Hence, the sum of outer-degrees of all dock nodes in each module is equal to $\theta_1\alpha_G p_G\beta_R p_R$, which implies $\theta_1\alpha_G p_G\beta_R p_R = r_2 \cdot t\mu x^2$. Since parameter r_2 is an integer, the value of $t\mu x^2$ should be a factor of $\theta_1\alpha_G p_G\beta_R p_R$; i.e., $t\mu x^2 \mid \theta_1\alpha_G p_G\beta_R p_R$.
- For design patterns with $d_1 = G$ and $d_2 =$ BIG(element), $\theta_1\alpha_R p_R$ GNs with outer-degree $\beta_G p_G$ are selected as dock nodes. Hence, the sum of outer-degrees of all dock nodes in each module is equal to $\theta_1 p_R\beta_G p_G$, which implies $\theta_1 p_R\beta_G p_G = r_2 \cdot t\mu x^2$. Since parameter r_2 is an integer, the value of $t\mu x^2$ should be a factor of $\theta_1 p_R\beta_G p_G$; i.e., $t\mu x^2 \mid \theta_1 p_R\beta_G p_G$.
- For design patterns with $d_2 =$ TDG(block), $\mu_1 v_1^2$ RNs with outer-degree $\beta_R p_R$ are selected as dock nodes. Hence, the sum of outer-degrees of all dock nodes in each module is equal to $\mu_1 v_1^2\beta_R p_R$, which implies $\mu_1 v_1^2\beta_R p_R = r_2 \cdot t\mu x^2$. Since parameter r_2 is an integer, the value of $t\mu x^2$ should be a factor of $\mu_1 v_1^2\beta_R p_R$; i.e., $t\mu x^2 \mid \mu_1 v_1^2\beta_R p_R$.
- For design patterns with $d_2 =$ TDG(element), $n_1 v_1(c_1' + 1)$ GNs with outer-degree $\beta_G p_G$ are selected as dock nodes. Hence, the sum of outer-degrees of all dock nodes in each module is equal to $n_1 v_1(c_1' + 1)\beta_G p_G$, which implies $n_1 v_1(c_1' + 1)\beta_G p_G = r_2 \cdot t\mu x^2$. Since parameter r_2 is an integer, the value of $t\mu x^2$ should be a factor of $n_1 v_1(c_1' + 1)\beta_G p_G$; i.e., $t\mu x^2 \mid n_1 v_1(c_1' + 1)\beta_G p_G$.

■ Note that each super-edge is a t-parallelized TD-(n, μ, x) interconnection. Since such a super-edge involves tx dock nodes in each module, the value of tx should be less than or equal to the number of dock nodes in each module. Under distinct design patterns, the detailed discussions are given as follows:

- For design patterns with $d_2 =$ CG(any), m_1 GNs are selected as dock nodes. Hence, we have $tx \le m_1$.
- For design patterns with $d_2 =$ CBG(any), m_1' GNs are selected as dock nodes. Hence, we have $tx \le m_1'$.
- For design patterns with $d_2 =$ BIG(block), $\theta_1\alpha_G p_G$ RNs are selected as dock nodes. Hence, we have $tx \le \theta_1\alpha_G p_G$.
- For design patterns with $d_2 =$ BIG(element), $\theta_1\alpha_R p_R$ GNs are selected as dock nodes. Hence, we have $tx \le \theta_1\alpha_R p_R$.

- For design patterns with $d_2 = \text{TDG(block)}$, $\mu_1 v_1^2$ RNs are selected as dock nodes. Hence, we have $tx \leq \mu_1 v_1^2$.
- For design patterns with $d_2 = \text{TDG(element)}$, $n_1 v_1 (c_1' + 1)$ GNs are selected as dock nodes. Hence, we have $tx \leq n_1 v_1 (c_1' + 1)$.

4.3.2 Parameters Under Each Design Pattern

In this subsection, by rearranging the results from our previous analysis, we summarize the parameters under each design pattern with $d_1 = G$, along with their relationships and the constraints to which they are subject.[3] The details are presented in the subsequent divisions (i.e., Sections 4.3.2.1–4.3.2.12).

In this process, we divide the parameters under each design pattern into two categories, namely *free* parameters and *bound* parameters. The former refer to the parameters that can be tuned freely over feasible ranges. Once their values are fixed, the values of the corresponding bound parameters are also determined. For ease of understanding, we take the parameters $\alpha_G, \beta_G, \gamma_G$ under the design patterns with $d_1 = G$ and $d_2 = \text{CG(any)}$ as an example for illustration. In this case, we take parameters α_G and γ_G as the free parameters and parameter β_G as the bound parameter. By equation (4.2) and the constraint $0 < \alpha_G, \beta_G, \gamma_G < 1$, the values of α_G and γ_G can be set within range (0, 1) such that $\alpha_G + \gamma_G < 1$. Given their values, the value of β_G is readily fixed as $\beta_G = 1 - \alpha_G - \gamma_G$. By distinguishing free parameters from bound parameters, it not only conduces to ease of analysis but also leads to systematic topology design with a lower complexity. In practice, network designers can take free parameters as convenient knobs and adjust their values flexibly to achieve expected performances. In Table 4.4, we provide a summary of free parameters under different design patterns with $d_1 = G$.

4.3.2.1 Design Pattern $< G, \text{CG(any)}, P >$

Under this design pattern, there are nine parameters, including the three parameters α_G, β_G, and γ_G describing the proportions of node degrees for distinct connections by GNs in modules, one parameter m_1 that denotes the number of GNs in the internal graph (a CG), the four parameters of ISNs and super-edges (i.e., x, w, l, and t), and one parameter m_2 that denotes the number of modules in the

3 The analysis for design patterns with $d_1 = H$ can be performed in a similar way by replacing parameters p_G, α_G, β_G, and γ_G with parameters p_H, α_H, β_H, and γ_H, respectively. The minor difference is that the value of $\gamma_H p_H$ should be a non-negative integer because $\gamma_H \geq 0$, while the value of $\gamma_G p_G$ must be a positive integer due to $\gamma_G > 0$.

Table 4.4 The summary of free parameters under each design pattern with $d_1 = G$. For each design pattern, the free parameters are divided into four groups, which represent the parameters of node degree coefficients, internal graph, super-edge, and skeleton graph, respectively.

d_2 \ d_3	P	B
CG(any)	$(\alpha_G, \gamma_G), (-), (t, x), (-)$	$(\alpha_G, \gamma_G), (-), (t, x, \mu), (\theta_2, k_2)$
CBG(any)	$(\alpha_G, \gamma_G), (-), (t, x), (-)$	$(\alpha_G, \gamma_G), (-), (t, x, \mu), (\theta_2, k_2)$
BIG(element)	$(\alpha_G, \gamma_G), (\theta_1, k_1), (t, x), (-)$	$(\alpha_G, \gamma_G), (\theta_1, k_1), (t, x, \mu), (\theta_2, k_2)$
BIG(block)	$(\alpha_R, \gamma_G), (\theta_1, k_1), (t, x), (-)$	$(\alpha_R, \gamma_G), (\theta_1, k_1), (t, x, \mu), (\theta_2, k_2)$
TDG(element)	$(\alpha_G, \gamma_G), (v_1, n_1), (t, x), (-)$	$(\alpha_G, \gamma_G), (v_1, n_1), (t, x, \mu), (\theta_2, k_2)$
TDG(block)	$(\alpha_R, \gamma_G), (v_1, n_1), (t, x), (-)$	$(\alpha_R, \gamma_G), (v_1, n_1), (t, x, \mu), (\theta_2, k_2)$

Note: The symbol (−) indicates that there are no free parameters in the group under the design pattern.

skeleton graph (a CG). The relationships among such parameters are summarized as follows:

$$\beta_G = 1 - \alpha_G - \gamma_G, \tag{4.7}$$

$$m_1 = \alpha_G p_G + 1, \tag{4.8}$$

$$w = x, \tag{4.9}$$

$$l = 1, \tag{4.10}$$

$$m_2 = \frac{(\alpha_G p_G + 1) \cdot (1 - \alpha_G - \gamma_G) \cdot p_G}{t x^2} + 1. \tag{4.11}$$

From the aforementioned relationships, we identify four free parameters, including parameters α_G, γ_G, t, and x. The remaining five parameters (i.e., β_G, m_1, w, l, m_2) are treated as bound parameters. Accordingly, the constraints to which the free parameters are subject can be summarized as follows:

- The values of $\alpha_G p_G, \gamma_G p_G$, and $(1 - \alpha_G - \gamma_G)p_G$ are all positive integers such that $\alpha_G, \gamma_G > 0$ and $\alpha_G + \gamma_G < 1$.
- The value of x should be a common factor of $\alpha_G p_G + 1$ and $(1 - \alpha_G - \gamma_G)p_G$.
- The value of t should be a factor of $\frac{(\alpha_G p_G + 1) \cdot (1 - \alpha_G - \gamma_G)p_G}{x^2}$ such that $t \leq \frac{\alpha_G p_G + 1}{x}$.

4.3.2.2 Design Pattern $< G$, CG(any), $B >$

Under this design pattern, there are 15 parameters, including the 3 parameters (i.e., α_G, β_G, and γ_G) describing the proportions of node degrees for distinct connections by GNs in modules, 1 parameter m_1 that denotes the number of GNs in the internal graph (a CG), the 6 parameters for ISNs and super-edges (i.e., x, w, l,

t, n, and μ), and the 5 parameters for the skeleton graph (a BIG-$(\theta_2, r_2, k_2, \lambda_2)$ with c_2 copies). The relationships among such parameters are given as follows:

$$\beta_G = 1 - \alpha_G - \gamma_G, \tag{4.12}$$

$$m_1 = \alpha_G p_G + 1, \tag{4.13}$$

$$n = k_2, \tag{4.14}$$

$$w = x, \tag{4.15}$$

$$l = 1, \tag{4.16}$$

$$c_2 = \frac{p_R}{tk_2} - 1, \tag{4.17}$$

$$r_2 = \frac{(\alpha_G p_G + 1)(1 - \alpha_G - \gamma_G)p_G}{t\mu x^2}, \tag{4.18}$$

$$\lambda_2 = \frac{(\alpha_G p_G + 1)(1 - \alpha_G - \gamma_G)p_G}{t\mu x^2} \cdot \frac{k_2 - 1}{\theta_2 k_2 - 1}. \tag{4.19}$$

From the aforementioned relationships, we identify seven free parameters (i.e., $\alpha_G, \gamma_G, t, x, \mu, \theta_2, k_2$) and eight bound parameters (i.e., $\beta_G, m_1, n, w, l, r_2, \lambda_2, c_2$). Accordingly, the constraints on the free parameters can be written as follows:

- The values of the free parameters should guarantee the existence of BIG-$(\theta_2, r_2, k_2, \lambda_2)$.
- The values of the free parameters should guarantee the existence of TD-(n, μ, x) interconnection.
- The values of $\alpha_G p_G$, $\gamma_G p_G$, and $(1 - \alpha_G - \gamma_G)p_G$ are all positive integers such that $\alpha_G, \gamma_G > 0$ and $\alpha_G + \gamma_G < 1$.
- The value of x should be a common factor of $\alpha_G p_G + 1$ and $(1 - \alpha_G - \gamma_G)p_G$.
- The value of t should be a common factor of p_R and $\frac{(\alpha_G p_G + 1) \cdot (1 - \alpha_G - \gamma_G)p_G}{x^2}$ with $t \le \frac{\alpha_G p_G + 1}{x}$.
- The value of μ should be a common factor of both $\frac{(1 - \alpha_G - \gamma_G)p_G \cdot (\alpha_G p_G + 1)}{tx^2}$ and $\frac{(1 - \alpha_G - \gamma_G)p_G}{x}$.
- The value of k_2 should be a factor of $\frac{p_R}{t}$ such that $k_2 \ge 2$.

4.3.2.3 Design Pattern $< G$, CBG(any), $P >$

Under this design pattern, there are nine parameters, including the three parameters α_G, β_G, and γ_G describing the proportions of node degrees for distinct connections by GNs in modules, one parameter m_1' that denotes the number of GNs in the internal graph (a CBG), the four parameters of ISNs and super-edges (i.e., x, w, l, and t), and one parameter m_2 that denotes the number of modules in the skeleton graph (a CG). The relationships among such parameters are summarized as follows:

$$\beta_G = 1 - \alpha_G - \gamma_G, \tag{4.20}$$

$$m_1' = 2\alpha_G p_G, \tag{4.21}$$

$$w = \begin{cases} 1, & \text{if } x = 1; \\ \lfloor \frac{x}{2} \rfloor, & \text{if } x > 1; \end{cases} \tag{4.22}$$

$$l = \begin{cases} 2, & \text{if } x = 1; \\ 1, & \text{if } x > 1; \end{cases} \tag{4.23}$$

$$m_2 = \frac{2\alpha_G p_G \cdot (1 - \alpha_G - \gamma_G) \cdot p_G}{tx^2} + 1. \tag{4.24}$$

From the aforementioned relationships, we identify four free parameters, including parameters α_G, γ_G, t, and x. The remaining five parameters (i.e., β_G, m_1', w, l, m_2) are treated as bound parameters. Accordingly, the constraints to which the free parameters are subject are summarized as follows:

- The values of $\alpha_G p_G, \gamma_G p_G$, and $(1 - \alpha_G - \gamma_G)p_G$ are all positive integers such that $\alpha_G, \gamma_G > 0$ and $\alpha_G + \gamma_G < 1$.
- The value of x should be a common factor of $2\alpha_G p_G$ and $(1 - \alpha_G - \gamma_G)p_G$.
- The value of t should be a factor of $\frac{2\alpha_G p_G \cdot (1 - \alpha_G - \gamma_G)p_G}{x^2}$ with $t \leq \frac{2\alpha_G p_G}{x}$.

4.3.2.4 Design Pattern $< G, \text{CBG(any)}, B >$

Under this design pattern, there are 15 parameters, including the 3 parameters (i.e., α_G, β_G, and γ_G) describing the proportions of node degrees for distinct connections by GNs in modules, 1 parameter m_1' that denotes the number of GNs in the internal graph (a CBG), the 6 parameters for ISNs and super-edges (i.e., x, w, l, t, n, and μ), and the 5 parameters for the skeleton graph (a BIG-($\theta_2, r_2, k_2, \lambda_2$) with c_2 copies). The relationships among such parameters are given as follows:

$$\beta_G = 1 - \alpha_G - \gamma_G, \tag{4.25}$$

$$m_1' = 2\alpha_G p_G, \tag{4.26}$$

$$n = k_2, \tag{4.27}$$

$$w = \begin{cases} 1, & \text{if } x = 1; \\ \lfloor \frac{x}{2} \rfloor, & \text{if } x > 1; \end{cases} \tag{4.28}$$

$$l = \begin{cases} 2, & \text{if } x = 1; \\ 1, & \text{if } x > 1; \end{cases} \tag{4.29}$$

$$c_2 = \frac{p_R}{tk_2} - 1, \tag{4.30}$$

$$r_2 = \frac{2\alpha_G p_G \cdot (1 - \alpha_G - \gamma_G)p_G}{t\mu x^2}, \tag{4.31}$$

$$\lambda_2 = \frac{2\alpha_G p_G \cdot (1 - \alpha_G - \gamma_G)p_G}{t\mu x^2} \cdot \frac{k_2 - 1}{\theta_2 k_2 - 1}. \tag{4.32}$$

From the aforementioned relationships, we identify seven free parameters (i.e., $\alpha_G, \gamma_G, t, x, \mu, \theta_2, k_2$) and eight bound parameters (i.e., $\beta_G, m_1', n, w, l, r_2, \lambda_2, c_2$). Accordingly, the constraints on the free parameters can be written as follows:

- The values of the free parameters should guarantee the existence of BIG-$(\theta_2, r_2, k_2, \lambda_2)$.
- The values of the free parameters should guarantee the existence of TD-(n, μ, x) interconnection.
- The values of $\alpha_G p_G$, $\gamma_G p_G$, and $(1 - \alpha_G - \gamma_G)p_G$ are all positive integers such that $\alpha_G, \gamma_G > 0$ and $\alpha_G + \gamma_G < 1$.
- The value of x should be a common factor of $2\alpha_G p_G$ and $(1 - \alpha_G - \gamma_G)p_G$.
- The value of t should be a common factor of p_R and $\frac{2\alpha_G p_G \cdot (1 - \alpha_G - \gamma_G)p_G}{x^2}$ such that $t \leq \frac{2\alpha_G p_G}{x}$.
- The value of μ should be a common factor of $\frac{(1 - \alpha_G - \gamma_G)p_G}{x}$ and $\frac{(1 - \alpha_G - \gamma_G)p_G \cdot 2\alpha_G p_G}{tx^2}$.
- The value of k_2 should be a factor of $\frac{p_R}{t}$ such that $k_2 \geq 2$.

4.3.2.5 Design Pattern $< G$, BIG(block), $P >$

Under this design pattern, there are 15 parameters, including the 3 parameters α_G, β_G, and γ_G describing the proportions of node degrees for distinct connections by GNs in modules, the 2 parameters α_R and β_R describing the proportions of node degrees for distinct connections by RNs in modules, the 5 parameters for the internal graph (a BIG-$(\theta_1, r_1, k_1, \lambda_1)$ with c_1 copies), the 4 parameters of ISNs and super-edges (i.e., x, w, l, and t), and 1 parameter m_2 that denotes the number of modules in the skeleton graph (a CG). The relationships among such parameters are summarized as follows:

$$\alpha_G = 1 - \gamma_G, \tag{4.33}$$

$$\beta_G = 0, \tag{4.34}$$

$$\beta_R = 1 - \alpha_R, \tag{4.35}$$

$$r_1 = (1 - \gamma_G)p_G, \tag{4.36}$$

$$\lambda_1 = (1 - \gamma_G)p_G \cdot \frac{k_1 - 1}{\theta_1 k_1 - 1}, \tag{4.37}$$

$$c_1 = \frac{\alpha_R p_R}{k_1} - 1, \tag{4.38}$$

$$w = \frac{x}{\theta_1}, \tag{4.39}$$

$$l = 1, \tag{4.40}$$

$$m_2 = \frac{\theta_1 \cdot (1 - \gamma_G)p_G \cdot (1 - \alpha_R)p_R}{tx^2} + 1. \tag{4.41}$$

From the aforementioned relationships, we identify six free parameters, including parameters α_R, γ_G, θ_1, k_1, t, and x. The remaining nine parameters (i.e., α_G, β_G, β_R, r_1, λ_1, c_1, w, l, m_2) are treated as bound parameters. Accordingly, the constraints to which the free parameters are subject can be summarized as follows:

- The values of the free parameters should guarantee the existence of BIG-$(\theta_1, r_1, k_1, \lambda_1)$.
- The values of the free parameters should guarantee that the BIBD-$(\theta_1 k_1, \theta_1 r_1, r_1, k_1, \lambda_1)$ is resolvable.
- The values of $\alpha_R p_R$ and $\gamma_G p_G$ are both positive integers with $0 < \alpha_R, \gamma_G < 1$.
- The value of k_1 should be a factor of $\alpha_R p_R$.
- The value of θ_1 should be a factor of x.
- The value of x should be a common factor of $\theta_1(1 - \gamma_G)p_G$ and $(1 - \alpha_R)p_R$.
- The value of t should be a factor of $\dfrac{\theta_1 \cdot (1 - \gamma_G)p_G \cdot (1 - \alpha_R)p_R}{x^2}$ with $t \leq \dfrac{\theta_1 \cdot (1 - \gamma_G)p_G}{x}$.

4.3.2.6 Design Pattern $< G$, BIG(block), $B >$

Under this design pattern, there are 21 parameters, including the 3 parameters (i.e., α_G, β_G, and γ_G) describing the proportions of node degrees for distinct connections by GNs in modules, the 2 parameters (i.e., α_R and β_R) describing the proportions of node degrees for distinct connections by RNs in modules, the 5 parameters for the internal graph (a BIG-$(\theta_1, r_1, k_1, \lambda_1)$ with c_1 copies), the 6 parameters for ISNs and super-edges (i.e., x, w, l, t, n, and μ), and the 5 parameters for the skeleton graph (a BIG-$(\theta_2, r_2, k_2, \lambda_2)$ with c_2 copies). The relationships among such parameters are given as follows:

$$\alpha_G = 1 - \gamma_G, \tag{4.42}$$

$$\beta_G = 0, \tag{4.43}$$

$$\beta_R = 1 - \alpha_R, \tag{4.44}$$

$$r_1 = (1 - \gamma_G)p_G, \tag{4.45}$$

$$\lambda_1 = (1-\gamma_G)p_G \cdot \frac{k_1-1}{\theta_1 k_1 - 1}, \tag{4.46}$$

$$c_1 = \frac{\alpha_R p_R}{k_1} - 1, \tag{4.47}$$

$$n = k_2, \tag{4.48}$$

$$w = \frac{x}{\theta_1}, \tag{4.49}$$

$$l = 1, \tag{4.50}$$

$$r_2 = \frac{\theta_1 \cdot (1-\gamma_G)p_G \cdot (1-\alpha_R)p_R}{t\mu x^2}, \tag{4.51}$$

$$\lambda_2 = \frac{\theta_1 \cdot (1-\gamma_G)p_G \cdot (1-\alpha_R)p_R}{t\mu x^2} \cdot \frac{k_2-1}{\theta_2 k_2 - 1}, \tag{4.52}$$

$$c_2 = \frac{p_R}{tk_2} - 1. \tag{4.53}$$

From the aforementioned relationships, we identify 9 free parameters (i.e., α_R, γ_G, θ_1, k_1, t, x, μ, θ_2, k_2) and 12 bound parameters (i.e., α_G, β_G, β_R, r_1, λ_1, c_1, n, w, l, r_2, λ_2, c_2). The constraints on the free parameters are summarized as follows:

- The values of the free parameters should guarantee the existence of BIG-$(\theta_1, r_1, k_1, \lambda_1)$.
- The values of the free parameters should guarantee that the BIBD-$(\theta_1 k_1, \theta_1 r_1, r_1, k_1, \lambda_1)$ is resolvable.
- The values of the free parameters should guarantee the existence of TD-(n, μ, x) interconnection.
- The values of the free parameters should guarantee the existence of BIG-$(\theta_2, r_2, k_2, \lambda_2)$.
- The values of $\alpha_R p_R$ and $\gamma_G p_G$ are both positive integers with $0 < \alpha_R, \gamma_G < 1$.
- The value of k_1 should be a factor of $\alpha_R p_R$.
- The value of θ_1 should be a factor of x.
- The value of x should be a common factor of $\theta_1(1-\gamma_G)p_G$ and $(1-\alpha_R)p_R$.
- The value of t should be a common factor of p_R and $\frac{(1-\alpha_R)p_R \cdot \theta_1(1-\gamma_G)p_G}{x^2}$ with $t \le \frac{\theta_1(1-\gamma_G)p_G}{x}$.
- The value of μ should be a common factor of $\frac{(1-\alpha_R)p_R \cdot \theta_1(1-\gamma_G)p_G}{tx^2}$ and $\frac{(1-\alpha_R)p_R}{x}$.
- The value of k_2 should be a factor of $\frac{p_R}{t}$.

4.3.2.7 Design Pattern < G, BIG(element), P >

Under this design pattern, there are 15 parameters, including the 3 parameters α_G, β_G, and γ_G describing the proportions of node degrees for distinct connections by GNs in modules, the 2 parameters α_R and β_R describing the proportions of node degrees for distinct connections by RNs in modules, the 5 parameters for the internal graph (a BIG-$(\theta_1, r_1, k_1, \lambda_1)$ with c_1 copies), the 4 parameters of ISNs and super-edges (i.e., x, w, l, and t), and 1 parameter m_2 that denotes the number of modules in the skeleton graph (a CG). The relationships among such parameters are summarized as follows:

$$\alpha_R = 1, \tag{4.54}$$

$$\beta_R = 0, \tag{4.55}$$

$$\beta_G = 1 - \alpha_G - \gamma_G, \tag{4.56}$$

$$r_1 = \alpha_G p_G, \tag{4.57}$$

$$\lambda_1 = \alpha_G p_G \cdot \frac{k_1 - 1}{\theta_1 k_1 - 1}, \tag{4.58}$$

$$c_1 = \frac{p_R}{k_1} - 1, \tag{4.59}$$

$$w = x, \tag{4.60}$$

$$l = 2, \tag{4.61}$$

$$m_2 = \frac{\theta_1 p_R \cdot (1 - \alpha_G - \gamma_G) p_G}{t x^2} + 1. \tag{4.62}$$

From the aforementioned relationships, we identify six free parameters, including parameters $\alpha_G, \gamma_G, \theta_1, k_1, t$, and x. The remaining nine parameters (i.e., $\alpha_R, \beta_R, \beta_G, r_1, \lambda_1, c_1, w, l, m_2$) are treated as bound parameters. Accordingly, the following are the constraints to which the free parameters are subject.

- The values of the free parameters should guarantee the existence of BIG-$(\theta_1, r_1, k_1, \lambda_1)$.
- The values of $\alpha_G p_G, \gamma_G p_G$, and $(1 - \alpha_G - \gamma_G) p_G$ are all positive integers such that $\alpha_G, \gamma_G > 0$ and $\alpha_G + \gamma_G < 1$.
- The value of k_1 should be a factor of p_R.
- The value of x should be a common factor of $\theta_1 p_R$ and $(1 - \alpha_G - \gamma_G) p_G$.
- The value of t should be a factor of $\frac{\theta_1 p_R \cdot (1 - \alpha_G - \gamma_G) p_G}{x^2}$ with $t \leq \frac{\theta_1 p_R}{x}$.

4.3.2.8 Design Pattern < G, BIG(element), B >

Under this design pattern, there are 21 parameters, including the 3 parameters (i.e., α_G, β_G, and γ_G) describing the proportions of node degrees for distinct connections by GNs in modules, the 2 parameters (i.e., α_R and β_R) describing the proportions of node degrees for distinct connections by RNs in modules, the 5 parameters for the internal graph (a BIG-$(\theta_1, r_1, k_1, \lambda_1)$ with c_1 copies), the

6 parameters for ISNs and super-edges (i.e., x, w, l, t, n, and μ), and the 5 parameters for the skeleton graph (a BIG-$(\theta_2, r_2, k_2, \lambda_2)$ with c_2 copies). The relationships among such parameters are given as follows:

$$\alpha_R = 1, \tag{4.63}$$

$$\beta_R = 0, \tag{4.64}$$

$$\beta_G = 1 - \alpha_G - \gamma_G, \tag{4.65}$$

$$r_1 = \alpha_G p_G, \tag{4.66}$$

$$\lambda_1 = \alpha_G p_G \cdot \frac{k_1 - 1}{\theta_1 k_1 - 1}, \tag{4.67}$$

$$c_1 = \frac{p_R}{k_1} - 1, \tag{4.68}$$

$$n = k_2, \tag{4.69}$$

$$w = x, \tag{4.70}$$

$$l = 2, \tag{4.71}$$

$$r_2 = \frac{\theta_1 p_R \cdot (1 - \alpha_G - \gamma_G) p_G}{t \mu x^2}, \tag{4.72}$$

$$\lambda_2 = \frac{\theta_1 p_R \cdot (1 - \alpha_G - \gamma_G) p_G}{t \mu x^2} \cdot \frac{k_2 - 1}{\theta_2 k_2 - 1}, \tag{4.73}$$

$$c_2 = \frac{p_R}{t k_2} - 1. \tag{4.74}$$

From the aforementioned relationships, we identify 9 free parameters (i.e., α_G, γ_G, θ_1, k_1, t, x, μ, θ_2, k_2) and 12 bound parameters (i.e., α_R, β_R, β_G, r_1, λ_1, c_1, n, w, l, r_2, λ_2, c_2). The constraints on the free parameters are summarized as follows:

- The values of the free parameters should guarantee the existence of BIG-$(\theta_1, r_1, k_1, \lambda_1)$.
- The values of the free parameters should guarantee the existence of TD-(n, μ, x) interconnection.
- The values of the free parameters should guarantee the existence of BIG-$(\theta_2, r_2, k_2, \lambda_2)$.
- The values of $\alpha_G p_G$, $\beta_G p_G$, and $\gamma_G p_G$ are all positive integers with $\alpha_G, \gamma_G > 0$ and $\alpha_G + \gamma_G < 1$.
- The value of k_1 should be a factor of p_R.
- The value of x should be a common factor of $\theta_1 p_R$ and $(1 - \alpha_G - \gamma_G) p_G$.
- The value of t should be a common factor of p_R and $\dfrac{\theta_1 p_R \cdot (1 - \alpha_G - \gamma_G) p_G}{x^2}$ with $t \leq \dfrac{\theta_1 p_R}{x}$.

- The value of μ should be a common factor of $\frac{\theta_1 p_R \cdot (1-\alpha_G-\gamma_G)p_G}{tx^2}$ and $\frac{(1-\alpha_G-\gamma_G)p_G}{x}$.
- The value of k_2 should be a factor of $\frac{p_R}{t}$.

4.3.2.9 Design Pattern $< G, \textbf{TDG(block)}, P >$

Under this design pattern, there are 14 parameters, including the 3 parameters α_G, β_G, and γ_G describing the proportions of node degrees for distinct connections by GNs in modules, the 2 parameters α_R and β_R describing the proportions of node degrees for distinct connections by RNs in modules, the 4 parameters for the internal graph (a TDG-(n, μ, υ) with c_1' copies), the 4 parameters of ISNs and super-edges (i.e., x, w, l, and t), and 1 parameter m_2 that denotes the number of modules in the skeleton graph (a CG). The relationships among such parameters are summarized as follows:

$$\alpha_G = 1 - \gamma_G, \tag{4.75}$$

$$\beta_G = 0, \tag{4.76}$$

$$\beta_R = 1 - \alpha_R, \tag{4.77}$$

$$\mu_1 = \frac{(1-\gamma_G)p_G}{\upsilon_1}, \tag{4.78}$$

$$c_1' = \frac{\alpha_R p_R}{n_1} - 1, \tag{4.79}$$

$$w = \frac{x}{\upsilon_1}, \tag{4.80}$$

$$l = 1, \tag{4.81}$$

$$m_2 = \frac{\upsilon_1 \cdot (1-\gamma_G)p_G \cdot \alpha_R p_R}{tx^2} + 1. \tag{4.82}$$

From the aforementioned relationships, we identify six free parameters, including parameters $\alpha_R, \gamma_G, n_1, \upsilon_1, t$, and x. The remaining eight parameters (i.e., $\alpha_G, \beta_G, \beta_R, \mu_1, c_1', w, l, m_2$) are treated as bound parameters. Accordingly, the constraints on the free parameters are summarized as follows:

- The values of the free parameters should guarantee the existence of TDG-(n_1, μ_1, υ_1).
- The values of the free parameters should guarantee that the TD-(n_1, μ_1, υ_1) is resolvable.
- The values of $\alpha_R p_R$ and $\gamma_G p_G$ are both positive integers with $0 < \alpha_R, \gamma_G < 1$.
- The value of n_1 should be a factor of $\alpha_R p_R$.
- The value of υ_1 should be a factor of $(1-\gamma_G)p_G$ and x.
- The value of x should be a common factor of $(1-\gamma_G)p_G \cdot \upsilon_1$ and $(1-\alpha_R)p_R$.

- The value of t should be a factor of $\frac{v_1 \cdot (1-\gamma_G)p_G \cdot (1-\alpha_R)p_R}{x^2}$ with $t \leq \frac{v_1 \cdot (1-\gamma_G)p_G}{x}$.

4.3.2.10 Design Pattern $< G$, TDG(block), $B >$

Under this design pattern, there are 20 parameters, including the 3 parameters (i.e., α_G, β_G, and γ_G) describing the proportions of node degrees for distinct connections by GNs in modules, the 2 parameters (i.e., α_R and β_R) describing the proportions of node degrees for distinct connections by RNs in modules, the 4 parameters for the internal graph (a TDG-(n_1, μ_1, v_1) with c_1' copies), the 6 parameters for ISNs and super-edges (i.e., x, w, l, t, n, and μ), and the 5 parameters for the skeleton graph (a BIG-$(\theta_2, r_2, k_2, \lambda_2)$ with c_2 copies). The relationships among such parameters are given as follows:

$$\alpha_G = 1 - \gamma_G, \tag{4.83}$$

$$\beta_G = 0, \tag{4.84}$$

$$\beta_R = 1 - \alpha_R, \tag{4.85}$$

$$\mu_1 = \frac{(1-\gamma_G)p_G}{v_1}, \tag{4.86}$$

$$c_1' = \frac{\alpha_R p_R}{n_1} - 1, \tag{4.87}$$

$$n = k_2, \tag{4.88}$$

$$w = \frac{x}{v_1}, \tag{4.89}$$

$$l = 1, \tag{4.90}$$

$$r_2 = \frac{v_1 \cdot (1-\gamma_G)p_G \cdot (1-\alpha_R)p_R}{t\mu x^2}, \tag{4.91}$$

$$\lambda_2 = \frac{v_1 \cdot (1-\gamma_G)p_G \cdot (1-\alpha_R)p_R}{t\mu x^2} \cdot \frac{k_2 - 1}{\theta_2 k_2 - 1}, \tag{4.92}$$

$$c_2 = \frac{p_R}{tk_2} - 1. \tag{4.93}$$

From the aforementioned relationships, we identify 9 free parameters (i.e., α_R, γ_G, n_1, v_1, t, x, μ, θ_2, k_2) and 12 bound parameters (i.e., α_G, β_G, β_R, μ_1, c_1', n, w, l, r_2, λ_2, c_2). Accordingly, the constraints on the free parameters are written as follows:

- The values of the free parameters should guarantee the existence of TDG-(n_1, μ_1, v_1).
- The values of the free parameters should guarantee that the TD-(n_1, μ_1, v_1) is resolvable.

- The values of the free parameters should guarantee the existence of TD-(n, μ, x) interconnection.
- The values of the free parameters should guarantee the existence of BIG-$(\theta_2, r_2, k_2, \lambda_2)$.
- The values of $\alpha_R p_R$ and $\gamma_G p_G$ are both positive integers with $0 < \alpha_R, \gamma_G < 1$.
- The value of n_1 should be a factor of $\alpha_R p_R$.
- The value of υ_1 should be a factor of $(1 - \gamma_G)p_G$ and x.
- The value of x should be a common factor of $\upsilon_1 \cdot (1 - \gamma_G)p_G$ and $(1 - \alpha_R)p_R$.
- The value of t should be a common factor of p_R and $\frac{\upsilon_1 \cdot (1 - \gamma_G)p_G \cdot (1 - \alpha_R)p_R}{x^2}$ with $t \leq \frac{\upsilon_1 \cdot (1 - \gamma_G)p_G}{x}$.
- The value of μ should be a common factor of $\frac{\upsilon_1 \cdot (1 - \gamma_G)p_G \cdot (1 - \alpha_R)p_R}{tx^2}$ and $\frac{(1 - \alpha_R)p_R}{x}$.
- The value of k_2 should be a factor of $\frac{p_R}{t}$.

4.3.2.11 Design Pattern $< G$, TDG(element), $P >$

Under this design pattern, there are 14 parameters, including the 3 parameters α_G, β_G, and γ_G describing the proportions of node degrees for distinct connections by GNs in modules, the 2 parameters α_R and β_R describing the proportions of node degrees for distinct connections by RNs in modules, the 4 parameters for the internal graph (a TDG-(n, μ, υ) with c_1' copies), the 4 parameters of ISNs and super-edges (i.e., x, w, l, and t), and 1 parameter m_2 that denotes the number of modules in the skeleton graph (a CG). The relationships among such parameters are summarized as follows:

$$\alpha_R = 1, \tag{4.94}$$

$$\beta_R = 0, \tag{4.95}$$

$$\beta_G = 1 - \alpha_G - \gamma_G, \tag{4.96}$$

$$\mu_1 = \frac{\alpha_G p_G}{\upsilon_1}, \tag{4.97}$$

$$c_1' = \frac{p_R}{n_1} - 1, \tag{4.98}$$

$$w = x - 1, \tag{4.99}$$

$$l = 2, \tag{4.100}$$

$$m_2 = \frac{\upsilon_1 \cdot (1 - \alpha_G - \gamma_G)p_G p_R}{tx^2} + 1. \tag{4.101}$$

From the aforementioned relationships, we identify six free parameters, including parameters $\alpha_G, \gamma_G, n_1, \upsilon_1, t$, and x. The remaining eight parameters (i.e., α_R, β_R,

$\beta_G, \mu_1, c_1', w, l, m_2$) are treated as bound parameters. Accordingly, the constraints on the free parameters are summarized as follows:

- The values of the free parameters should guarantee the existence of TDG-(n_1, μ_1, v_1).
- The values of the free parameters should guarantee that the TD-(n_1, μ_1, v_1) is resolvable.
- The values of $\alpha_G p_G, \gamma_G p_G$, and $(1 - \alpha_G - \gamma_G)p_G$ are all positive integers with $\alpha_G, \gamma_G > 0$ and $\alpha_G + \gamma_G < 1$.
- The value of n_1 should be a common factor of p_R and x.
- The value of v_1 should be a factor of $\alpha_G p_G$.
- The value of x should be a common factor of $v_1 p_R$ and $(1 - \alpha_G - \gamma_G)p_G$.
- The value of t should be a factor of $\frac{v_1 \cdot (1 - \alpha_G - \gamma_G)p_G p_R}{x^2}$ with $t \leq \frac{v_1 p_R}{x}$.

4.3.2.12 Design Pattern $< G$, TDG(element), $B >$

Under this design pattern, there are 20 parameters, including the 3 parameters (i.e., α_G, β_G, and γ_G) describing the proportions of node degrees for distinct connections by GNs in modules, the 2 parameters (i.e., α_R and β_R) describing the proportions of node degrees for distinct connections by RNs in modules, the 4 parameters for the internal graph (a TDG-(n_1, μ_1, v_1) with c_1' copies), the 6 parameters for ISNs and super-edges (i.e., x, w, l, t, n, and μ), and the 5 parameters for the skeleton graph (a BIG-$(\theta_2, r_2, k_2, \lambda_2)$ with c_2 copies). The relationships among such parameters are given as follows:

$$\alpha_R = 1, \tag{4.102}$$

$$\beta_R = 0, \tag{4.103}$$

$$\beta_G = 1 - \alpha_G - \gamma_G, \tag{4.104}$$

$$\mu_1 = \frac{\alpha_G p_G}{v_1}, \tag{4.105}$$

$$c_1' = \frac{p_R}{n_1} - 1, \tag{4.106}$$

$$n = k_2, \tag{4.107}$$

$$w = x - 1, \tag{4.108}$$

$$l = 2, \tag{4.109}$$

$$r_2 = \frac{v_1 \cdot (1 - \alpha_G - \gamma_G)p_G p_R}{t\mu x^2}, \tag{4.110}$$

$$\lambda_2 = \frac{v_1 \cdot (1 - \alpha_G - \gamma_G)p_G p_R}{t\mu x^2} \cdot \frac{k_2 - 1}{\theta_2 k_2 - 1}, \tag{4.111}$$

$$c_2 = \frac{p_R}{tk_2} - 1. \tag{4.112}$$

From the aforementioned relationships, we identify 9 free parameters (i.e., α_G, γ_G, n_1, υ_1, t, x, μ, θ_2, k_2) and 11 bound parameters (i.e., α_R, β_R, β_G, μ_1, c_1', n, w, l, r_2, λ_2, c_2). The constraints on the free parameters can be written as follows:

- The values of the free parameters should guarantee the existence of TDG-(n_1, μ_1, υ_1).
- The values of the free parameters should guarantee the existence of TD-(n, μ, x) interconnection.
- The values of the free parameters should guarantee the existence of BIG-(θ_2, r_2, k_2, λ_2).
- The values of $\alpha_G p_G$, $\gamma_G p_G$, $(1 - \alpha_G - \gamma_G)p_G$ are all positive integers such that $\alpha_G, \gamma_G > 0$ and $\alpha_G + \gamma_G < 1$.
- The value of n_1 should be a common factor of p_R and x.
- The value of υ_1 should be a factor of $\alpha_G p_G$.
- The value of x should be a common factor of $\upsilon_1 p_R$ and $(1 - \alpha_G - \gamma_G)p_G$.
- The value of t should be a common factor of p_R and $\frac{\upsilon_1 \cdot (1 - \alpha_G - \gamma_G)p_G p_R}{x^2}$ such that $t \leq \frac{\upsilon_1 p_R}{x}$.
- The value of μ should be a common factor of $\frac{\upsilon_1 \cdot (1 - \alpha_G - \gamma_G)p_G p_R}{tx^2}$ and $\frac{(1 - \alpha_G - \gamma_G)p_G}{x}$.
- The value of k_2 should be a factor of $\frac{p_R}{t}$.

4.4 Performance Metrics of Design Patterns

In the previous section, we summarized the parameters under the design patterns, as well as the relationships and constraints they need to satisfy. Based on such results, in this section, we demonstrate how the performance metrics[4] under each design pattern are characterized quantitatively in terms of the corresponding parameters. The analysis shall serve as the basis of our subsequent discussions on the guidelines for systematic topology design in Sections 4.5 and 4.6.

4.4.1 Scalability

By Definition 2.4, the scalability of a network topology is defined as the total number of its nodes with computing function (i.e., CNs and HNs). Under the assumptions made in Section 4.1, scalability is equal to the product of two terms, denoted by symbols S_1 and S_2, respectively.

4 Recall that there are five performance metrics considered in this monograph, including scalability, cost, latency, throughput, and fault-tolerance.

- The term S_1 represents the attachment number of each module. In particular, for design patterns with $d_1 = G$, the value of S_1 equals the number of CNs attached to GNs in each module. For design patterns with $d_1 = H$, its value equals the number of CNs attached to HNs plus the number of HNs in each module.
- The term S_2 represents the number of modules in the topology.

For the aforementioned notations, we present the following two lemmas to characterize their values under different design patterns. The proofs are given in Sections A.4.1 and A.4.2, respectively.

Lemma 4.1 *Let term S_1 denote the number of nodes with computing function attached to each module. The value of S_1 is determined by the selected options of terms d_1 and d_2. Specifically,*

- *for design patterns with d_2 =CG(any), we have $S_1 = \gamma_G p_G \cdot (\alpha_G p_G + 1)$ if $d_1 = G$ or $S_1 = (\gamma_H p_H + 1) \cdot (\alpha_H p_H + 1)$ if $d_1 = H$;*
- *for design patterns with d_2 =CBG(any), we have $S_1 = \gamma_G p_G \cdot 2\alpha_G p_G$ if $d_1 = G$ or $S_1 = (\gamma_H p_H + 1) \cdot 2\alpha_H p_H$ if $d_1 = H$;*
- *for design patterns with d_2 =BIG(block), we have $S_1 = \gamma_G p_G \cdot \theta_1 \alpha_R p_R$ if $d_1 = G$ or $S_1 = (\gamma_H p_H + 1) \cdot \theta_1 \alpha_R p_R$ if $d_1 = H$;*
- *for design patterns with d_2 =BIG(element), we have $S_1 = \gamma_G p_G \cdot \theta_1 p_R$ if $d_1 = G$ or $S_1 = (\gamma_H p_H + 1) \cdot \theta_1 p_R$ if $d_1 = H$;*
- *for design patterns with d_2 =TDG(block), we have $S_1 = \gamma_G p_G \cdot \upsilon_1 \alpha_R p_R$ if $d_1 = G$ or $S_1 = (\gamma_H p_H + 1) \cdot \upsilon_1 \alpha_R p_R$ if $d_1 = H$;*
- *for design patterns with d_2 =TDG(element), we have $S_1 = \gamma_G p_G \cdot \upsilon_1 p_R$ if $d_1 = G$ or $S_1 = (\gamma_H p_H + 1) \cdot \upsilon_1 p_R$ if $d_1 = H$.*

Lemma 4.2 *The value of S_2 (i.e., the number of modules in the topology) is determined by the selected option of term d_3. Specifically,*

- *if $d_3 = P$, we have $S_2 = m_2$;*
- *if $d_3 = B$, we have $S_2 = \theta_2 p_R / t$.*

Based on the two lemmas, the scalabilities under design patterns with $d_1 = G$ and with $d_1 = H$ are summarized in Tables 4.5 and 4.6, respectively.

Besides, the two lemmas also inform quantitative effects of choosing different options for terms d_1, d_2, and d_3 on the scalability metric, respectively.

- When option $d_1 = G$ turns into $d_1 = H$, the terms $\alpha_G p_G$, $\beta_G p_G$, and $\gamma_G p_G$ in scalability are changed into $\alpha_H p_H$, $\beta_H p_H$, and $\gamma_H p_H + 1$, respectively.
- The effect of term d_2 on scalability is the multiplication of S_1.
- The effect of term d_3 on scalability is the multiplication of S_2.

Table 4.5 A summary of scalabilities under design patterns with $d_1 = G$.

d_2 \ d_3	P	B
CG(any)	$\gamma_G p_G \cdot (\alpha_G p_G + 1) \cdot m_2$	$\gamma_G p_G \cdot (\alpha_G p_G + 1) \cdot (\theta_2 p_R / t)$
CBG(any)	$\gamma_G p_G \cdot (2\alpha_G p_G) \cdot m_2$	$\gamma_G p_G \cdot (2\alpha_G p_G) \cdot (\theta_2 p_R / t)$
BIG(block)	$\gamma_G p_G \cdot (\theta_1 \alpha_R p_R) \cdot m_2$	$\gamma_G p_G \cdot (\theta_1 \alpha_R p_R) \cdot (\theta_2 p_R / t)$
BIG(element)	$\gamma_G p_G \cdot (\theta_1 p_R) \cdot m_2$	$\gamma_G p_G \cdot (\theta_1 p_R) \cdot (\theta_2 p_R / t)$
TDG(block)	$\gamma_G p_G \cdot (\upsilon_1 \alpha_R p_R) \cdot m_2$	$\gamma_G p_G \cdot (\upsilon_1 \alpha_R p_R) \cdot (\theta_2 p_R / t)$
TDG(element)	$\gamma_G p_G \cdot (\upsilon_1 p_R) \cdot m_2$	$\gamma_G p_G \cdot (\upsilon_1 p_R) \cdot (\theta_2 p_R / t)$

Note: Parameters p_G and p_R denote the node degrees of GNs and RNs, respectively. Parameters α_G and γ_G denote the node degree proportions for GNs. Parameter α_R denotes the node degree proportion for RNs.
Parameters θ_1 and υ_1 are the parameters of internal graph options BIG-$(\theta_1, r_1, k_1, \lambda_1)$ and TDG-(n_1, μ_1, υ_1), respectively.
Parameters θ_2 and m_2 are the parameters of skeleton graph options BIG-$(\theta_2, r_2, k_2, \lambda_2)$ and CG with m_2 nodes, respectively.
Parameter t is the parameter of t-parallelized TD-(n, μ, x) interconnection for super-edge under the bipartite mode.

Table 4.6 A summary of scalabilities under design patterns with $d_1 = H$.

d_2 \ d_3	P	B
CG(any)	$(\gamma_H p_H + 1) \cdot (\alpha_H p_H + 1) \cdot m_2$	$(\gamma_H p_H + 1) \cdot (\alpha_H p_H + 1) \cdot (\theta_2 p_R / t)$
CBG(any)	$(\gamma_H p_H + 1) \cdot (2\alpha_H p_H) \cdot m_2$	$(\gamma_H p_H + 1) \cdot (2\alpha_H p_H) \cdot (\theta_2 p_R / t)$
BIG(block)	$(\gamma_H p_H + 1) \cdot (\theta_1 \alpha_R p_R) \cdot m_2$	$(\gamma_H p_H + 1) \cdot (\theta_1 \alpha_R p_R) \cdot (\theta_2 p_R / t)$
BIG(element)	$(\gamma_H p_H + 1) \cdot (\theta_1 p_R) \cdot m_2$	$(\gamma_H p_H + 1) \cdot (\theta_1 p_R) \cdot (\theta_2 p_R / t)$
TDG(block)	$(\gamma_H p_H + 1) \cdot (\upsilon_1 \alpha_R p_R) \cdot m_2$	$(\gamma_H p_H + 1) \cdot (\upsilon_1 \alpha_R p_R) \cdot (\theta_2 p_R / t)$
TDG(element)	$(\gamma_H p_H + 1) \cdot (\upsilon_1 p_R) \cdot m_2$	$(\gamma_H p_H + 1) \cdot (\upsilon_1 p_R) \cdot (\theta_2 p_R / t)$

Note: Parameters p_H and p_R denote the node degrees of HNs and RNs, respectively. Parameters α_H and γ_H denote the node degree proportions for GNs. Parameter α_R denotes the node degree proportion for RNs.
Parameters θ_1 and υ_1 are the parameters of internal graph options BIG-$(\theta_1, r_1, k_1, \lambda_1)$ and TDG-(n_1, μ_1, υ_1), respectively.
Parameters θ_2 and m_2 are the parameters of skeleton graph options BIG-$(\theta_2, r_2, k_2, \lambda_2)$ and CG with m_2 nodes, respectively.
Parameter t is the parameter of t-parallelized TD-(n, μ, x) interconnection for super-edge under the bipartite mode.

4.4.2 Cost

Throughout this monograph, we employ the metric of apportionment coefficient to characterize the cost of a network (denoted by C). By Definition 2.5, the apportionment coefficient C is defined as $(N_1 + \epsilon N_3)/(N_2 + N_3)$, where ϵ is a constant and parameters N_1, N_2, and N_3 denote the number of GNs and RNs, the number of CNs, and the number of HNs in the topology, respectively. Note that all GNs and HNs are contained in modules, while each RN is contained in either a module or an RN cluster. Correspondingly, the value of apportionment coefficient C is equal to the sum of the following two terms (denoted by symbols C_1 and C_2).

- The term C_1 is the ratio of the number of GNs (or ϵ times the number of HNs) and RNs in modules to the number of CNs and HNs.
- The term C_2 is the ratio of the number of RNs in RN clusters to the number of CNs and HNs.

Based on the aforementioned notations, we have the following two lemmas to characterize the values of such two terms. The proofs are given in Sections A.4.3 and A.4.4, respectively.

Lemma 4.3 *Let term C_1 denote the ratio of the number of GNs (or ϵ times the number of HNs) and RNs in modules to the number of CNs and HNs. Then the value of C_1 can be characterized by the ratio of the number of RNs in each module to the number of GNs and HNs in each module (denoted by symbol R_1). Such a characterization is determined by the selected option of term d_1. Specifically,*

- *if $d_1 = G$, we have $C_1 = (1 + R_1)/\gamma_G p_G$;*
- *if $d_1 = H$, we have $C_1 = (\epsilon + R_1)/(\gamma_H p_H + 1)$.*

The value of ratio R_1 is determined by the selected options of terms d_1 and d_2. Specifically,

- *for design patterns with $d_2 =$ CG(any) or $d_2 =$ CBG(any), we have $R_1 = 0$;*
- *for design patterns with $d_2 =$ BIG(block) or $d_2 =$ TDG(block), we have $R_1 = \alpha_G p_G/\alpha_R p_R$ if $d_1 = G$ or $R_1 = \alpha_H p_H/\alpha_R p_R$ if $d_1 = H$;*
- *for design patterns with $d_2 =$ BIG(element) or $d_2 =$ TDG(element), we have $R_1 = \alpha_G p_G/p_R$ if $d_1 = G$ or $R_1 = \alpha_H p_H/p_R$ if $d_1 = H$.*

Lemma 4.4 *Let R_2 denote the ratio of the total outer-degree of all dock nodes in each module to the number of GNs and HNs in each module. Then the value of term C_2 (*i.e.*, the ratio of the number of RNs in modules to the number of CNs and HNs) is determined by the selected options of terms d_1 and d_3. Specifically,*

- *for design patterns with $d_3 = P$, we have $C_2 = 0$;*
- *for design patterns with $d_3 = B$, we have $C_2 = R_2/\gamma_G p_G p_R$ if $d_1 = G$ or $C_2 = R_2/(\gamma_H p_H + 1)p_R$ if $d_1 = H$.*

The value of ratio R_2 is determined by the selected options of terms d_1 and d_2. Specifically,

- *for design patterns with* $d_2 = CG(any)$, $d_2 = CBG(any)$, $d_2 = BIG(element)$, *or* $d_2 = TDG(element)$, *we have* $R_2 = \beta_G p_G$ *if* $d_1 = G$ *or* $R_2 = \beta_H p_H$ *if* $d_1 = H$;
- *for design patterns with* $d_2 = BIG(block)$ *or* $d_2 = TDG(block)$, *we have* $R_2 = \beta_R \alpha_G p_G / \alpha_R$ *if* $d_1 = G$ *or* $R_2 = \beta_R \alpha_H p_H / \alpha_R$ *if* $d_1 = H$.

Based on the two lemmas, the costs (i.e., the apportionment coefficients) under design patterns with $d_1 = G$ and with $d_1 = H$ are summarized in Tables 4.7 and 4.8, respectively.

Besides, the two lemmas also inform the quantitative effects of the terms d_1, d_2, and d_3 on the cost metric, respectively.

- When option $d_1 = G$ turns into $d_1 = H$, the terms $\alpha_G p_G$ and $\beta_G p_G$ are changed into $\alpha_H p_H$ and $\beta_H p_H$, respectively. The term $\gamma_G p_G$ is changed into $(\gamma_H p_H + 1)/\epsilon$ or $(\gamma_H p_H + 1)$.
- The effect on cost of term d_2 is the addition of C_1.
- The effect on cost of term d_3 is the addition of C_2.

4.4.3 Latency

By Definition 2.6, the latency of a network is defined as the maximum distance between two nodes with computing function in the framework. Let symbol l' denote the maximum distance between two GNs (or two HNs) in the network.

Table 4.7 A summary of costs (apportionment coefficients) under design patterns with $d_1 = G$.

d_2 \ d_3	**P**	**B**
CG(any)	$\frac{1}{\gamma_G p_G}$	$\frac{1}{\gamma_G p_G} + \frac{\beta_G}{\gamma_G p_R}$
CBG(any)	$\frac{1}{\gamma_G p_G}$	$\frac{1}{\gamma_G p_G} + \frac{\beta_G}{\gamma_G p_R}$
BIG(block)	$\frac{1}{\gamma_G p_G} + \frac{\alpha_G}{\gamma_G \alpha_R p_R}$	$\frac{1}{\gamma_G p_G} + \frac{\alpha_G}{\gamma_G \alpha_R p_R} + \frac{\alpha_G \beta_R}{\gamma_G \alpha_R p_R}$
BIG(element)	$\frac{1}{\gamma_G p_G} + \frac{\alpha_G}{\gamma_G p_R}$	$\frac{1}{\gamma_G p_G} + \frac{\alpha_G}{\gamma_G p_R} + \frac{\beta_G}{\gamma_G p_R}$
TDG(block)	$\frac{1}{\gamma_G p_G} + \frac{\alpha_G}{\gamma_G \alpha_R p_R}$	$\frac{1}{\gamma_G p_G} + \frac{\alpha_G}{\gamma_G \alpha_R p_R} + \frac{\alpha_G \beta_R}{\gamma_G \alpha_R p_R}$
TDG(element)	$\frac{1}{\gamma_G p_G} + \frac{\alpha_G}{\gamma_G p_R}$	$\frac{1}{\gamma_G p_G} + \frac{\alpha_G}{\gamma_G p_R} + \frac{\beta_G}{\gamma_G p_R}$

Note: Parameters p_G and p_R denote the node degrees of GNs and RNs, respectively. Parameters α_G, β_G, and γ_G are the parameters of node degree proportions for GNs. The parameters α_R and β_R are the parameters of node degree proportions for RNs.

Table 4.8 A summary of costs (apportionment coefficients) under design patterns with $d_1 = H$.

d_2 \ d_3	**P**	**B**
CG(any)	$\frac{\epsilon}{\gamma_H p_H + 1}$	$\frac{\epsilon}{\gamma_H p_H + 1} + \frac{\beta_H p_H}{(\gamma_H p_H + 1)p_R}$
CBG(any)	$\frac{\epsilon}{\gamma_H p_H + 1}$	$\frac{\epsilon}{\gamma_H p_H + 1} + \frac{\beta_H p_H}{(\gamma_H p_H + 1)p_R}$
BIG(block)	$\frac{\epsilon}{\gamma_H p_H + 1} + \frac{\alpha_H p_H}{(\gamma_H p_H + 1)\alpha_R p_R}$	$\frac{\epsilon}{\gamma_H p_H + 1} + \frac{\alpha_H p_H}{(\gamma_H p_H + 1)\alpha_R p_R} + \frac{\alpha_H p_H \beta_R}{(\gamma_H p_H + 1)\alpha_R p_R}$
BIG(element)	$\frac{\epsilon}{\gamma_H p_H + 1} + \frac{\alpha_H p_H}{(\gamma_H p_H + 1)p_R}$	$\frac{\epsilon}{\gamma_H p_H + 1} + \frac{\alpha_H p_H}{(\gamma_H p_H + 1)p_R} + \frac{\beta_H p_H}{(\gamma_H p_H + 1)p_R}$
TDG(block)	$\frac{\epsilon}{\gamma_H p_H + 1} + \frac{\alpha_H p_H}{(\gamma_H p_H + 1)\alpha_R p_R}$	$\frac{\epsilon}{\gamma_H p_H + 1} + \frac{\alpha_H p_H}{(\gamma_H p_H + 1)\alpha_R p_R} + \frac{\alpha_H p_H \beta_R}{(\gamma_H p_H + 1)\alpha_R p_R}$
TDG(element)	$\frac{\epsilon}{\gamma_H p_H + 1} + \frac{\alpha_H p_H}{(\gamma_H p_H + 1)p_R}$	$\frac{\epsilon}{\gamma_H p_H + 1} + \frac{\alpha_H p_H}{(\gamma_H p_H + 1)p_R} + \frac{\beta_H p_H}{(\gamma_H p_H + 1)p_R}$

Note: Parameters p_H and p_R denote the node degrees of HNs and RNs, respectively.
Parameters α_H, β_H, and γ_H are the parameters of node degree proportions for HNs. The parameters α_R and β_R are the parameters of node degree proportions for RNs.

Then the relationship between latency L and the value of l' is determined by the option of term d_1, as illustrated by the following lemma. The proof is given in Section A.4.5.

Lemma 4.5 *Given the value of l' (i.e., the maximum distance between two GNs or two HNs in the topology), the latency L is determined by the selected option of term d_1. Specifically,*

- *for design patterns with $d_1 = G$ or $d_1 = H$ and $\gamma_H > 0$, we have $L = l' + 2$;*
- *for design patterns with $d_1 = H$ and $\gamma_H = 0$, we have $L = l'$.*

The values of l' under different design patterns are summarized by the following 12 lemmas. Their proofs are presented in Sections A.4.6–A.4.17.

Lemma 4.6 *For design patterns with $d_2 = CG(any)$ and $d_3 = P$, the feasible values of l' (i.e., the maximum distance between two GNs or two HNs) include 1, 2, and 3. Particularly, equation (4.113) provides several conditions on low latency, where N_D denotes the number of dock nodes in each module.*

$$\begin{cases} l' = 1, & if\ and\ only\ if\ x = N_D; \\ l' = 2, & if\ tx = N_D\ and\ t > 1. \end{cases} \tag{4.113}$$

Lemma 4.7 *For design patterns with $d_2 = CG(any)$ and $d_3 = B$, the feasible values of l' (i.e., the maximum distance between two GNs or two HNs) include 2, 3, and 4. Particularly, equation (4.114) provides several conditions on low latency, where N_D denotes the number of dock nodes in each module.*

$$\begin{cases} l' = 2, & if\ x = N_D; \\ l' = 2, & if\ tx = N_D\ and\ c_2 = 0; \\ l' \leq 3, & if\ \theta_2 = 1; \\ l' \leq 3, & if\ tx = N_D. \end{cases} \tag{4.114}$$

Lemma 4.8 *For design patterns with $d_2 = CBG(any)$ and $d_3 = P$, the feasible values of l' (i.e., the maximum distance between two GNs or two HNs) include 2, 3, 4, and 5. Particularly, when any super-edge between two modules uses dock nodes in distinct node partitions of each module (CBG), equation (4.115) provides several conditions on low latency, where N_D denotes the number of dock nodes in each module.*

$$\begin{cases} l' = 2, & if\ tx = N_D\ and\ x \geq 2; \\ l' \leq 3, & if\ x \geq 2; \\ l' \leq 3, & if\ x = 1\ and\ t = N_D; \\ l' \leq 4, & if\ x = 1\ and\ t \geq 2. \end{cases} \tag{4.115}$$

Lemma 4.9 *For design patterns with d_2 = CBG(any) and d_3 = B, the feasible values of l' (i.e., the maximum distance between two GNs or two HNs) include 2, 3, 4, and 6. Particularly, equation (4.116) provides several conditions on low latency, where N_D denotes the number of dock nodes in each module.*

$$\begin{cases} l' = 2, & if\ x = 1\ and\ t = N_D; \\ l' = 2, & if\ tx = N_D, x \geq 2,\ and\ c_2 = 0; \\ l' \leq 3, & if\ \theta_2 = 1\ and\ x \geq 2; \\ l' \leq 4, & if\ tx = N_D\ and\ x \geq 2; \\ l' \leq 4, & if\ \theta_2 = 1. \end{cases} \tag{4.116}$$

Lemma 4.10 *For design patterns with d_2 = BIG(block) and d_3 = P, the value of l' (i.e., the maximum distance between two GNs or two HNs) is equal to* 3.

Lemma 4.11 *For design patterns with d_2 = BIG(block) and d_3 = B, the value of l' (i.e., the maximum distance between two GNs or two HNs) is equal to* 4.

Lemma 4.12 *For design patterns with d_2 = BIG(element) and d_3 = P, the feasible values of l' (i.e., the maximum distance between two GNs or two HNs) include 2, 3, 4, and 5. Particularly, equation (4.117) provides several conditions on low latency, where N_D denotes the number of dock nodes in each module.*

$$\begin{cases} l' = 2, & if\ x = N_D; \\ l' \leq 3, & if\ tx = N_D; \\ l' \leq 4, & if\ m_2 = 3. \end{cases} \tag{4.117}$$

Lemma 4.13 *For design patterns with d_2 = BIG(element) and d_3 = B, the feasible values of l' (i.e., the maximum distance between two GNs or two HNs) include 2, 4, and 6. Particularly, equation (4.118) provides several conditions on low latency, where N_D denotes the number of dock nodes in each module.*

$$\begin{cases} l' = 2, & if\ tx = N_D\ and\ c_2 = 0; \\ l' = 2, & if\ x = 1\ and\ t = N_D; \\ l' \leq 4, & if\ tx = N_D; \\ l' \leq 4, & if\ \theta_2 = 1. \end{cases} \tag{4.118}$$

Lemma 4.14 *For design patterns with d_2 = TDG(block) and $d_3 = P$, the value of l' (i.e., the maximum distance between two GNs or two HNs) is equal to* 4.

Lemma 4.15 *For design patterns with d_2 = TDG(block) and $d_3 = B$, the feasible values of l' (i.e., the maximum distance between two GNs or two HNs) include* 4 *and* 6. *Particularly, the value of l' is equal to* 4 *if $c_2 = 0$.*

Lemma 4.16 *For design patterns with d_2 = TDG(element) and $d_3 = P$, the feasible values of l' (i.e., the maximum distance between two GNs or two HNs) include* 2, 3, 4, *and* 5. *Particularly, equation* (4.119) *provides several conditions on low latency, where N_D denotes the number of dock nodes in each module.*

$$\begin{cases} l' = 2, & if\ x = N_D; \\ l' \leq 4, & if\ tx = N_D. \end{cases} \tag{4.119}$$

Lemma 4.17 *For design patterns with d_2 = TDG(element) and $d_3 = B$, the feasible values of l' (i.e., the maximum distance between two GNs or two HNs) include* 2, 4, *and* 6. *Particularly, equation* (4.120) *provides several conditions on low latency, where N_D denotes the number of dock nodes in each module.*

$$\begin{cases} l' = 2, & if\ x = p_R, t = v_1, and\ c_2 = 0; \\ l' \leq 4, & if\ tx = N_D\ and\ c_2 = 0; \\ l' \leq 4, & if\ \theta_2 = 1\ and\ c_2 = 0. \end{cases} \tag{4.120}$$

Based on the aforementioned discussions, the latencies under different design patterns are summarized in Tables 4.9 and 4.10, respectively.

4.4.4 Throughput

Recall that in the framework, the throughput of a network is characterized by two metrics, namely its intra-module throughput and its inter-module throughput. Their values under different design patterns are discussed as follows, respectively.

Table 4.9 A summary of latencies under design patterns with $d_1 = G$ or $d_1 = H$ and $\gamma_H > 0$.

d_2 \ d_3	P	B
CG(any)	3, 4, 5	4, 5, 6
CBG(any)	4, 5, 6, 7	4, 5, 6, 8
BIG(block)	5	6
BIG(element)	4, 5, 6, 7	4, 6, 8
TDG(block)	6	6, 8
TDG(element)	4, 5, 6, 7	4, 6, 8

Table 4.10 A summary of latencies under design patterns with $d_1 = H$ and $\gamma_H = 0$.

d_2 \ d_3	P	B
CG(any)	1, 2, 3	2, 3, 4
CBG(any)	2, 3, 4, 5	2, 3, 4, 6
BIG(block)	3	4
BIG(element)	2, 3, 4, 5	2, 4, 6
TDG(block)	4	4, 6
TDG(element)	2, 3, 4, 5	2, 4, 6

Case 1: Intra-module throughput.

The following lemma investigates the intra-module throughput under different design patterns. The proof is given in Section A.4.18.

Lemma 4.18 *The intra-module throughput T_1 is equal to α_G/γ_G if $d_1 = G$ or $\alpha_H p_H/(\gamma_H p_H + 1)$ if $d_1 = H$.*

Case 2: Inter-module throughput.

The value of inter-module throughput T_2 is specified in the following lemma and proved in Section A.4.19.

Lemma 4.19 *Let symbol R_3 denote the ratio of the total outer-degree of all dock nodes in each module to the attachment number of each module. Then the inter-module throughput T_2 is dependent on the option of term d_3. Specifically,*

- *if $d_3 = P$, we have $T_2 = R_3/(m_2 - 1)$;*
- *if $d_3 = B$, we have $T_2 = R_3 \cdot \lambda_2/r_2$,*

where the value of parameter R_3 is determined by the selected options of terms d_1 and d_2 as follows:

- *for design patterns with $d_2 = $ CG(any), $d_2 = $CBG(any), $d_2 = $ BIG(element), or $d_2 = $ TDG(element), we have $R_3 = \beta_G/\gamma_G$ if $d_1 = G$ or $R_3 = (\beta_H p_H)/(\gamma_H p_H + 1)$ if $d_1 = H$.*
- *for design patterns with $d_2 = $ BIG(block) or $d_2 = $ TDG(block), we have $R_3 = (\beta_R \alpha_G)/(\alpha_R \gamma_G)$ if $d_1 = G$ or $R_3 = (\beta_R \alpha_H p_H)/[\alpha_R \cdot (\gamma_H p_H + 1)]$ if $d_1 = H$.*

Based on our analysis, the throughput under design patterns with $d_1 = G$ and $d_1 = H$ are summarized in Tables 4.11 and 4.12, respectively.

By Lemmas 4.18 and 4.19, the quantitative effects of terms d_1, d_2, and d_3 on (intra-module and inter-module) throughputs are informed as follows, respectively.

- When option $d_1 = G$ turns into $d_1 = H$, the terms $\alpha_G p_G$, $\beta_G p_G$, and $\gamma_G p_G$ are changed into $\alpha_H p_H$, $\beta_H p_H$, and $\gamma_H p_H + 1$, respectively.
- The effect on intra-module throughput of term d_2 is the value of T_1 itself. The effect on inter-module throughput of term d_2 is the multiplication of R_3.
- The effect on inter-module throughput of term d_3 is the multiplication of $1/(m_2 - 1)$ if $d_3 = P$ or the multiplication of λ_2/r_2 if $d_3 = B$.

Table 4.11 A summary of throughputs under design patterns with $d_1 = G$.

d_2 \ d_3	P	B
CG(any)	$T_1 = \frac{\alpha_G}{\gamma_G}, T_2 = \frac{\beta_G}{\gamma_G} \cdot \frac{1}{m_2 - 1}$	$T_1 = \frac{\alpha_G}{\gamma_G}, T_2 = \frac{\beta_G}{\gamma_G} \cdot \frac{\lambda_2}{r_2}$
CBG(any)	$T_1 = \frac{\alpha_G}{\gamma_G}, T_2 = \frac{\beta_G}{\gamma_G} \cdot \frac{1}{m_2 - 1}$	$T_1 = \frac{\alpha_G}{\gamma_G}, T_2 = \frac{\beta_G}{\gamma_G} \cdot \frac{\lambda_2}{r_2}$
BIG(block)	$T_1 = \frac{\alpha_G}{\gamma_G}, T_2 = \frac{\beta_R \alpha_G}{\alpha_R \gamma_G} \cdot \frac{1}{m_2 - 1}$	$T_1 = \frac{\alpha_G}{\gamma_G}, T_2 = \frac{\beta_R \alpha_G}{\alpha_R \gamma_G} \cdot \frac{\lambda_2}{r_2}$
BIG(element)	$T_1 = \frac{\alpha_G}{\gamma_G}, T_2 = \frac{\beta_G}{\gamma_G} \cdot \frac{1}{m_2 - 1}$	$T_1 = \frac{\alpha_G}{\gamma_G}, T_2 = \frac{\beta_G}{\gamma_G} \cdot \frac{\lambda_2}{r_2}$
TDG(block)	$T_1 = \frac{\alpha_G}{\gamma_G}, T_2 = \frac{\beta_R \alpha_G}{\alpha_R \gamma_G} \cdot \frac{1}{m_2 - 1}$	$T_1 = \frac{\alpha_G}{\gamma_G}, T_2 = \frac{\beta_R \alpha_G}{\alpha_R \gamma_G} \cdot \frac{\lambda_2}{r_2}$
TDG(element)	$T_1 = \frac{\alpha_G}{\gamma_G}, T_2 = \frac{\beta_G}{\gamma_G} \cdot \frac{1}{m_2 - 1}$	$T_1 = \frac{\alpha_G}{\gamma_G}, T_2 = \frac{\beta_G}{\gamma_G} \cdot \frac{\lambda_2}{r_2}$

Note: Parameters α_G, β_G, and γ_G are the parameters of node degree proportions for GNs.
Parameter α_R is the parameter of node degree proportion for RNs.
Parameters λ_2, r_2, and m_2 are the parameters of skeleton graph options BIG-$(\theta_2, r_2, k_2, \lambda_2)$ and CG with m_2 nodes, respectively.

Table 4.12 A summary of throughputs under design patterns with $d_1 = H$.

d_2 \ d_3	P	B
CG(any)	$T_1 = \frac{\alpha_H p_H}{\gamma_H p_H + 1}, T_2 = \frac{\beta_H p_H}{\gamma_H p_H + 1} \cdot \frac{1}{m_2 - 1}$	$T_1 = \frac{\alpha_H p_H}{\gamma_H p_H + 1}, T_2 = \frac{\beta_H p_H}{\gamma_H p_H + 1} \cdot \frac{\lambda_2}{r_2}$
CBG(any)	$T_1 = \frac{\alpha_H p_H}{\gamma_H p_H + 1}, T_2 = \frac{\beta_H p_H}{\gamma_H p_H + 1} \cdot \frac{1}{m_2 - 1}$	$T_1 = \frac{\alpha_H p_H}{\gamma_H p_H + 1}, T_2 = \frac{\beta_H p_H}{\gamma_H p_H + 1} \cdot \frac{\lambda_2}{r_2}$
BIG(block)	$T_1 = \frac{\alpha_H p_H}{\gamma_H p_H + 1}, T_2 = \frac{\beta_R}{\alpha_R} \cdot \frac{\alpha_H p_H}{\gamma_H p_H + 1} \cdot \frac{1}{m_2 - 1}$	$T_1 = \frac{\alpha_H p_H}{\gamma_H p_H + 1}, T_2 = \frac{\beta_R}{\alpha_R} \cdot \frac{\alpha_H p_H}{\gamma_H p_H + 1} \cdot \frac{\lambda_2}{r_2}$
BIG(element)	$T_1 = \frac{\alpha_H p_H}{\gamma_H p_H + 1}, T_2 = \frac{\beta_H p_H}{\gamma_H p_H + 1} \cdot \frac{1}{m_2 - 1}$	$T_1 = \frac{\alpha_H p_H}{\gamma_H p_H + 1}, T_2 = \frac{\beta_H p_H}{\gamma_H p_H + 1} \cdot \frac{\lambda_2}{r_2}$
TDG(block)	$T_1 = \frac{\alpha_H p_H}{\gamma_H p_H + 1}, T_2 = \frac{\beta_R}{\alpha_R} \cdot \frac{\alpha_H p_H}{\gamma_H p_H + 1} \cdot \frac{1}{m_2 - 1}$	$T_1 = \frac{\alpha_H p_H}{\gamma_H p_H + 1}, T_2 = \frac{\beta_R}{\alpha_R} \cdot \frac{\alpha_H p_H}{\gamma_H p_H + 1} \cdot \frac{\lambda_2}{r_2}$
TDG(element)	$T_1 = \frac{\alpha_H p_H}{\gamma_H p_H + 1}, T_2 = \frac{\beta_H p_H}{\gamma_H p_H + 1} \cdot \frac{1}{m_2 - 1}$	$T_1 = \frac{\alpha_H p_H}{\gamma_H p_H + 1}, T_2 = \frac{\beta_H p_H}{\gamma_H p_H + 1} \cdot \frac{\lambda_2}{r_2}$

Note: Parameter p_H denotes the node degree of each HN.
1 Parameters α_H, β_H, and γ_H denote the node degree proportions for HNs. Parameter α_R is the parameter of node degree proportion for RNs.
Parameters λ_2, r_2, and m_2 are the parameters of skeleton graph options BIG-(θ_2, r_2, k_2, λ_2) and CG with m_2 nodes, respectively.

4.4.5 Fault-tolerance

The fault-tolerance F of a network topology (see Definition 2.9) under different design patterns is specified by the following lemma. The proof is given in Section A.4.20. This lemma shows that the effect of term d_3 on fault-tolerance is the value of F itself.

Lemma 4.20 *The value of fault-tolerance F is determined by the selected option of term d_3. Specifically,*

- *if $d_3 = P$, we have $F = w$;*
- *if $d_3 = B$, we have $F = t \cdot [\frac{\theta_2 - \frac{1}{k_2}}{\theta_2 - \frac{t}{p_R}} \cdot (w-1) + 1]$.*

4.5 Guidelines on Selection of Design Patterns

In this section, we provide a number of guidelines for the selection of design patterns in systematic topology design. To this end, in Section 4.5.1, we compare different options for each term of the design patterns based on the previous analysis of their effects on the performance metrics. Then in Section 4.5.2, we elucidate the guidelines by demonstrating the option selections and design patterns adopted by existing topology designs.

4.5.1 Analysis on Option Selection for Each Term

To provide guidelines on the selection of design patterns to achieve distinct design goals, we compare the performance metrics under different options of terms d_1, d_2, and d_3 in Sections 4.5.1.1–4.5.1.3, respectively.

4.5.1.1 Comparison Among Options of Term d_1

Recall that there are two options for term d_1. The consequent performance metrics are compared as follows:

- *Scalability*. As the option for term d_1 is switched from G to H, the values of $\alpha_G p_G$, $\beta_G p_G$, and $\gamma_G p_G$ in the scalability metric would turn into $\alpha_H p_H$, $\beta_H p_H$, and $\gamma_H p_H + 1$, respectively. In particular, when $p_G = p_H$, $\alpha_G = \alpha_H$, $\beta_G = \beta_H$, and $\gamma_G = \gamma_H$, the option $d_1 = H$ would lead to a higher scalability.
- *Cost*. As the option for term d_1 is switched from G to H, in the cost metric, we would have the value of $\alpha_G p_G$ turn into $\alpha_H p_H$, the value of $\beta_G p_G$ into $\beta_H p_H$, and the value of $\gamma_G p_G$ into $\gamma_H p_H + 1$ or $(\gamma_H p_H + 1)/\epsilon$. In particular, when $p_G = p_H$, $\alpha_G = \alpha_H$, $\beta_G = \beta_H$, $\gamma_G = \gamma_H$, and $\epsilon \leq 1$, the option $d_1 = H$ would lead to a lower cost.

- *Latency*. As the option for term d_1 is switched from G to H, the latency metric would be reduced by two if $\gamma_H = 0$ or remain the same otherwise.
- *Throughput*. As the option for term d_1 is switched from G to H, in the throughput metric, we would have the value of $\alpha_G p_G$ turn into $\alpha_H p_H$, the value of $\beta_G p_G$ into $\beta_H p_H$, and the value of $\gamma_G p_G$ into $\gamma_H p_H + 1$, respectively. In particular, when $p_G = p_H, \alpha_G = \alpha_H, \beta_G = \beta_H$, and $\gamma_G = \gamma_H$, the option $d_1 = H$ would lead to a lower throughput.

In summary, compared to the option $d_1 = G$, the option $d_1 = H$ would result in a higher scalability, a lower cost, and a lower throughput under the same parameter settings of node degree and their proportions. Moreover, the option $d_1 = H$ would incur a lower latency provided that $\gamma_H = 0$ (i.e., when no CNs exist in a topology with $d_1 = H$).[5] Considering that most existing topologies adopt the option $d_1 = G$, in subsequent analysis, we fix the option $d_1 = G$ and compare the options of terms d_2 and d_3, respectively. We also assume that $p_R = p_G = p$, as switches with identical port numbers are common in practical systems (Al-Fares et al. 2008; Walraed-Sullivan et al. 2013; Liu et al. 2013; Ahn et al. 2009; Cheng and Chuang 1994; Kim et al. 2007).

4.5.1.2 Comparison Among Options of Term d_2

By fixing $d_1 = G$ and $d_3 = P$, we summarize the quantitative effects of the options of term d_2 on the performance metrics in Table 4.13.[6] For ease of description, we use symbols $S(d_2)$, $C(d_2)$, $T_1(d_2)$, and $T_2(d_2)$ to denote the effects of term d_2 on scalability, cost, intra-module throughput, and inter-module throughput, respectively.

From Table 4.13, we notice that the cost and throughput metrics are both affected by the parameter settings of node degree proportions (i.e., $\alpha_R, \beta_R, \alpha_G, \beta_G$, and γ_G). Besides such parameters, the scalability metric is also affected by parameters of the internal graph (e.g., the parameter θ_1 of BIG or the parameter υ_1 of TDG). In contrast, the latency metric is not specified by any of the parameters. In the following comparison, we divide the performance metrics into three groups (i.e., the first for latency, the second for cost and throughput, the third for scalability).

- *Comparison of latency*. For each option of term d_2, the latency metric ranges within a set of integers. To compare the latencies of two options, we consider the minimum and maximum values, respectively. For two options A and B, if the minimum value (resp. the maximum value) of the range A is smaller than or equal to the minimum value (resp. the maximum value) of the range B, we say the latency of option A is lower than or equal to the latency of option B.

5 In fact, such settings have been widely adopted in existing topologies with $d_1 = H$, e.g., the well-known DCell Guo et al. (2008) and BCube Guo et al. (2009).

6 Note that we omit the analysis on the case with $d_3 = B$ because the results are similar.

Hence, the option A is a better option for low latency compared with option B, denoted by $A \geq B$. Based on such a criterion, together with the results in Table 4.13, we have the following two preference lists of the options over low latency. They can be the guideline for option selection when latency is a key concern.

$$\text{CG(any)} \geq \text{CBG(any)} = \text{BIG(element)} = \text{TDG(element)}; \tag{4.121}$$

$$\text{CG(any)} \geq \text{BIG(block)} \geq \text{TDG(block)}. \tag{4.122}$$

Note that compared with BIG(block) and TDG(block), the latencies of CBG(any), BIG(element), and TDG(element) are lower only under some special conditions (see Lemmas 4.8–4.17). Hence, it is suggested to either select BIG(block) and TDG(block), or CBG(any), BIG(element), TDG(element) satisfying those conditions in pursuit of a low latency.

- *Comparison of cost and throughput.* Next, we analyze the trade-offs between the cost and throughput metrics under each option of term d_2.
  - For design patterns with $d_2 = \text{CG(any)}$ or $d_2 = \text{CBG(any)}$, the values of α_G, β_G, and γ_G satisfy $\alpha_G + \beta_G + \gamma_G = 1$. As a result, it follows that

$$C(d_2) \cdot p = T_1(d_2) + T_2(d_2) + 1. \tag{4.123}$$

  - For design patterns with $d_2 = \text{BIG(element)}$ or $d_2 = \text{TDG(element)}$, the values of α_G, β_G, and γ_G satisfy $\alpha_G + \beta_G + \gamma_G = 1$. As a result, it follows that

$$C(d_2) \cdot p = 2T_1(d_2) + T_2(d_2) + 1. \tag{4.124}$$

  - For design patterns with $d_2 = \text{BIG(block)}$ or $d_2 = \text{TDG(block)}$, the values of α_R, β_R, α_G, β_G, and γ_G satisfy $\alpha_R + \beta_R = 1$ and $\alpha_G + \gamma_G = 1$. As a result, it follows that

$$C(d_2) \cdot p = 2T_1(d_2) + T_2(d_2) + 1. \tag{4.125}$$

The aforementioned trade-offs lead to the following conclusions.

  - Adopting CG or CBG as the internal graph has the same effect on the trade-off between cost and throughput.
  - No matter which types of nodes are selected as dock nodes (i.e., element nodes or block nodes), the adoption of BIG or TDG as internal graph has the same effect on the trade-off between cost and throughput.
  - Compared with BIG and TDG, employing CG and CBG as the internal graph leads to a better trade-off between the effects on cost and throughput. In particular, given the fixed values of $T_1(d_2)$ and $T_2(d_2)$, CG and CBG have lower values of $C(d_2)$ than BIG and TDG; given the fixed values of $C(d_2)$ and $T_1(d_2)$ (resp. $T_2(d_2)$), CG and CBG have higher values of $T_2(d_2)$ (resp. $T_1(d_2)$) than BIG and TDG.

Considering the trade-off between low cost and high throughput, the CG and CBG are better options than BIG and TDG for the internal graph.

- *Comparison of scalability.* Among the parameters that affect the scalability (i.e., parameters $\alpha_G, \gamma_G, \theta_1$, and υ_1), parameter γ_G has constant effects on the trade-off between scalability and other performance metrics (see Table 4.13). To be specific, the value of γ_G is proportional to the value of scalability and inversely proportional to the values of cost and throughput. To compare the effects of term d_2 on the scalability metric, we consider the values of $S(d_2)/\gamma_G$ under its six options. Note that each of the options jointly involves the selections of internal graph and dock node type. First, we analyze the effects of internal graphs when the type of all dock nodes is fixed as GN, and then compare the scalabilities incurred by different selections of dock node type.
  - *Analysis on the selection of internal graph.* By fixing GNs as dock nodes, we compare the values of $S(d_2)/\gamma_G$ under four options of term d_2, including CG(any), CBG(any), BIG(element), and TDG(element), as they all employ GNs as dock nodes. For two options A and B, if the minimum value (resp. the maximum value) of $S(d_2)/\gamma_G$ of option A is smaller than or equal to that of $S(d_2)/\gamma_G$ of option B, we say the scalability of option A is lower than or equal to that of option B (denoted by $A \leq B$). Accordingly, we have the following preference list over the options with respect to the scalability metric. It serves as guideline on the selection of internal graph when scalability is the key concern for topology design.

$$\text{CG(any)} \leq \text{CBG(any)} \leq \text{BIG(element)} \leq \text{TDG(element)}. \tag{4.126}$$

    The reason is given as follows:
    - *Comparison between CG(any) and CBG(any).* Since the value of $\alpha_G p$ is a positive integer, we have $\alpha_G p \geq 1$. This implies that

$$\alpha_G p + 1 \leq 2\alpha_G p. \tag{4.127}$$

      Meanwhile, the values of $S(d_2)/\gamma_G$ satisfy the following equation.

$$\frac{S(d_2)}{\gamma_G} = \begin{cases} (\alpha_G p + 1)p, & \text{if } d_2 = \text{CG(any)}; \\ 2\alpha_G p^2, & \text{if } d_2 = \text{CBG(any)}. \end{cases} \tag{4.128}$$

      As a result, the option $d_2 =$ CBG(any) leads to a higher scalability than the option $d_2 =$ CG(any); i.e., CG(any) $\leq$ CBG(any).
    - *Comparison between CBG(any) and BIG(element).* Such a comparison is conducted upon the minimum values and the maximum values of parameters $2\alpha_G$ and θ_1, respectively. Regarding the minimum values, we have $\theta_1 \geq 1$ by Definition 1.23 and $\alpha_G \geq 1/p$ because $\alpha_G p$ is a positive integer. Hence, the minimum value of $2\alpha_G$ is less than or equal to that of θ_1. For the maximum values, we have $\alpha_G < 1$ and accordingly $2\alpha_G < 2$. By

Table 4.13 A summary of the effects of term d_2 on the performance metrics when $d_1 = G$ and $d_3 = P$.

Effects \ Option	CG(any)	CBG(any)	BIG(block)	BIG(element)	TDG(block)	TDG(element)
$S(d_2)$	$\times(\alpha_G p + 1)\gamma_G p$	$\times 2\alpha_G \gamma_G p^2$	$\times \theta_1 \alpha_R \gamma_G p^2$	$\times \theta_1 \gamma_G p^2$	$\times \upsilon_1 \alpha_R \gamma_G p^2$	$\times \upsilon_1 \gamma_G p^2$
$C(d_2)$	$+\frac{1}{\gamma_G p}$	$+\frac{1}{\gamma_G p}$	$+\frac{\alpha_R + \alpha_G}{\alpha_R \gamma_G p}$	$+\frac{1 + \alpha_G}{\gamma_G p}$	$+\frac{\alpha_R + \alpha_G}{\alpha_R \gamma_G p}$	$+\frac{1 + \alpha_G}{\gamma_G p}$
L	3, 4, 5	4, 5, 6, 7	5	4, 5, 6, 7	6	4, 5, 6, 7
$T_1(d_2)$	$\frac{\alpha_G}{\gamma_G}$	$\frac{\alpha_G}{\gamma_G}$	$\frac{\alpha_G}{\gamma_G}$	$\frac{\alpha_G}{\gamma_G}$	$\frac{\alpha_G}{\gamma_G}$	$\frac{\alpha_G}{\gamma_G}$
$T_2(d_2)$	$\times\frac{\beta_G}{\gamma_G}$	$\times\frac{\beta_G}{\gamma_G}$	$\times\frac{\beta_R \alpha_G}{\alpha_R \gamma_G}$	$\times\frac{\beta_G}{\gamma_G}$	$\times\frac{\beta_R \alpha_G}{\alpha_R \gamma_G}$	$\times\frac{\beta_G}{\gamma_G}$

Note: The symbol L denotes the latency metric.
Symbols $S(d_2)$, $C(d_2)$, $T_1(d_2)$, and $T_2(d_2)$ denote the effects of term d_2 on scalability, cost, intra-module throughput, and inter-module throughput, respectively.
Symbols $\times$ and $+$ denote the effects of multiplication and addition incurred by the options on the performance metrics, respectively.

Lemma 1.7, we know that the value of θ_1 can be greater than two. Hence, the maximum value of $2\alpha_G$ is less than that of θ_1. Meanwhile, the values of $S(d_2)/\gamma_G$ is subject to the following equation.

$$\frac{S(d_2)}{\gamma_G} = \begin{cases} 2\alpha_G p^2, & \text{if } d_2 = \text{CBG(any)}; \\ \theta_1 p^2, & \text{if } d_2 = \text{BIG(element)}. \end{cases} \tag{4.129}$$

Therefore, the option $d_2 = \text{BIG(element)}$ conduces to a higher scalability compared to the option $d_2 = \text{CBG(any)}$; i.e., CBG(any) $\leq$ BIG(element).

- *Comparison between BIG(element) and TDG(element).* We conduct the comparison by considering the minimum values and the maximum values of parameters θ_1 and υ_1, respectively. Regarding the minimum values, by Definitions 1.23 and 1.24, we know that the parameters of BIG-$(\theta_1, r_1, k_1, \lambda_1)$ and TDG-(n_1, μ_1, υ_1) satisfy $\theta_1 \geq 1$ and $\upsilon_1 \geq 2$, respectively. Hence, the minimum value of θ_1 is less than that of υ_1. When it comes to the maximum values, since $\mu_1\upsilon_1 = \alpha_G p_G$ and $\mu_1 \geq 1$, the maximum value of υ_1 ($\upsilon_1 = \alpha_G p_G$) is achieved in the case with $\mu_1 = 1$. Recall that $\frac{r_1}{\lambda_1} = \frac{(\theta_1 k_1 - 1)}{k_1 - 1}$ and $\lambda_1 \geq 1$. Then given that $\theta_1 > 1$, we have

$$\theta_1 = \frac{\theta_1 k_1 - \theta_1}{k_1 - 1} < \frac{\theta_1 k_1 - 1}{k_1 - 1} = \frac{r_1}{\lambda_1} \leq r_1 = \alpha_G p_G. \tag{4.130}$$

This implies that the maximum value of υ_1 is greater than the maximum value of θ_1. Meanwhile, the values of $S(d_2)/\gamma_G$ are subject to the following equation.

$$\frac{S(d_2)}{\gamma_G} = \begin{cases} \theta_1 p^2, & \text{if } d_2 = \text{BIG(element)}; \\ \upsilon_1 p^2, & \text{if } d_2 = \text{TDG(element)}. \end{cases} \tag{4.131}$$

In conclusion, TDG(element) conduces to a higher scalability compared to the option BIG(element); i.e., BIG(element) $\leq$ TDG(element).

- *Analysis on the selection of dock node type.* Provided that BIG or TDG is selected as the internal graph, the next step is to decide which type of nodes should be selected as dock nodes (i.e., element nodes or block nodes). In either case, parameters θ_1 and υ_1 have the same effect on the scalability metric. The only difference lies in that there is an extra multiplication factor α_R when block nodes are selected as dock nodes. Since $0 < \alpha_R < 1$, then selecting element nodes as dock nodes conduces a higher scalability.

The aforementioned discussions lead to guidelines for the option selection of term d_2.

- If low cost and high throughput are both the design goals, it is suggested to select CG(any) or CBG(any) as the option of term d_2. The choice is determined by the relative preference of high scalability or low latency.
  - If low latency is more desirable, it is suggested to select CG(any).
  - If high scalability is more desirable, it is suggested to select CBG(any).
- Otherwise, the selection of options for term d_2 is determined by the requirements on scalability and latency.
  - If low latency is more desirable, CG(any) is a prior choice. Regarding the other options, designers need to satisfy the special conditions (see Lemmas 4.8–4.17) to achieve a low latency.
  - If high scalability is more desirable, the priority order of internal graphs is CG≤CBG≤BIG≤TDG. When BIG or TDG is selected as the internal graph, element nodes can be selected as dock nodes to achieve a higher scalability.

4.5.1.3 Comparison for Options of Term d_3

By fixing $d_1 = G$, we first summarize the quantitative effects under the two options of term d_3 on the performance metrics in Table 4.14. For ease of description, we use symbols $S(d_3)$, $C(d_3)$, $T_2(d_3)$, and $F(d_3)$ to denote the effects of term d_3 on scalability, cost, inter-module throughput, and fault-tolerance, respectively.

In the following, we divide the performance metrics into three groups for comparison, the first group for cost and latency, the second group for scalability and inter-module throughput, and the third for fault-tolerance.

Table 4.14 A summary of the quantitative effects on the performance metrics under the options of term d_3 with $d_1 = G$.

Effect \ Option	P	B
$S(d_3)$	$\times m_2$	$\times \theta_2 p/t$
$C(d_3)$	$+0$	$+\frac{\beta_G}{\gamma_G p}$ or $+\frac{\alpha_G \beta_R}{\gamma_G \alpha_R p}$
L	3 – 7	4 – 8
$T_2(d_3)$	$\times \frac{1}{m_2 - 1}$	$\times \frac{k_2 - 1}{\theta_2 k_2 - 1}$
$F(d_3)$	$= w$	$= t \cdot [\frac{\theta_2 - \frac{1}{k_2}}{\theta_2 - \frac{t}{p}} \cdot (w - 1) + 1]$

Note: Symbol L denotes the latency metric, which ranges over the set of feasible values under all options of term d_2.

Symbols $S(d_3)$, $C(d_3)$, $T_2(d_3)$, and $F(d_3)$ denote the effects of term d_3 on scalability, cost, intra-module throughput, and inter-module throughput, respectively.

Symbols $\times$ and $+$ denote the multiplicative and additive effects on the performance metrics.

- *Comparison of cost and latency.* Compared to option $d_3 = P$, the option $d_3 = B$ leads to a higher cost and a higher latency.
- *Comparison of scalability and inter-module throughput.* From Table 4.14, we see that there exists a trade-off between scalability and inter-module throughput. Under option $d_3 = P$, the trade-off is determined by parameter m_2. A larger value of m_2 leads to a higher scalability at the cost of a lower inter-module throughput. Under option $d_3 = B$, the trade-off is determined by parameter θ_2. A larger value of θ_2 leads to a higher scalability at the cost of a lower inter-module throughput. To compare the effects of the two options, we fix their effects on one metric and investigate which option leads to a larger value of the other metric.
  - *Comparison of scalability.* We fix the effect on inter-module throughput and compare which option leads to a higher scalability. From Table 4.14, the effect of option $d_3 = P$ on scalability (denoted by $S(d_3 = P)$) is equal to m_2, while the effect of option $d_3 = B$ on scalability (denoted by $S(d_3 = B)$) is equal to $\theta_2 p/t$. When the effects of two options on inter-module throughput are fixed (i.e., $1/(m_2 - 1) = (k_2 - 1)/(\theta_2 k_2 - 1)$), we have

$$S(d_3 = B) = \frac{\theta_2 p}{t} \tag{4.132}$$

$$= \frac{(k_2 - 1)(m_2 - 1) + 1}{k_2} \cdot \frac{p}{t} \tag{4.133}$$

$$= (m_2 - 1 - \frac{m_2 - 2}{k_2}) \cdot \frac{p}{t} \tag{4.134}$$

$$\geq (m_2 - 1 - \frac{m_2 - 2}{2}) \cdot \frac{p}{t} \tag{4.135}$$

$$= \frac{m_2}{2} \cdot \frac{p}{t} \tag{4.136}$$

$$\geq m_2 = S(d_3 = P). \tag{4.137}$$

  Note that inequation (4.135) holds because k_2 is a parameter of BIG-$(\theta_2, r_2, k_2, \lambda_2)$ with $k_2 \geq 2$ (see Definition 1.23). As for inequation (4.137), it holds because tk_2 is a factor of p (hence $tk_2 \leq p$ and $p/t \geq k_2 \geq 2$).
  - *Comparison of inter-module throughput.* We fix the effect on the scalability metric and investigate which option leads to a higher inter-module throughput. From Table 4.14, the effect of option $d_3 = P$ on inter-module throughput (denoted by $T_2(d_3 = P)$) is equal to $1/(m_2 - 1)$, while the effect of option $d_3 = B$ on inter-module throughput (denoted by $T_2(d_3 = B)$) is equal to $(k_2 - 1)/(\theta_2 k_2 - 1)$. When the effects of two options on scalability are fixed (i.e., $m_2 = \theta_2 p/t$), we have

$$T_2(d_3 = P) = \frac{1}{m_2 - 1} \tag{4.138}$$

$$= \frac{1}{\theta_2 p/t - 1} \tag{4.139}$$

$$\leq \frac{1}{\theta_2 k_2 - 1} \tag{4.140}$$

$$\leq \frac{k_2 - 1}{\theta_2 k_2 - 1} \tag{4.141}$$

$$= T_2(d_3 = B). \tag{4.142}$$

Note that in equation (4.140) holds because tk_2 is a factor of p (hence $tk_2 \leq p$), while inequation (4.141) holds because k_2 is a parameter of BIG-$(\theta_2, r_2, k_2, \lambda_2)$ with $k_2 \geq 2$ (see Definition 1.23). Hence, given the fixed effect on the scalability metric, the option $d_3 = B$ leads to a higher inter-module throughput than the option $d_3 = P$.

In summary, the option $d_3 = B$ leads to a better trade-off between scalability and inter-module throughput than the option $d_3 = P$.

- *Comparison of fault-tolerance.* Let symbols $F(d_3 = P)$ and $F(d_3 = B)$ denote the fault-tolerance of design patterns with $d_3 = P$ and $d_3 = B$, respectively. Then by Lemma 4.20, we have the following two equations.

$$F(d_3 = P) = w, \tag{4.143}$$

$$F(d_3 = B) = t \cdot \left[\frac{\theta_2 - \frac{1}{k_2}}{\theta_2 - \frac{t}{p}} \cdot (w - 1) + 1 \right]. \tag{4.144}$$

These equations imply that a larger value of w incurs larger values of both $F(d_3 = P)$ and $F(d_3 = B)$. Meanwhile, the value of $F(d_3 = B)$ becomes larger as the value of t increases. Recall that parameter t should be a positive integer with tk_2 as a factor of p. Hence, the value of t is bounded within range $[1, p/k_2]$. To compare the values of $F(d_3 = P)$ and $F(d_3 = B)$, we fix the value of w and conduct the comparison under the minimum value and the maximum value of parameter t, respectively.

  - When $t = 1$, the value of k_2 is a factor of p and accordingly smaller than or equal to the value of p (i.e., $k_2 \leq p$). Hence, we have

$$F(d_3 = B) = \frac{\theta_2 - \frac{1}{k_2}}{\theta_2 - \frac{1}{p}} \cdot (w - 1) + 1 \leq w = F(d_3 = P). \tag{4.145}$$

  - When $t = p/k_2$, we have

$$F(d_3 = B) = t \cdot [\frac{\theta_2 - \frac{1}{k_2}}{\theta_2 - \frac{p}{pk_2}} \cdot (w - 1) + 1] = tw \geq w = F(d_3 = P). \tag{4.146}$$

Given the same value of w, the minimum value of $F(d_3 = B)$ is smaller than or equal to the value of $F(d_3 = P)$, while the maximum value of $F(d_3 = B)$ is greater than or equal to the value of $F(d_3 = P)$. Hence, there is no obvious superiority for fault-tolerance between the two options of term d_3. Nonetheless, the option B has a greater effect to improve fault-tolerance as the value of parameter t increases.

The discussion provides guidelines for the option selection of term d_3, as summarized as follows:

- If low cost or low latency are more desirable than high inter-module throughput or high scalability, it is suggested to select the option $d_3 = P$.
- If high inter-module throughput or high scalability are more desirable than low cost or low latency, it is suggested to select the option $d_3 = B$.
- Otherwise, it is suggested to select option $d_3 = B$ as such an option has a greater effect to improve fault-tolerance.

4.5.2 Demonstrations by Existing Topologies

To demonstrate the guidelines on selection of design patterns, we take existing topologies listed in Section 2.5 as examples to reveal the rationales of design pattern selection behind them. In the following, we first group them by their implicitly selected design patterns and lay out their positions within the design space in Figure 4.2. Then in the following two remarks, by comparing their selections on design patterns and performance differences, we show that existing topologies implicitly follow the guidelines on design pattern selection to satisfy the performance requirements of their application scenarios.

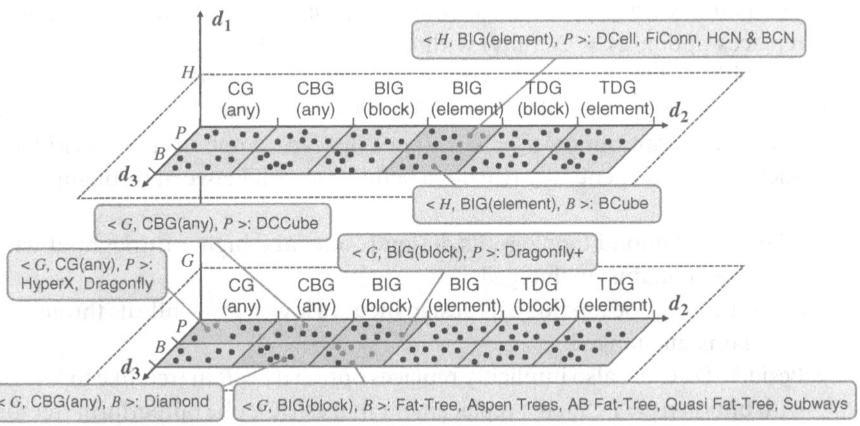

Figure 4.2 An overview of the locations of existing topologies (mentioned in Section 2.5) in the topology design space. Each dot symbolically represents a concrete topology design. The areas of design patterns containing such topologies are highlighted.

- *Design pattern* $< G$, CG(any), $P >$: HyperX and Dragonfly.
- *Design pattern* $< G$, CBG(any), $P >$: DCCube.
- *Design pattern* $< G$, CBG(any), $B >$: Diamond.
- *Design pattern* $< G$, BIG(block), $P >$: Dragonfly+.
- *Design pattern* $< G$, BIG(block), $B >$: Fat-Tree, Aspen Trees, AB Fat-Tree, Quasi Fat-Tree, and Subways.
- *Design pattern* $< H$, BIG(element), $P >$: DCell, FiConn, and HCN & BCN.
- *Design pattern* $< H$, BIG(element), $B >$: BCube.

Remark 4.2 Based on the earlier results, we find that:

- On one hand, existing topologies for supercomputing (e.g., HyperX and Dragonfly) implicitly employ options $d_2 =$ CG(any) and $d_3 = P$ in their designs. Such options conduce to a low latency, which match the demand of supercomputing as they are required to perform a large number of calculations within the limited time.
- On the other hand, existing topologies of data center networks (e.g., tree-like topologies) implicitly employ options $d_2 =$ BIG(block) and $d_3 = B$. Such options would lead to a high scalability, which match the demand of data center networks as they need to employ huge amounts of devices with computing functions to provide services.

We also find that there exist other topologies achieving distinct degrees of performance compromise between the two types of topologies discussed earlier. For example:

- Compared with HyperX and Dragonfly, Dragonfly+ topology implicitly replaces option $d_2 =$ CG(any) with option $d_2 =$ BIG(block). As a result, the scalability is improved at the cost of a higher latency and a higher cost.
- Compared with tree-like topologies, Diamond and DCCube topologies implicitly adopt option $d_2 =$ CBG(any) instead of option $d_2 =$ BIG(block). Such a selection generally reaches a better trade-off between throughput and cost.
  - For the Diamond topology, its design results in a higher throughput while the cost remains unchanged.
  - For the DCCube topology, its design incurs a lower cost while its throughput remains unchanged.

  Besides, DCCube also implicitly replaces option $d_3 = B$ in tree-like topologies with option $d_3 = P$. Such a replacement makes DCCube outperform tree-like topologies like Fat-Tree in terms of a lower latency and a lower cost.

Remark 4.3 A significant division for topology designs of large-scale computer networks is whether or not servers are also involved to provide routing function. From the perspective of the framework, this distinguishes the topologies under design patterns with $d_1 = G$ (i.e., topologies that contain no HNs, aka switch-centric topologies) from those with $d_1 = H$ (i.e., topologies that contain HNs, aka server-centric topologies). Our discussions on their option selection are presented as follows:

- In existing server-centric topologies (e.g., DCell and BCube), HNs are not directly connected to any CNs (i.e., $\gamma_H = 0$). These topologies adopt such a setting because it conduces to a lower latency (see Lemma 4.5).
- Given that BIG with copies is selected as the internal graph, existing switch-centric topologies (e.g., Fat-Tree) select block nodes as dock nodes, while existing server-centric topologies (e.g., BCube and DCell) select element nodes as dock nodes. Based on our analysis in Section 4.5.1, it turns out that the option selection adopted by the latter outperforms the former in terms of a higher scalability. Meanwhile, a higher latency may also be incurred, which can be mitigated under particular parameter settings. Such a finding also inspires potential improvement over existing switch-centric topologies. In particular, under the design patterns with options $d_1 = G$ and $d_2 =$ BIG(element), we can devise a novel switch-centric topology achieving a higher scalability than the classical Fat-Tree without compromising other performance metrics. This will be elaborated in Section 4.6.

4.6 Guidelines on Determination of Parameter Values

Following the selection of a design pattern, the values of its parameters should be determined to generate a concrete topology. In this section, we provide a scheme as the guidelines for parameter determination. Such a scheme would guide designers to derive a parameter setting satisfying their needs based on the analytical results on parameters and performances. To demonstrate this scheme, we not only reverse-engineer existing topologies under the implicit performance targets behind their parameter settings, but also devise a few concrete topologies with individual performance requirements. Next, we first introduce the scheme in Section 4.6.1, and then demonstrate it in Section 4.6.2.

4.6.1 Parameter Determination Scheme

Given a design pattern, parameter determination scheme aims to lead designers to search for the parameter settings based on some preset target performance. Note that the parameter settings of a design pattern can affect all five performance metrics simultaneously. Hence, optimizing such metrics requires the consideration of trade-offs among them, resulting in prohibitively high design complexity.

To reduce design complexity, the parameter determination scheme would prioritize these metrics, followed by considering how to meet the designers' demands for each metric in turn. In this way, the top-ranked metrics are given higher priority for consideration. In the following, we present the details of the scheme, where design pattern $< H, \text{BIG(element)}, P >$ with $p_H = 2$ and $p_R = p$ is used as an example for illustration.

As shown in Figure 4.3, the scheme consists of three stages, which are introduced as follows, respectively.

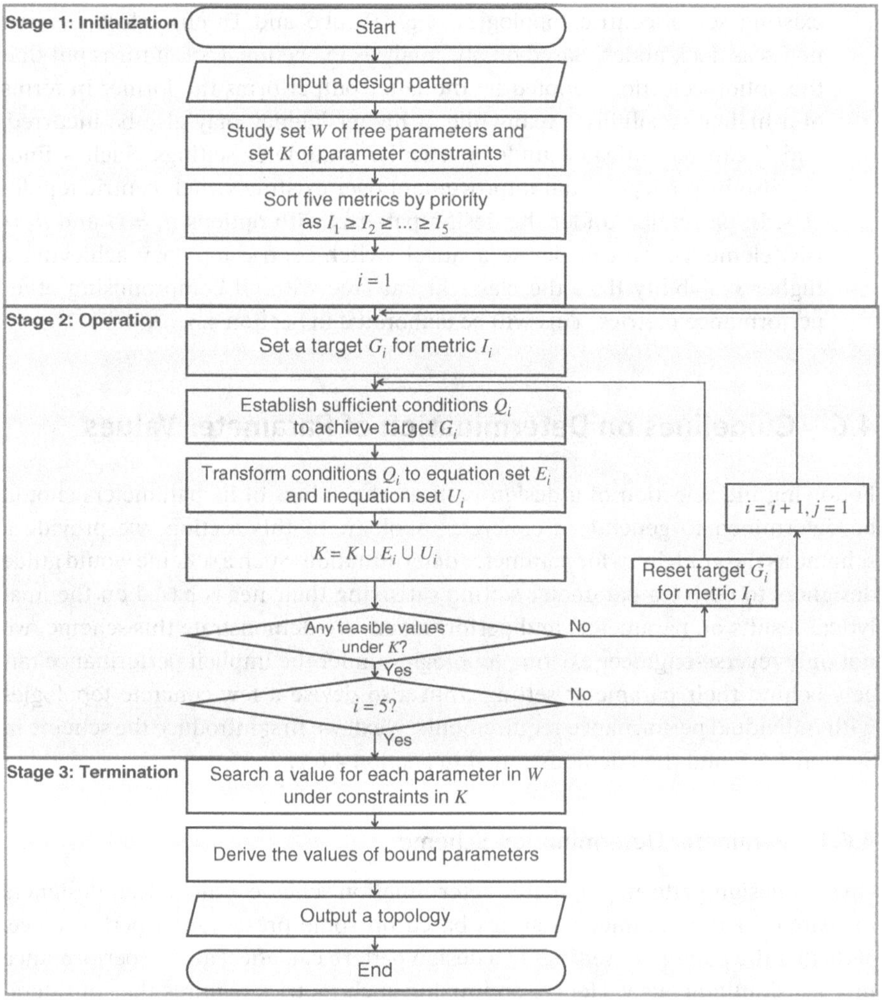

Figure 4.3 The flowchart of parameter determination scheme.

- **Stage 1: Initialization.** Given a design pattern as input, this stage consists of the following four steps.

Step 1 Study the free parameters of the design pattern. The set of free parameters is denoted by W. For example, the free parameters of design pattern $< H$, BIG(element) $, P >$ include α_H, γ_H, θ_1, k_1, t, and x.

Step 2 Study the constraints on free parameters based on the results in Section 4.3.2. The set of constraints is denoted by symbol K. Regarding design pattern $< H$, BIG(element) $, P >$, its free parameters are constrained as follows:

- The values of the free parameters should guarantee the existence of BIG-$(\theta_1, r_1, k_1, \lambda_1)$.
- The values of $\alpha_H p_H$, $\gamma_H p_H$, and $(1 - \alpha_H - \gamma_H) p_H$ are all non-negative integers such that $\alpha_H > 0$, $\gamma_H \geq 0$, and $\alpha_H + \gamma_H < 1$.
- The value of k_1 should be a factor of p.
- The value of x should be a common factor of $\theta_1 p$ and $(1 - \alpha_H - \gamma_H) p_H$.
- The value of t should be a factor of $\dfrac{\theta_1 p \cdot (1 - \alpha_H - \gamma_H) p_H}{x^2}$ with $t \leq \dfrac{\theta_1 p}{x}$.

When $p_H = 2$, to satisfy these constraints, we have $\alpha_H = 0.5$, $\gamma_H = 0$, $x = 1$, and $\theta_1 = 1$. The reasons are given as follows:

- To satisfy the second constraint, the only feasible values of α_H and γ_H are $\alpha_H = 0.5$ and $\gamma_H = 0$.
- To satisfy the third constraint, the value of x is a factor of $(1 - \alpha_H - \gamma_H) p_H = 1$, leading to $x = 1$. Meanwhile, $x = 1$ also guarantees that the third constraint is satisfied.
- By Definition 1.23, the existence of BIG-$(\theta_1, r_1, k_1, \lambda_1)$ implies $\theta_1, \lambda_1 \geq 1$. Since parameters satisfy $\lambda_1 = \alpha_H p_H \cdot (k_1 - 1)/(\theta_1 k_1 - 1) = (k_1 - 1)/(\theta_1 k_1 - 1)$, we have $\lambda_1 \leq 1$, where the equality holds if and only if $\theta_1 = 1$. Hence, to satisfy the first constraint, the only feasible value of θ_1 is $\theta_1 = 1$. Meanwhile, according to Lemma 1.7, $\theta_1 = 1$ also guarantees the first constraint if and only if $k_1 \geq 2$.

As a result, the set K of constraints only includes the following constraints.

- The value of k_1 should be a factor of p with $k_1 \geq 2$.
- The value of t should be a factor of p.
- $\alpha_H = 0.5$, $\gamma_H = 0$, $x = 1$, and $\theta_1 = 1$.

Step 3 Sort the five performance metrics by priority. The five performance metrics are denoted by I_1, I_2, I_3, I_4, I_5 from the highest priority to the lowest priority (i.e., $I_1 \geq I_2 \geq I_3 \geq I_4 \geq I_5$). For example, designers can set the priority of performance metrics as scalability ≥ cost ≥ latency ≥ throughput ≥ fault-tolerance.

Step 4 Initialize a global counter i to one (i.e., $i = 1$).

- **Stage 2: Operation.** In this stage, designers sequentially consider how to satisfy their demands on each performance metric. The basic idea is to constrained the free parameters to achieve each preset target. Specifically, for each performance metric I_i ($i \leq 5$), designers conduct the following seven steps in this stage.

Step 1 : Set a target G_i for metric I_i. The target could be a specific value (e.g., latency equal to 3), the potential extreme value (e.g., scalability equal to its largest value), a range of values (e.g., cost less than 1), or even no requirements. For the first metric I_1 in the example described earlier (i.e., scalability), designers can set target G_1 as achieving the largest value of scalability and set targets for other performance metrics as no requirements.

Step 2 : Establish sufficient conditions Q_i to achieve target G_i for metric I_i. In the example described earlier, there are no requirements for other metrics except scalability. Hence, we use scalability to illustrate the following steps. According to the performance analysis results in Table 4.6, the value of scalability is given by

$$S = (\gamma_H p_H + 1) \cdot (\theta_1 p_R) \cdot \left[\frac{\theta_1 p_R \cdot (1 - \alpha_H - \gamma_H) p_H}{tx^2} + 1 \right] = p(\frac{p}{t} + 1). \tag{4.147}$$

This equation indicates that a larger value of t leads to a smaller value of scalability. To achieve the target G_1 (i.e., achieving the largest value of scalability), the sufficient condition Q_1 is to set t as its smallest value.

Step 3 : Utilize equation set E_i and inequation set U_i to describe condition Q_i. Note that parameter t should be a factor of p. Hence, the smallest value of t is one. As a result, equation set E_1 contains one equation $t = 1$, while inequation set U_1 is empty.

Step 4 : Add equations in set E_i and inequations in set U_i to the set K of parameter constraints (i.e., $K = K \cup E_i \cup U_i$). In the example described earlier, since there is one equation $t = 1$ in set E_1 and no inequation in set U_1, the set K now contains the following constraints.

- The value of k_1 should be a factor of p with $k_1 \geq 2$.
- The value of t should be a factor of p.
- $\alpha_H = 0.5$, $\gamma_H = 0$, $t = x = 1$, and $\theta_1 = 1$.

Note that given $t = 1$, it is already guaranteed that the value of t is a factor of p. Hence, the constraints in set K can be reduced as follows:

- The value of k_1 should be a factor of p with $k_1 \geq 2$.
- $\alpha_H = 0.5, \gamma_H = 0, t = x = 1$, and $\theta_1 = 1$.

Step 5 : Check whether there exist feasible values of free parameters under the constraints in K. If so, designers can execute the seven steps for next performance metric by setting $i = i + 1$ when $i < 5$ or go to the next stage when $i = 5$. Otherwise, designers need to re-execute these seven steps by resetting target G_i for metric I_i. In the example described earlier, there are six free parameters (i.e., $\alpha_H, \gamma_H, \theta_1, k_1, t$, and x), and the values of $\alpha_H, \gamma_H, \theta_1, t$, and x are fixed as $\alpha_H = 0.5$, $\gamma_H = 0$, $\theta_1 = 1$, $t = 1$, and $x = 1$. We observe that the only undetermined free parameter is k_1, which should be a factor of p with $k_1 \geq 2$. For any value of $p \geq 2$, there exists a feasible value of free parameter k_1, i.e., $k_1 = p$. Hence, designers can execute the seven steps for other metrics. Recall that there are no requirements for other metrics and thus no extra constraints. As a result, designers can enter the next stage with constraints in set K remaining unchanged.

- **Stage 3: Termination.** With the operation stage completed, designers have determined the constraint for each free parameter to achieve the target performance. In this stage, designers conduct the following two steps to devise a concrete topology.

Step 1 : Search for a feasible value of each free parameter to satisfy its constraints in set K. Recall that there are six free parameters in the example, including $\alpha_H, \gamma_H, \theta_1, k_1, t$, and x. Regarding parameter k_1, since it should be a factor of p with $k_1 \geq 2$, we set $k_1 = p$ such that this constraint is satisfied for any value of $p \geq 2$. As for other free parameters, their values are fixed as $\alpha_H = 0.5, \gamma_H = 0, \theta_1 = 1, t = 1$, and $x = 1$ under the constraints.

Step 2 : Derive the values of bound parameters based on the values of free parameters. In this example, the values of bound parameters are derived by replacing parameters α_G, β_G, γ_G, and p_G in equations (4.54)–(4.62) with parameters α_H, β_H, γ_H, and p_H, respectively. Among these bound parameters, the values of parameters α_R, β_R, and l are not determined by the values of free parameters, but fixed as $\alpha_R = 1$, $\beta_R = 0$, and $l = 2$. The values of other bound parameters are derived as follows:

$$\beta_H = 1 - \alpha_H - \gamma_H = 0.5, \tag{4.148}$$

$$r_1 = \alpha_H p_H = 1, \tag{4.149}$$

$$\lambda_1 = \alpha_H p_H \cdot \frac{k_1 - 1}{\theta_1 k_1 - 1} = 1, \tag{4.150}$$

$$c_1 = \frac{p_R}{k_1} - 1 = 0, \tag{4.151}$$

$$w = x = 1, \tag{4.152}$$

$$m_2 = \frac{\theta_1 p_R \cdot (1 - \alpha_H - \gamma_H) p_H}{t x^2} + 1 = p + 1. \tag{4.153}$$

Through these two steps, given port number of switches as p (i.e., $p_R = p$), the devised topology is described as follows:

- *Intra-module design.* The internal graph is a BIG-$(1, 1, p, 1)$ where element nodes and block nodes are viewed as HNs and RNs, respectively.
- *Connection design.* In each module, all HNs are selected as dock nodes. Each dock node serves as an ISN-$(1, 1, 2)$. Then dock node interconnection is constructed under the plain mode. Specifically, each composite-edge between any two ISNs is designed as an all-to-all interconnection.
- *Inter-module design.* The skeleton graph is a CG with $p + 1$ nodes. Each node corresponds to a module. Each super-edge between two modules is instantiated as a 1-parallelized all-to-all interconnection.
- *Finalization.* This step is omitted as the topology contains no CNs.

Actually, this topology is the level-1 DCell topology (see Figure 4.4). Hence, this example can be regarded as the reverse-engineering process of a level-1 DCell topology via the scheme. Such a process reveals that designers implicitly achieve the highest scalability for level-1 DCell topology design.

4.6.2 Demonstrations by Various Design Patterns

Based on the parameter determination scheme, designers can devise concrete topologies according to their performance demands in a systematic fashion. In the following, we illustrate such a process under five representative design patterns with $d_1 = G$ and $p_R = p_G = p$ in Sections 4.6.2.1–4.6.2.5, respectively. Under each design pattern, we use some specific target as an example to demonstrate how to generate a required topology. Particularly, if a classical topology exists in some design pattern, we employ the implicit performance target behind such a topology and generate it in a reverse-engineering fashion. Otherwise, we set a new performance target to forward-engineer a novel topology.

4.6.2.1 Design Pattern $< G, \mathrm{CG(any)}, P >$

Under design pattern $< G, \mathrm{CG(any)}, P >$, there exists a class of topologies called HyperX, whose parameter settings satisfy $x = 1$ and $t = \alpha_G p + 1$. In the following, we adopt the parameter determination scheme to reverse-engineer HyperX topologies.

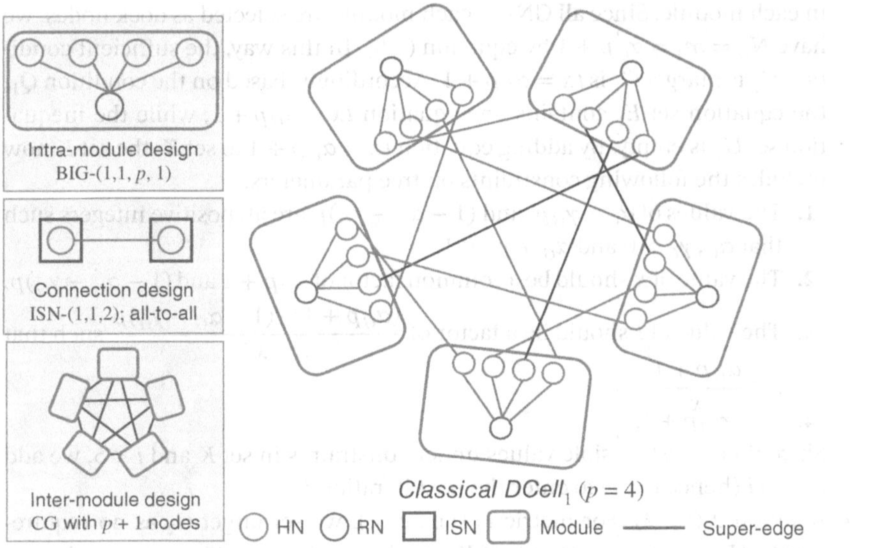

Figure 4.4 An instance of a DCell_1 topology under design pattern $< H, \text{CG(any)}, P >$ with $p = 4$.

Stage 1: Initialization. Given design pattern $< G, \text{CG(any)}, P >$ as input, based on results in Section 4.3.2.1, we can study its free parameters and parameter constraints. Particularly, under this design pattern, the set W of free parameters contains parameters α_G, γ_G, t, and x. Meanwhile, the set K of parameter constraints contains the following constraints.

1. The values of $\alpha_G p$, $\gamma_G p$, and $(1 - \alpha_G - \gamma_G)p$ are all positive integers such that $\alpha_G, \gamma_G > 0$ and $\alpha_G + \gamma_G < 1$.
2. The value of x should be a common factor of $\alpha_G p + 1$ and $(1 - \alpha_G - \gamma_G)p$.
3. The value of t should be a factor of $\dfrac{(\alpha_G p + 1) \cdot (1 - \alpha_G - \gamma_G)p}{x^2}$ such that $t \leq \dfrac{\alpha_G p + 1}{x}$.

Then we sort the five performance metrics by priority as I_1 (latency) $\geq I_2$ (cost) $\geq I_3$ (scalability) $\geq I_4$ (throughput) $\geq I_5$ (fault-tolerance). After initializing global counter i to one (i.e., $i = 1$), we enter the next stage.

Stage 2: Operation. This stage consists of five iterations controlled by the global counter i. In the following, we conduct each iteration.

- Iteration 1 ($i = 1$). For metric I_1 (i.e., latency), we set target G_1 as avoiding the worst latency 5 (i.e., latency $L \leq 4$). By Lemmas 4.5 and 4.6, the value of L is no more than four if $tx = N_D$, where N_D denotes the number of dock nodes in each module. Since all GNs in each module are selected as dock nodes, we have $N_D = m_1 = \alpha_G p + 1$ by equation (4.8). In this way, the sufficient condition Q_1 for target G_1 is $tx = \alpha_G p + 1$. Accordingly, based on the condition Q_1, the equation set E_1 contains one equation $tx = \alpha_G p + 1$, while the inequation set U_1 is empty. By adding equation $tx = \alpha_G p + 1$ to set K, the set K now includes the following constraints on free parameters.

1. The values of $\alpha_G p, \gamma_G p$, and $(1 - \alpha_G - \gamma_G)p$ are all positive integers such that $\alpha_G, \gamma_G > 0$ and $\alpha_G + \gamma_G < 1$.
2. The value of x should be a common factor of $\alpha_G p + 1$ and $(1 - \alpha_G - \gamma_G)p$.
3. The value of t should be a factor of $\frac{(\alpha_G p + 1) \cdot (1 - \alpha_G - \gamma_G)p}{x^2}$ such that $t \leq \frac{\alpha_G p + 1}{x}$.
4. $tx = \alpha_G p + 1$.

Since there exist feasible values under constraints in set K and $i < 5$, we add one to i (hence $i = 2$) to enter the next iteration.[7]

- Iteration 2 ($i = 2$). For metric I_2 (i.e., cost), we set target G_2 as no requirements. Hence, the constraint set K remains unchanged. Since there exist feasible values under constraints in set K and $i < 5$, we add one to i (hence $i = 3$) to enter the next iteration.
- Iteration 3 ($i = 3$). For metric I_3 (i.e., scalability), we set target G_3 as no requirements. Hence, the constraint set K remains unchanged. Since there exist feasible values under constraints in set K and $i < 5$, we add one to i (hence $i = 4$) to enter the next iteration.
- Iteration 4 ($i = 4$). For metrics I_4 (i.e., throughput), we set target G_4 as no requirements. Hence, the constraint set K remains unchanged. Since there exist feasible values under constraints in set K and $i < 5$, we add one to i (hence $i = 5$) to enter the next iteration.
- Iteration 5 ($i = 5$). For metrics I_5 (i.e., fault-tolerance), we set target G_5 as no requirements. Hence, the constraint set K remains unchanged. Since there exist feasible values under constraints in set K and $i = 5$, we jump out of these iterations and enter the next stage.

Stage 3: Termination. First, we need to search for a feasible value for each free parameter in W under constraints in K. The constraints in set K are listed as follows for reference.

1. The values of $\alpha_G p, \gamma_G p$, and $(1 - \alpha_G - \gamma_G)p$ are all positive integers such that $\alpha_G, \gamma_G > 0$ and $\alpha_G + \gamma_G < 1$.
2. The value of x should be a common factor of $\alpha_G p + 1$ and $(1 - \alpha_G - \gamma_G)p$.

---

7 The process of checking whether there exist feasible values under constraints in set K is omitted, as the devised topology in the termination stage is an example with the feasible values.

3. The value of t should be a factor of $\frac{(\alpha_G p+1)\cdot(1-\alpha_G-\gamma_G)p}{x^2}$ such that $t \leq \frac{\alpha_G p+1}{x}$.
4. $tx = \alpha_G p + 1$.

Next, we search for feasible values of free parameters (i.e., α_G, γ_G, x, and t) under the given constraints.

- Parameters α_G and γ_G. Note that the values of $\alpha_G p$, $\gamma_G p$, and $(1-\alpha_G-\gamma_G)p$ are all positive integers such that $\alpha_G, \gamma_G > 0$ and $\alpha_G + \gamma_G < 1$. To satisfy this constraint, we set $\alpha_G = a/p$ and $\gamma_G = b/p$, where a and b are positive integers with $a + b < p$.
- Parameters x and t. Note that the value of x should be a common factor of $\alpha_G p + 1$ and $(1-\alpha_G-\gamma_G)p$. Meanwhile, the value of $t = (\alpha_G p + 1)/x$ should be a factor of $\frac{(\alpha_G p+1)\cdot(1-\alpha_G-\gamma_G)p}{x^2}$. We can observe that $x = 1$ is a feasible value for any values of α_G, γ_G, and p. Hence, we set $x = 1$ and accordingly $t = \alpha_G p + 1 = a + 1$.

Based on the preceding discussions, given two positive integers a and b with $a + b < p$, the values of free parameters are set as $t = a + 1$, $x = 1$, $\alpha_G = a/p$, and $\gamma_G = b/p$. Given the fixed values of free parameters, the values of bound parameters can be derived by equations (4.7)–(4.11). Particularly, the value of parameter l is not determined by the values of free parameters, but fixed as $l = 1$. The values of other bound parameters are derived as follows:

$$\beta_G = 1 - \alpha_G - \gamma_G = \frac{p-a-b}{p}, \tag{4.154}$$

$$m_1 = \alpha_G p + 1 = a + 1, \tag{4.155}$$

$$w = x = 1, \tag{4.156}$$

$$m_2 = \frac{(\alpha_G p+1)\cdot(1-\alpha_G-\gamma_G)\cdot p}{tx^2} + 1 = p - a - b + 1. \tag{4.157}$$

Under the preceding parameter settings, given the port number of switches as p, the complete design of devised topologies can be summarized as follows, where a and b are positive integers with $a + b < p$.

- *Intra-module design*. A CG with $a + 1$ GNs is employed as the internal graph.
- *Connection design*. In each module, all GNs are selected as dock nodes. Each dock node forms an ISN-(1, 1, 1). Dock nodes in distinct modules are all connected under the plain mode. Specifically, two ISNs-(1, 1, 1) are interconnected by a composite edge which is an all-to-all interconnection.
- *Inter-module design*. A CG with $p - a - b + 1$ modules is employed as the skeleton graph to interconnect different modules. The super-edge between

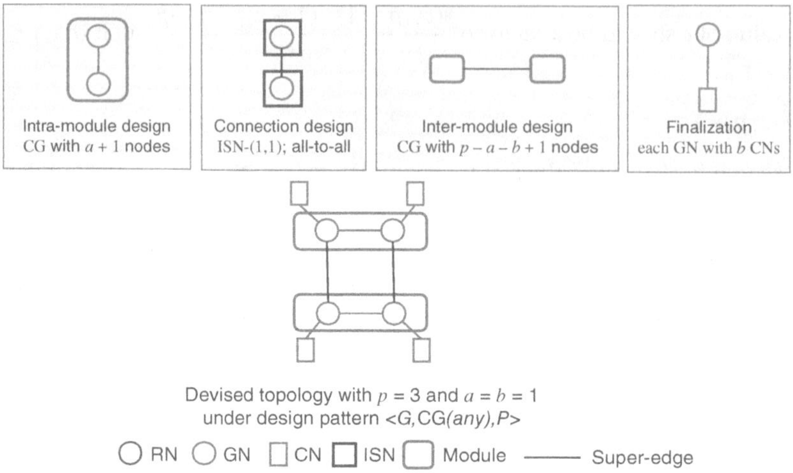

Figure 4.5 The complete design of a devised topology under design pattern $< G$, CG(any), $P >$ with $p = 3$ and $a = b = 1$. This topology is actually a concrete HyperX topology.

two modules in the CG is instantiated as a $(a + 1)$-parallelized all-to-all interconnection.

- *Finalization.* Each GN in the modules is directly connected to b CNs.

An example of these topologies with $p = 3$ and $a = b = 1$ is provided in Figure 4.5. The preceding parameter settings indicate that these topologies are the classical HyperX topologies (see Section 2.5.5). In other words, HyperX topologies can be reverse-engineered under design pattern $< G$, CG(any), $P >$ to achieve latency $L \leq 4$.

4.6.2.2 Design Pattern $< G$, BIG(block), $B >$

Based on Section 3.1.1, we can learn that tree-like topologies—one of the most widely adopted and well-suited classes of topologies in data center networks and AI infrastructures—constitute a class of special cases of design pattern $< G$, BIG(block), $B >$. In fact, design pattern $< G$, BIG(block), $B >$ specifies the design options for each step of the design procedure for tree-like topologies, ensuring fundamental structural and performance properties of the resulting topologies. At the same time, it offers a broader parameter space beyond tree-like topologies, which enables more customized performance trade-off and parameter setting schemes to accommodate diverse performance requirements of various application scenarios.

Under design pattern $< G$, BIG(block), $B >$, the classical Fat-Tree topology with port number p is a special topology with $\alpha_R = \gamma_G = 0.5$, $\theta_1 = \theta_2 = x = t = 1$, $\mu = k_1 = p/2$, and $k_2 = p$. In the following, we adopt the three stages in the parameter determination scheme to reverse-engineer the Fat-Tree topology.

Stage 1: Initialization. Given design pattern $< G, \text{BIG(block)}, B >$ as input, based on results in Section 4.3.2.6, we can study its free parameters and parameter constraints. Particularly, under this design pattern, the set W contains nine parameters, including $\alpha_R, \gamma_G, \theta_1, k_1, t, x, \mu, \theta_2$, and k_2. Meanwhile, the set K of parameter constraints contains the following constraints.

1. The values of the free parameters should guarantee the existence of BIG-$(\theta_1, r_1, k_1, \lambda_1)$.
2. The values of the free parameters should guarantee that the BIBD-$(\theta_1 k_1, \theta_1 r_1, r_1, k_1, \lambda_1)$ is resolvable.
3. The values of the free parameters should guarantee the existence of TD-(n, μ, x) interconnection.
4. The values of the free parameters should guarantee the existence of BIG-$(\theta_2, r_2, k_2, \lambda_2)$.
5. The values of $\alpha_R p$ and $\gamma_G p$ are both positive integers with $0 < \alpha_R, \gamma_G < 1$.
6. The value of k_1 should be a factor of $\alpha_R p$.
7. The value of θ_1 should be a factor of x.
8. The value of x should be a common factor of $\theta_1(1-\gamma_G)p$ and $(1-\alpha_R)p$.
9. The value of t should be a common factor of p and $\dfrac{(1-\alpha_R)\cdot\theta_1(1-\gamma_G)p^2}{x^2}$ with $t \leq \dfrac{\theta_1(1-\gamma_G)p}{x}$.
10. The value of μ should be a common factor of $\dfrac{(1-\alpha_R)\cdot\theta_1(1-\gamma_G)p^2}{tx^2}$ and $\dfrac{(1-\alpha_R)p}{x}$.
11. The value of k_2 should be a factor of $\dfrac{p}{t}$.

Then we sort the five performance metrics by priority as I_1 (throughput) $\geq I_2$ (cost) $\geq I_3$ (scalability) $\geq I_4$ (fault-tolerance) $\geq I_5$ (latency). After initializing global counter i to one (i.e., $i = 1$), we enter the next stage.

Stage 2: Operation. This stage consists of five iterations controlled by the global counter i. In the following, we conduct each iteration.

- Iteration 1 ($i = 1$). For metric I_1 (i.e., throughput), we set target G_1 as $T_1 = T_2 = 1$, where T_1 and T_2 denote the values of intra-module throughput and inter-module throughput, respectively. By substituting $\beta_R = 1 - \alpha_R$, $\alpha_G = 1 - \gamma_G$, and $\lambda_2/r_2 = (k_2 - 1)/(\theta_2 k_2 - 1)$ (see equations 4.42, 4.44, 4.51, and 4.52) into analysis results on throughput in Table 4.11, it follows that

$$T_1 = \frac{\alpha_G}{\gamma_G} = \frac{1-\gamma_G}{\gamma_G}, \tag{4.158}$$

$$T_2 = \frac{\beta_R \alpha_G}{\alpha_R \gamma_G} \cdot \frac{\lambda_2}{r_2} = \frac{(1-\alpha_R)(1-\gamma_G)}{\alpha_R \gamma_G} \cdot \frac{k_2 - 1}{\theta_2 k_2 - 1}. \tag{4.159}$$

Hence, the sufficient (also necessary) condition Q_1 on $T_1 = T_2 = 1$ is $\gamma_G = 0.5$ and $(1 - \alpha_R)/\alpha_R = (\theta_2 k_2 - 1)/(k_2 - 1)$. Based on this condition, the equation set E_1 includes two equations $\gamma_G = 0.5$ and $(1 - \alpha_R)/\alpha_R = (\theta_2 k_2 - 1)/(k_2 - 1)$, while the inequation set U_1 is empty. By adding these two equations to set K, the set K now includes the following constraints on free parameters.

1. The values of the free parameters should guarantee the existence of BIG-$(\theta_1, r_1, k_1, \lambda_1)$.
2. The values of the free parameters should guarantee that the BIBD-$(\theta_1 k_1, \theta_1 r_1, r_1, k_1, \lambda_1)$ is resolvable.
3. The values of the free parameters should guarantee the existence of TD-(n, μ, x) interconnection.
4. The values of the free parameters should guarantee the existence of BIG-$(\theta_2, r_2, k_2, \lambda_2)$.
5. The values of $\alpha_R p$ and $\gamma_G p$ are both positive integers with $0 < \alpha_R, \gamma_G < 1$.
6. The value of k_1 should be a factor of $\alpha_R p$.
7. The value of θ_1 should be a factor of x.
8. The value of x should be a common factor of $\theta_1(1 - \gamma_G)p$ and $(1 - \alpha_R)p$.
9. The value of t should be a common factor of p and $\dfrac{(1 - \alpha_R) \cdot \theta_1(1 - \gamma_G)p^2}{x^2}$ with $t \leq \dfrac{\theta_1(1 - \gamma_G)p}{x}$.
10. The value of μ should be a common factor of $\dfrac{(1 - \alpha_R) \cdot \theta_1(1 - \gamma_G)p^2}{tx^2}$ and $\dfrac{(1 - \alpha_R)p}{x}$.
11. The value of k_2 should be a factor of $\dfrac{p}{t}$.
12. $\gamma_G = 0.5$ and $(1 - \alpha_R)/\alpha_R = (\theta_2 k_2 - 1)/(k_2 - 1)$.

Since there exist feasible values under constraints in set K and $i < 5$, we add one to i (hence $i = 2$) to enter the next iteration.[8]

- Iteration 2 ($i = 2$). For metric I_2 (i.e., cost), we set target G_2 as achieving the lowest cost. By substituting $\beta_R = 1 - \alpha_R$ and $\alpha_G = 1 - \gamma_G$ (see equations 4.42 and 4.44) into analysis results on cost in Table 4.7, it follows that

$$C = \frac{1}{\gamma_G p} + \frac{\alpha_G}{\gamma_G \alpha_R p} + \frac{\alpha_G \beta_R}{\gamma_G \alpha_R p} = \frac{1}{\gamma_G p} + \frac{1 - \gamma_G}{\gamma_G \alpha_R p} + \frac{(1 - \gamma_G)(1 - \alpha_R)}{\gamma_G \alpha_R p}. \tag{4.160}$$

8 The process of checking whether there exist feasible values under constraints in set K is omitted, as the devised topology in the termination stage is an example with the feasible values.

Note that we have $\gamma_G = 0.5$ and $\alpha_R = (\theta_2 k_2 - 1)/(\theta_2 k_2 + k_2 - 2)$ based on the discussions on metric I_1. By substituting these two equations into equation (4.160), we have

$$C = \frac{1}{\gamma_G p} + \frac{1-\gamma_G}{\gamma_G \alpha_R p} + \frac{(1-\gamma_G)(1-\alpha_R)}{\gamma_G \alpha_R p} = \frac{3}{p} + \frac{2}{p} \cdot \frac{\theta_2 k_2 - 1}{k_2 - 1}. \quad (4.161)$$

By this equation, we can observe that a smaller value of θ_2 leads to a lower cost. Note that the parameter θ_2 of BIG-$(\theta_2, r_2, k_2, \lambda_2)$ is greater than or equal to one. Therefore, the lowest cost is achieved when $\theta_2 = 1$, leading to

$$C = \frac{3}{p} + \frac{2}{p} \cdot \frac{\theta_2 k_2 - 1}{k_2 - 1} \geq \frac{5}{p}. \quad (4.162)$$

In other words, the sufficient (also necessary) condition Q_2 on the lowest cost is $\theta_2 = 1$. Accordingly, the equation set E_2 contains one equation $\theta_2 = 1$, while the inequation set U_2 is empty. By adding equation $\theta_2 = 1$ to set K, the set K now includes the following constraints on free parameters.

1. The values of the free parameters should guarantee the existence of BIG-$(\theta_1, r_1, k_1, \lambda_1)$.
2. The values of the free parameters should guarantee that the BIBD-$(\theta_1 k_1, \theta_1 r_1, r_1, k_1, \lambda_1)$ is resolvable.
3. The values of the free parameters should guarantee the existence of TD-(n, μ, x) interconnection.
4. The values of the free parameters should guarantee the existence of BIG-$(\theta_2, r_2, k_2, \lambda_2)$.
5. The values of $\alpha_R p$ and $\gamma_G p$ are both positive integers with $0 < \alpha_R, \gamma_G < 1$.
6. The value of k_1 should be a factor of $\alpha_R p$.
7. The value of θ_1 should be a factor of x.
8. The value of x should be a common factor of $\theta_1(1-\gamma_G)p$ and $(1-\alpha_R)p$.
9. The value of t should be a common factor of p and $\dfrac{(1-\alpha_R)\cdot\theta_1(1-\gamma_G)p^2}{x^2}$ with $t \leq \dfrac{\theta_1(1-\gamma_G)p}{x}$.
10. The value of μ should be a common factor of $\dfrac{(1-\alpha_R)\cdot\theta_1(1-\gamma_G)p^2}{tx^2}$ and $\dfrac{(1-\alpha_R)p}{x}$.
11. The value of k_2 should be a factor of $\dfrac{p}{t}$.
12. $\gamma_G = 0.5$ and $(1-\alpha_R)/\alpha_R = (\theta_2 k_2 - 1)/(k_2 - 1)$.
13. $\theta_2 = 1$.

Since there exist feasible values under constraints in set K and $i < 5$, we add one to i (hence $i = 3$) to enter the next iteration.

- Iteration 3 ($i = 3$). For metric I_3 (i.e., scalability), we set target G_3 as achieving the highest scalability. According to analysis results on scalability (see Table 4.5), it follows that

$$S = \gamma_G p \cdot \theta_1 \alpha_R p \cdot (\theta_2 p/t) = \frac{\theta_1 \theta_2 \gamma_G \alpha_R}{t} \cdot p^3. \tag{4.163}$$

Note that the discussions on metrics I_1 and I_2 require $\gamma_G = 0.5$, $\theta_2 = 1$, and $\alpha_R = (\theta_2 k_2 - 1)/(\theta_2 k_2 + k_2 - 2) = 0.5$. By substituting these three equations into equation (4.163), we have

$$S = \frac{\theta_1 \theta_2 \gamma_G \alpha_R}{t} \cdot p^3 = \frac{\theta_1}{4t} p^3. \tag{4.164}$$

By this equation, we can observe that a smaller value of t leads to a higher scalability. Under the constraints in set K, parameter t should be a common factor of p and $(\theta_1 p^2)/(4x^2)$ with $t \le \frac{\theta_1 p}{2x}$. This implies that t is a positive integer (hence $t \ge 1$). Therefore, the highest scalability is achieved when $t = 1$. Meanwhile, parameter θ_1 should guarantee the existence of BIG-$(\theta_1, r_1, k_1, \lambda_1)$. Hence, although a larger value of θ_1 leads to a higher scalability, we set $\theta_1 = 1$ to guarantee the universal existence of BIG-$(\theta_1, r_1, k_1, \lambda_1)$. Therefore, under the consideration of imposing the fewest restrictions on port number, the sufficient condition Q_3 on the highest scalability is $t = 1$ and $\theta_1 = 1$. Based on this condition, the equation set E_3 contains two equations $t = 1$ and $\theta_1 = 1$, while the inequation set U_3 is empty. By adding equations $t = 1$ and $\theta_1 = 1$ to set K, the set K now includes the following constraints on free parameters.

1. The values of the free parameters should guarantee the existence of BIG-$(\theta_1, r_1, k_1, \lambda_1)$.
2. The values of the free parameters should guarantee that the BIBD-$(\theta_1 k_1, \theta_1 r_1, r_1, k_1, \lambda_1)$ is resolvable.
3. The values of the free parameters should guarantee the existence of TD-(n, μ, x) interconnection.
4. The values of the free parameters should guarantee the existence of BIG-$(\theta_2, r_2, k_2, \lambda_2)$.
5. The values of $\alpha_R p$ and $\gamma_G p$ are both positive integers with $0 < \alpha_R, \gamma_G < 1$.
6. The value of k_1 should be a factor of $\alpha_R p$.
7. The value of θ_1 should be a factor of x.
8. The value of x should be a common factor of $\theta_1(1 - \gamma_G)p$ and $(1 - \alpha_R)p$.
9. The value of t should be a common factor of p and $\frac{(1 - \alpha_R) \cdot \theta_1(1 - \gamma_G)p^2}{x^2}$ with $t \le \frac{\theta_1(1 - \gamma_G)p}{x}$.
10. The value of μ should be a common factor of $\frac{(1 - \alpha_R) \cdot \theta_1(1 - \gamma_G)p^2}{tx^2}$ and $\frac{(1 - \alpha_R)p}{x}$.
11. The value of k_2 should be a factor of $\frac{p}{t}$.
12. $\gamma_G = 0.5$ and $(1 - \alpha_R)/\alpha_R = (\theta_2 k_2 - 1)/(k_2 - 1)$.

13. $\theta_2 = 1$.
14. $t = 1$ and $\theta_1 = 1$.

Since there exist feasible values under constraints in set K and $i < 5$, we add one to i (hence $i = 4$) to enter the next iteration.

- Iteration 4 ($i = 4$). For metrics I_4 (i.e., fault-tolerance), we set target G_4 as no requirements. Hence, the set K remains unchanged. Since there exist feasible values under constraints in set K and $i < 5$, we add one to i (hence $i = 5$) to enter the next iteration.
- Iteration 5 ($i = 5$). For metrics I_5 (i.e., latency), we set target G_5 as no requirements. Hence, the set K remains unchanged. Since there exist feasible values under constraints in set K and $i = 5$, we jump out of these iterations and enter the next stage.

Stage 3: Termination. First, we need to search for a feasible value for each free parameter in W under constraints in K. The constraints in set K are listed as follows for reference.

1. The values of the free parameters should guarantee the existence of BIG-$(\theta_1, r_1, k_1, \lambda_1)$.
2. The values of the free parameters should guarantee that the BIBD-$(\theta_1 k_1, \theta_1 r_1, r_1, k_1, \lambda_1)$ is resolvable.
3. The values of the free parameters should guarantee the existence of TD-(n, μ, x) interconnection.
4. The values of the free parameters should guarantee the existence of BIG-$(\theta_2, r_2, k_2, \lambda_2)$.
5. The values of $\alpha_R p$ and $\gamma_G p$ are both positive integers with $0 < \alpha_R, \gamma_G < 1$.
6. The value of k_1 should be a factor of $\alpha_R p$.
7. The value of θ_1 should be a factor of x.
8. The value of x should be a common factor of $\theta_1(1-\gamma_G)p$ and $(1-\alpha_R)p$.
9. The value of t should be a common factor of p and $\dfrac{(1-\alpha_R)\cdot\theta_1(1-\gamma_G)p^2}{x^2}$ with $t \le \dfrac{\theta_1(1-\gamma_G)p}{x}$.
10. The value of μ should be a common factor of $\dfrac{(1-\alpha_R)\cdot\theta_1(1-\gamma_G)p^2}{tx^2}$ and $\dfrac{(1-\alpha_R)p}{x}$.
11. The value of k_2 should be a factor of $\dfrac{p}{t}$.
12. $\gamma_G = 0.5$ and $(1-\alpha_R)/\alpha_R = (\theta_2 k_2 - 1)/(k_2 - 1)$.
13. $\theta_2 = 1$.
14. $t = 1$ and $\theta_1 = 1$.

Next, we search for feasible values of free parameters (i.e., $\alpha_R, \gamma_G, \theta_1, k_1, t, x, \mu, \theta_2$, and k_2) under these constraints.

- Parameters α_R and γ_G. According to the constraints in set K, the values of α_R and γ_G are fixed as $\alpha_R = \gamma_G = 0.5$. Note that the values of $\alpha_R p$ and $\gamma_G p$ should be both positive integers with $0 < \alpha_R, \gamma_G < 1$. As a result, the values of $0.5p$ should be a positive integer. To satisfy this constraint, we set $p = 2q$, where q is a positive integer.
- Parameters θ_1 and k_1. The constraints in set K fix the value of θ_1 as $\theta_1 = 1$, which can guarantee the existence of BIG-($\theta_1, r_1, k_1, \lambda_1$) if and only if $k_1 \geq 2$ by Lemma 1.7. Note that the value of k_1 should also be a factor of $p/2$. Hence, we set $k_1 = p/2 = q$ and $q \geq 2$ to satisfy these constraints.
- Parameter t. The constraints in set K fix parameter t as $t = 1$.
- Parameters x. Note that the value of x should be a factor of $p/2$ and guarantee the existence of TD-(n, μ, x) interconnection. By Lemma 2.1, TD-(n, μ, x) interconnection exists for any integer $n \geq 2$ if $x = 1$. Meanwhile, we can observe that $x = 1$ is a factor of $p/2$. Hence, we set $x = 1$ to satisfy all constraints on parameter x.
- Parameter μ. Note that the value of μ should be a common factor of $\frac{p^2}{4}$ and $\frac{p}{2}$. To satisfy this constraint, we set $\mu = p/2 = q$. In this way, parameter μ is a common factor of $\frac{p^2}{4}$ and $\frac{p}{2}$ for any value of p.
- Parameters θ_2 and k_2. According to the constraints, the value of θ_2 is fixed as $\theta_2 = 1$, which can guarantee the existence of BIG-($\theta_2, r_2, k_2, \lambda_2$) if and only if $k_2 \geq 2$ by Lemma 1.7. Besides, the parameter k_2 should also be a factor of p/t (also equal p as we set $t = 1$). Hence, we set $k_2 = p = 2q$ to satisfy the preceding constraints.

Based on the aforementioned discussions, given $p = 2q$ with q as a positive integer, the values of free parameters are set as $\alpha_R = \gamma_G = 0.5$, $\theta_1 = 1$, $k_1 = q$, $t = 1$, $x = 1$, $\mu = q$, $\theta_2 = 1$, and $k_2 = 2q$. Given the fixed values of free parameters, the values of bound parameters can be derived by equations (4.42)–(4.53). Particularly, the values of parameters β_G and l are not determined by the values of free parameters, but fixed as $\beta_G = 0$ and $l = 1$. The values of other bound parameters are derived as follows:

$$\alpha_G = 1 - \gamma_G = 0.5, \tag{4.165}$$

$$\beta_R = 1 - \alpha_R = 0.5, \tag{4.166}$$

$$r_1 = (1 - \gamma_G)p = \frac{p}{2}, \tag{4.167}$$

$$\lambda_1 = (1 - \gamma_G)p \cdot \frac{k_1 - 1}{\theta_1 k_1 - 1} = \frac{p}{2}, \tag{4.168}$$

$$c_1 = \frac{\alpha_R p}{k_1} - 1 = 0, \tag{4.169}$$

$$n = k_2 = p, \tag{4.170}$$

$$w = x = 1, \tag{4.171}$$

$$r_2 = \frac{\theta_1(1 - \gamma_G)p^2}{t\mu x^2} = \frac{p}{2}, \tag{4.172}$$

$$\lambda_2 = \frac{\theta_1(1 - \gamma_G)p^2}{t\mu x^2} \cdot \frac{k_2 - 1}{\theta_2 k_2 - 1} = \frac{p}{2}, \tag{4.173}$$

$$c_2 = \frac{p}{tk_2} - 1 = 0. \tag{4.174}$$

Under the preceding parameter settings, given port number of switches as $p = 2q$, where q is an integer greater than one, the complete design of devised topologies can be summarized as follows:

- *Intra-module design.* A BIG-$(1, q, q, q)$ is employed as the internal graph for each module, where element nodes are GNs and block nodes are RNs.
- *Connection design.* In each module, all RNs are selected as dock nodes. Each dock node forms an ISNs-$(1, 1, 1)$. Dock nodes in distinct modules are all connected under the bipartite mode. Specifically, $2q$ ISNs-$(1, 1, 1)$ are interconnected by a composite edge which is a TD-$(2q, q, 1)$ interconnection.
- *Inter-module design.* A BIG-$(1, q, 2q, q)$ is employed as the skeleton graph to interconnect different modules. For each RN cluster, it is incident to $2q$ super-edges. These super-edges are instantiated as a 1-parallelized TD-$(2q, q, 1)$ interconnection.
- *Finalization.* Each GN in the modules is directly connected to q CNs.

An example of this topology with $p = 4$ and $q = 2$ is provided in Figure 4.6. The preceding parameter settings indicate that the devised topology is the Fat-Tree topology (see Section 2.5.1). In other words, Fat-Tree topology can be reverse-engineered under design pattern $< G, \mathrm{BIG(block)}, B >$ using the targets of achieving throughput $T_1 = T_2 = 1$, achieving the lowest cost, and achieving the highest scalability.

4.6.2.3 Design Pattern $< G, \mathrm{BIG(element)}, B >$

Under design pattern $< G, \mathrm{BIG(element)}, B >$, since there are no existing topologies, we exploit the performance target of achieving the same latency and throughput as Fat-Tree topology to generate a topology of this design pattern. The generation process is conducted under the guidance of the three-stage scheme.

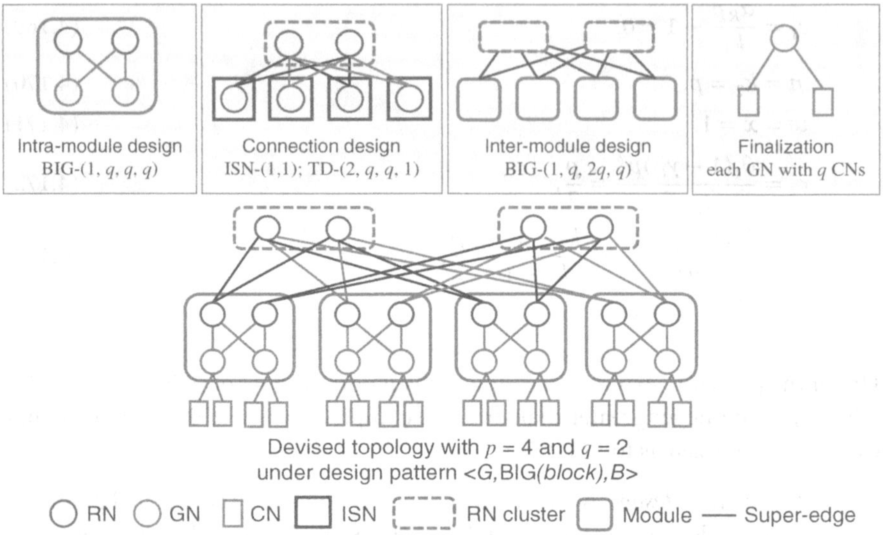

Figure 4.6 The complete design of a devised topology under design pattern $< G$, BIG(block), $B >$ with $p = 4$ and $q = 2$. This topology is actually a concrete Fat-Tree topology.

Stage 1: Initialization. Given design pattern $< G$, BIG(element), $B >$ as input, based on results in Section 4.3.2.8, we can study its free parameters and parameter constraints. Particularly, under this design pattern, the set W contains nine free parameters, including α_G, γ_G, θ_1, k_1, t, x, μ, θ_2, and k_2. Meanwhile, the set K contains the following constraints.

1. The values of the free parameters should guarantee the existence of BIG-$(\theta_1, r_1, k_1, \lambda_1)$.
2. The values of the free parameters should guarantee the existence of TD-(n, μ, x) interconnection.
3. The values of the free parameters should guarantee the existence of BIG-$(\theta_2, r_2, k_2, \lambda_2)$.
4. The values of $\alpha_G p$, $\gamma_G p$, $(1 - \alpha_G - \gamma_G)p$ are all positive integers such that $\alpha_G, \gamma_G > 0$ and $\alpha_G + \gamma_G < 1$.
5. The value of k_1 should be a factor of p.
6. The value of x should be a common factor of $\theta_1 p$ and $(1 - \alpha_G - \gamma_G)p$.
7. The value of t should be a common factor of p and $\dfrac{\theta_1 \cdot (1 - \alpha_G - \gamma_G)p^2}{x^2}$ such that $t \leq \dfrac{\theta_1 p}{x}$.

8. The value of μ should be a common factor of $\frac{\theta_1 \cdot (1-\alpha_G-\gamma_G)p^2}{tx^2}$ and $\frac{(1-\alpha_G-\gamma_G)p}{x}$.

9. The value of k_2 should be a factor of $\frac{p}{t}$.

Then we sort the five performance metrics by priority as I_1 (latency) $\geq I_2$ (throughput) $\geq I_3$ (scalability) $\geq I_4$ (fault-tolerance) $\geq I_5$ (cost). After initializing global counter i to one (i.e., $i = 1$), we enter the next stage.

Stage 2: Operation. This stage consists of five iterations controlled by the global counter i. In the following, we conduct each iteration.

- Iteration 1 ($i = 1$). For metric I_1 (i.e., latency), we set target G_1 as latency less than or equal to 6. According to analysis results on latency (see Lemmas 4.5 and 4.13), we learn that the value of latency is less than or equal to 6 if $\theta_2 = 1$. Hence, we set $\theta_2 = 1$ as the sufficient condition Q_1 on target G_1. Based on this condition, the equation set E_1 contains one equation $\theta_2 = 1$, while the inequation set U_1 is empty. By adding equation $\theta_2 = 1$ to set K, the set K now includes the following constraints on free parameters.
  1. The values of the free parameters should guarantee the existence of BIG-$(\theta_1, r_1, k_1, \lambda_1)$.
  2. The values of the free parameters should guarantee the existence of TD-(n, μ, x) interconnection.
  3. The values of the free parameters should guarantee the existence of BIG-$(\theta_2, r_2, k_2, \lambda_2)$.
  4. The values of $\alpha_G p, \gamma_G p, (1-\alpha_G-\gamma_G)p$ are all positive integers such that $\alpha_G, \gamma_G > 0$ and $\alpha_G + \gamma_G < 1$.
  5. The value of k_1 should be a factor of p.
  6. The value of x should be a common factor of $\theta_1 p$ and $(1-\alpha_G-\gamma_G)p$.
  7. The value of t should be a common factor of p and $\frac{\theta_1 \cdot (1-\alpha_G-\gamma_G)p^2}{x^2}$ such that $t \leq \frac{\theta_1 p}{x}$.
  8. The value of μ should be a common factor of $\frac{\theta_1 \cdot (1-\alpha_G-\gamma_G)p^2}{tx^2}$ and $\frac{(1-\alpha_G-\gamma_G)p}{x}$.
  9. The value of k_2 should be a factor of $\frac{p}{t}$.
  10. $\theta_2 = 1$.

Since there exist feasible values under constraints in set K and $i < 5$, we add one to i (hence $i = 2$) to enter the next iteration.[9]

- Iteration 2 ($i = 2$). For metric I_2 (i.e., throughput), we set target G_2 as $T_1 = T_2 = 1$, where T_1 and T_2 denote the values of intra-module throughput and inter-module throughput, respectively. By substituting $\beta_G = 1 - \alpha_G - \gamma_G$, and $\lambda_2/r_2 = (k_2 - 1)/(\theta_2 k_2 - 1)$ (see equations 4.65, 4.71, and 4.72) into analysis results on throughput in Table 4.11, it follows that

$$T_1 = \frac{\alpha_G}{\gamma_G}, \tag{4.175}$$

$$T_2 = \frac{\beta_G}{\gamma_G} \cdot \frac{\lambda_2}{r_2} = \frac{1 - \alpha_G - \gamma_G}{\gamma_G} \cdot \frac{k_2 - 1}{\theta_2 k_2 - 1}. \tag{4.176}$$

Note that we have $\theta_2 = 1$ based on the discussions on metric I_1. By substituting equation $\theta_2 = 1$ into equation (4.176), we have

$$T_2 = \frac{1 - \alpha_G - \gamma_G}{\gamma_G} \cdot \frac{k_2 - 1}{\theta_2 k_2 - 1} = \frac{1 - \alpha_G - \gamma_G}{\gamma_G}. \tag{4.177}$$

Hence, the sufficient (also necessary) condition Q_2 on $T_1 = T_2 = 1$ is $\gamma_G = \alpha_G = 1/3$. Based on this condition, the equation set E_2 contains one equation $\gamma_G = \alpha_G = 1/3$, while the inequation set U_2 is empty. By adding equation $\gamma_G = \alpha_G = 1/3$ to set K, the set K now includes the following constraints on free parameters.

1. The values of the free parameters should guarantee the existence of BIG-$(\theta_1, r_1, k_1, \lambda_1)$.
2. The values of the free parameters should guarantee the existence of TD-(n, μ, x) interconnection.
3. The values of the free parameters should guarantee the existence of BIG-$(\theta_2, r_2, k_2, \lambda_2)$.
4. The values of $\alpha_G p, \gamma_G p, (1 - \alpha_G - \gamma_G)p$ are all positive integers such that $\alpha_G, \gamma_G > 0$ and $\alpha_G + \gamma_G < 1$.
5. The value of k_1 should be a factor of p.
6. The value of x should be a common factor of $\theta_1 p$ and $(1 - \alpha_G - \gamma_G)p$.
7. The value of t should be a common factor of p and $\frac{\theta_1 \cdot (1 - \alpha_G - \gamma_G)p^2}{x^2}$ such that $t \leq \frac{\theta_1 p}{x}$.
8. The value of μ should be a common factor of $\frac{\theta_1 \cdot (1 - \alpha_G - \gamma_G)p^2}{tx^2}$ and $\frac{(1 - \alpha_G - \gamma_G)p}{x}$.

9 The process of checking whether there exist feasible values under constraints in set K is omitted, as the devised topology in the termination stage is an example with the feasible values.

9. The value of k_2 should be a factor of $\frac{p}{t}$.
10. $\theta_2 = 1$.
11. $\gamma_G = \alpha_G = 1/3$.

Since there exist feasible values under constraints in set K and $i < 5$, we add one to i (hence $i = 3$) to enter the next iteration.

- Iteration 3 ($i = 3$). For metric I_3 (i.e., scalability), we set target G_3 as achieving the highest scalability. According to analysis results on scalability in Table 4.5, it follows that

$$S = \gamma_G p \cdot \theta_1 p \cdot (\theta_2 p/t) = \frac{\theta_1\theta_2\gamma_G}{t} \cdot p^3. \tag{4.178}$$

Note that we have $\theta_2 = 1$ and $\gamma_G = 1/3$ based on the discussions on metrics I_1 and I_2. By substituting these two equations into equation (4.178), we have

$$S = \frac{\theta_1\theta_2\gamma_G}{t} \cdot p^3 = \frac{\theta_1}{3t} \cdot p^3. \tag{4.179}$$

By this equation, we can observe that a smaller value of t leads to a higher scalability. Under the constraints in set K, t should be a common factor of p and $\frac{\theta_1 \cdot p^2}{3x^2}$. This implies that t is a positive integer (hence $t \geq 1$). Therefore, the highest scalability is achieved when $t = 1$. Meanwhile, parameter θ_1 should guarantee the existence of BIG-$(\theta_1, r_1, k_1, \lambda_1)$. Hence, although a larger value of θ_1 leads to a higher scalability, we set $\theta_1 = 1$ to guarantee the universal existence of BIG-$(\theta_1, r_1, k_1, \lambda_1)$. Therefore, under the consideration of imposing the fewest restrictions on port number, the sufficient condition Q_3 on the highest scalability is $t = 1$ and $\theta_1 = 1$. Based on this condition, the equation set E_3 contains two equations $t = 1$ and $\theta_1 = 1$, while the inequation set U_3 is empty. By adding equations $t = 1$ and $\theta_1 = 1$ to set K, the set K now includes the following constraints on free parameters.

1. The values of the free parameters should guarantee the existence of BIG-$(\theta_1, r_1, k_1, \lambda_1)$.
2. The values of the free parameters should guarantee the existence of TD-(n, μ, x) interconnection.
3. The values of the free parameters should guarantee the existence of BIG-$(\theta_2, r_2, k_2, \lambda_2)$.
4. The values of $\alpha_G p, \gamma_G p, (1 - \alpha_G - \gamma_G)p$ are all positive integers such that $\alpha_G, \gamma_G > 0$ and $\alpha_G + \gamma_G < 1$.
5. The value of k_1 should be a factor of p.
6. The value of x should be a common factor of $\theta_1 p$ and $(1 - \alpha_G - \gamma_G)p$.
7. The value of t should be a common factor of p and $\frac{\theta_1 \cdot (1 - \alpha_G - \gamma_G)p^2}{x^2}$ such that $t \leq \frac{\theta_1 p}{x}$.

8. The value of μ should be a common factor of $\frac{\theta_1 \cdot (1-\alpha_G-\gamma_G)p^2}{tx^2}$ and $\frac{(1-\alpha_G-\gamma_G)p}{x}$.
9. The value of k_2 should be a factor of $\frac{p}{t}$.
10. $\theta_2 = 1$.
11. $\gamma_G = \alpha_G = 1/3$.
12. $t = 1$ and $\theta_1 = 1$.

Since there exist feasible values under constraints in set K and $i < 5$, we add one to i (hence $i = 4$) to enter the next iteration.

- Iteration 4 ($i = 4$). For metrics I_4 (i.e., fault-tolerance), we set target G_4 as no requirements. Hence, the set K remains unchanged. Since there exist feasible values under constraints in set K and $i < 5$, we add one to i (hence $i = 5$) to enter the next iteration.
- Iteration 5 ($i = 5$). For metrics I_5 (i.e., cost), we set target G_5 as no requirements. Hence, the set K remains unchanged. Since there exist feasible values under constraints in set K and $i = 5$, we jump out of these iterations and enter the next stage.

Stage 3: Termination. First, we need to search for a feasible value for each free parameter in W under constraints in K. The constraints in set K are listed as follows for reference.

1. The values of the free parameters should guarantee the existence of BIG-$(\theta_1, r_1, k_1, \lambda_1)$.
2. The values of the free parameters should guarantee the existence of TD-(n, μ, x) interconnection.
3. The values of the free parameters should guarantee the existence of BIG-$(\theta_2, r_2, k_2, \lambda_2)$.
4. The values of $\alpha_G p$, $\gamma_G p$, $(1-\alpha_G-\gamma_G)p$ are all positive integers such that $\alpha_G, \gamma_G > 0$ and $\alpha_G + \gamma_G < 1$.
5. The value of k_1 should be a factor of p.
6. The value of x should be a common factor of $\theta_1 p$ and $(1-\alpha_G-\gamma_G)p$.
7. The value of t should be a common factor of p and $\frac{\theta_1 \cdot (1-\alpha_G-\gamma_G)p^2}{x^2}$ such that $t \le \frac{\theta_1 p}{x}$.
8. The value of μ should be a common factor of $\frac{\theta_1 \cdot (1-\alpha_G-\gamma_G)p^2}{tx^2}$ and $\frac{(1-\alpha_G-\gamma_G)p}{x}$.
9. The value of k_2 should be a factor of $\frac{p}{t}$.
10. $\theta_2 = 1$.

11. $\gamma_G = \alpha_G = 1/3$.
12. $t = 1$ and $\theta_1 = 1$.

Next, we search for feasible values of free parameters (i.e., α_G, γ_G, θ_1, k_1, t, x, μ, θ_2, and k_2) under these constraints.

- Parameters α_G and γ_G. According to the constraints in set K, the values of α_G and γ_G are fixed as $\alpha_G = \gamma_G = 1/3$. Note that the values of $\alpha_G p$, $\gamma_G p$, $(1 - \alpha_G - \gamma_G)p$ are all positive integers such that $\alpha_G, \gamma_G > 0$ and $\alpha_G + \gamma_G < 1$. As a result, the values of $p/3$ should be a positive integer. To satisfy this constraint, we set $p = 3q$, where q is a positive integer.
- Parameter θ_1 and k_1. The constraints in set K fix the value of θ_1 as $\theta_1 = 1$, which can guarantee the existence of BIG-$(\theta_1, r_1, k_1, \lambda_1)$ if and only if $k_1 \geq 2$ by Lemma 1.7. Hence, we set $k_1 = p = 3q$ to satisfy these constraints.
- Parameter t. The constraints in set K fix parameter t as $t = 1$.
- Parameter x. Note that the value of x should not only be a common factor of p and $p/3$, but also guarantee the existence of TD-(n, μ, x) interconnection. By Lemma 2.1, TD-(n, μ, x) interconnection exists for any integer $n \geq 2$ if $x = 1$. Meanwhile, we can observe that $x = 1$ is a common factor of p and $p/3$. Hence, we set $x = 1$ to satisfy all constraints on parameter x.
- Parameter μ. Note that the value of μ should be a common factor of $\frac{p^2}{3}$ and $\frac{p}{3}$. To satisfy this constraint, we set $\mu = p/3 = q$. In this way, parameter μ is a common factor of $\frac{p^2}{3}$ and $\frac{p}{3}$ for any value of p.
- Parameters θ_2 and k_2. According to the constraints, the value of θ_2 is fixed as $\theta_2 = 1$, which can guarantee the existence of BIG-$(\theta_2, r_2, k_2, \lambda_2)$ if and only if $k_2 \geq 2$ by Lemma 1.7. Besides, the parameter k_2 should be a factor of p with $k_2 \geq 2$. Hence, we set $k_2 = p = 3q$ to satisfy the preceding constraints.

Based on the aforementioned discussions, given $p = 3q$ with q as a positive integer, the values of free parameters are set as $\alpha_G = \gamma_G = 1/3, \theta_1 = 1, k_1 = 3q, t = 1, x = 1, \mu = q, \theta_2 = 1$, and $k_2 = 3q$. Given the fixed values of free parameters, the values of bound parameters can be derived by equations (4.63)–(4.74). Particularly, the values of parameters α_R, β_R, and l are not determined by the values of free parameters, but fixed as $\alpha_R = 1, \beta_R = 0$, and $l = 2$. The values of other bound parameters are derived as follows:

$$\beta_G = 1 - \alpha_G - \gamma_G = 1/3, \tag{4.180}$$

$$r_1 = \alpha_G p = \frac{p}{3}, \tag{4.181}$$

$$\lambda_1 = \alpha_G p \cdot \frac{k_1 - 1}{\theta_1 k_1 - 1} = \frac{p}{3}, \tag{4.182}$$

$$c_1 = \frac{p}{k_1} - 1 = \frac{p}{3} - 1, \tag{4.183}$$

$$n = k_2 = p, \tag{4.184}$$

$$w = x = 1, \tag{4.185}$$

$$r_2 = \frac{\theta_1(1 - \alpha_G - \gamma_G)p^2}{t\mu x^2} = p, \tag{4.186}$$

$$\lambda_2 = \frac{\theta_1(1 - \alpha_G - \gamma_G)p^2}{t\mu x^2} \cdot \frac{k_2 - 1}{\theta_2 k_2 - 1} = p, \tag{4.187}$$

$$c_2 = \frac{p}{tk_2} - 1 = 0. \tag{4.188}$$

Under the preceding parameter settings, given the port number of switches as $p = 3q$, where q is a positive integer, the complete design of devised topologies can be summarized as follows:

- *Intra-module design*. A BIG-$(1, q, 3q, q)$ is employed as the internal graph for each module, where element nodes are GNs and block nodes are RNs.
- *Connection design*. In each module, all GNs are selected as dock nodes. Each dock node forms an ISNs-$(1, 1, 2)$. Dock nodes in distinct modules are all connected under the bipartite mode. Specifically, $3q$ ISNs-$(1, 1, 2)$ are interconnected by a composite edge which is a TD-$(3q, q, 1)$ interconnection.
- *Inter-module design*. A BIG-$(1, 3q, 3q, 3q)$ is employed as the skeleton graph to interconnect different modules. For each RN cluster, it is incident to $3q$ super-edges. These super-edges are instantiated as a 1-parallelized TD-$(3q, q, 1)$ interconnection.
- *Finalization*. Each GN in the modules is directly connected to q CNs.

An example of this topology with $p = 3$ and $q = 1$ is provided in Figure 4.7.

4.6.2.4 Design Pattern < G, TDG(block), B >

Under design pattern < G, TDG(block), B >, since there are no existing topologies, we exploit the similar target behind Fat-Tree (the existing topology of design pattern < G, BIG(block), B >) to generate a topology of this design pattern. Such a generation is conducted under the guidance of the three-stage scheme.

Stage 1: Initialization. Given design pattern < G, TDG(block), B > as input, based on results in Section 4.3.2.10, we can study its free parameters and parameter constraints. Particularly, under this design pattern, the set W contains nine free parameters, including α_R, γ_G, v_1, n_1, t, x, μ, θ_2, and k_2. Meanwhile, the set K contains the following constraints.

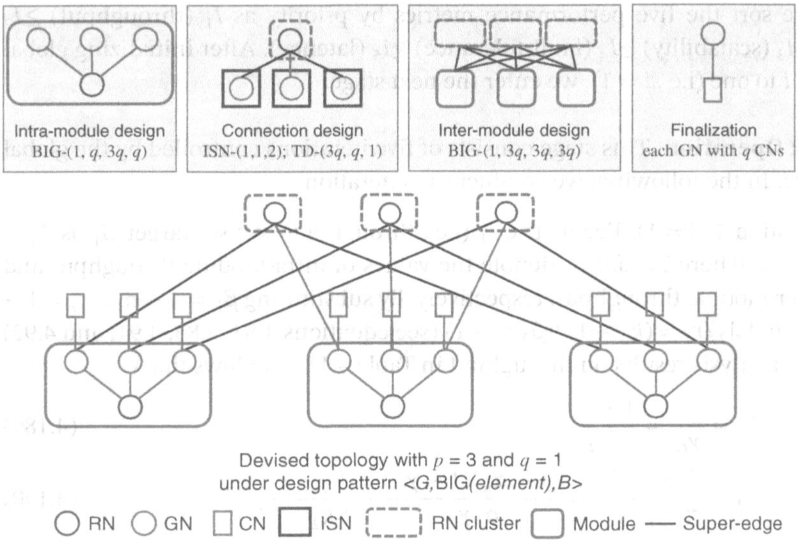

Figure 4.7 The complete design of a devised topology under design pattern $< G$, BIG(element), $B >$ with $p = 3$ and $q = 1$.

1. The values of the free parameters should guarantee the existence of TDG-(n_1, μ_1, v_1).
2. The values of the free parameters should guarantee that the TD-(n_1, μ_1, v_1) is resolvable.
3. The values of the free parameters should guarantee the existence of TD-(n, μ, x) interconnection.
4. The values of the free parameters should guarantee the existence of BIG-$(\theta_2, r_2, k_2, \lambda_2)$.
5. The values of $\alpha_R p$ and $\gamma_G p$ are both positive integers with $0 < \alpha_R, \gamma_G < 1$.
6. The value of n_1 should be a factor of $\alpha_R p$.
7. The value of v_1 should be a factor of $(1 - \gamma_G)p$ and x.
8. The value of x should be a common factor of $v_1 \cdot (1 - \gamma_G)p$ and $(1 - \alpha_R)p$.
9. The value of t should be a common factor of p and $\dfrac{v_1 \cdot (1 - \gamma_G)p \cdot (1 - \alpha_R)p}{x^2}$ with $t \leq \dfrac{v_1 \cdot (1 - \gamma_G)p}{x}$.
10. The value of μ should be a common factor of $\dfrac{v_1 \cdot (1 - \gamma_G)p \cdot (1 - \alpha_R)p}{tx^2}$ and $\dfrac{(1 - \alpha_R)p}{x}$.
11. The value of k_2 should be a factor of $\dfrac{p}{t}$.

Then we sort the five performance metrics by priority as I_1 (throughput) $\geq I_2$ (cost) $\geq I_3$ (scalability) $\geq I_4$ (fault-tolerance) $\geq I_5$ (latency). After initializing global counter i to one (i.e., $i = 1$), we enter the next stage.

Stage 2: Operation. This stage consists of five iterations controlled by the global counter i. In the following, we conduct each iteration.

- Iteration 1 ($i = 1$). For metric I_1 (i.e., throughput), we set target G_1 as $T_1 = T_2 = 1$, where T_1 and T_2 denote the values of intra-module throughput and inter-module throughput, respectively. By substituting $\beta_R = 1 - \alpha_R$, $\alpha_G = 1 - \gamma_G$, and $\lambda_2/r_2 = (k_2 - 1)/(\theta_2 k_2 - 1)$ (see equations 4.83, 4.85, 4.91, and 4.92) into analysis results on throughput in Table 4.11, it follows that

$$T_1 = \frac{\alpha_G}{\gamma_G} = \frac{1-\gamma_G}{\gamma_G}, \tag{4.189}$$

$$T_2 = \frac{\beta_R \alpha_G}{\alpha_R \gamma_G} \cdot \frac{\lambda_2}{r_2} = \frac{(1-\alpha_R)(1-\gamma_G)}{\alpha_R \gamma_G} \cdot \frac{k_2 - 1}{\theta_2 k_2 - 1}. \tag{4.190}$$

  Hence, the sufficient (also necessary) condition Q_1 on $T_1 = T_2 = 1$ is $\gamma_G = 0.5$ and $(1 - \alpha_R)/\alpha_R = (\theta_2 k_2 - 1)/(k_2 - 1)$. Based on this condition, the equation set E_1 contains two equations $\gamma_G = 0.5$ and $(1 - \alpha_R)/\alpha_R = (\theta_2 k_2 - 1)/(k_2 - 1)$, while the inequation set U_1 is empty. By adding equations $\gamma_G = 0.5$ and $(1 - \alpha_R)/\alpha_R = (\theta_2 k_2 - 1)/(k_2 - 1)$ to set K, the set K now includes the following constraints.

  1. The values of the free parameters should guarantee the existence of TDG-(n_1, μ_1, υ_1).
  2. The values of the free parameters should guarantee that the TD-(n_1, μ_1, υ_1) is resolvable.
  3. The values of the free parameters should guarantee the existence of TD-(n, μ, x) interconnection.
  4. The values of the free parameters should guarantee the existence of BIG-$(\theta_2, r_2, k_2, \lambda_2)$.
  5. The values of $\alpha_R p$ and $\gamma_G p$ are both positive integers with $0 < \alpha_R, \gamma_G < 1$.
  6. The value of n_1 should be a factor of $\alpha_R p$.
  7. The value of υ_1 should be a factor of $(1 - \gamma_G)p$ and x.
  8. The value of x should be a common factor of $\upsilon_1 \cdot (1 - \gamma_G)p$ and $(1 - \alpha_R)p$.
  9. The value of t should be a common factor of p and $\dfrac{\upsilon_1 \cdot (1-\gamma_G)p \cdot (1-\alpha_R)p}{x^2}$ with $t \leq \dfrac{\upsilon_1 \cdot (1-\gamma_G)p}{x}$.
  10. The value of μ should be a common factor of $\dfrac{\upsilon_1 \cdot (1-\gamma_G)p \cdot (1-\alpha_R)p}{tx^2}$ and $\dfrac{(1-\alpha_R)p}{x}$.

11. The value of k_2 should be a factor of $\frac{p}{t}$.
12. $\gamma_G = 0.5$ and $(1 - \alpha_R)/\alpha_R = (\theta_2 k_2 - 1)/(k_2 - 1)$ to set K.

Since there exist feasible values under constraints in set K and $i < 5$, we add one to i (hence $i = 2$) to enter the next iteration.[10]

- Iteration 2 ($i = 2$). For metric I_2 (i.e., cost), we set target G_2 as achieving the lowest cost. By substituting $\beta_R = 1 - \alpha_R$ and $\alpha_G = 1 - \gamma_G$ (see equations 4.83 and 4.85) into analysis results on cost in Table 4.7, it follows that

$$C = \frac{1}{\gamma_G p} + \frac{\alpha_G}{\gamma_G \alpha_R p} + \frac{\alpha_G \beta_R}{\gamma_G \alpha_R p} = \frac{1}{\gamma_G p} + \frac{1 - \gamma_G}{\gamma_G \alpha_R p} + \frac{(1 - \gamma_G)(1 - \alpha_R)}{\gamma_G \alpha_R p}. \tag{4.191}$$

Note that we have $\gamma_G = 0.5$ and $\alpha_R = (\theta_2 k_2 - 1)/(\theta_2 k_2 + k_2 - 2)$ based on the discussions on metric I_1. By substituting these two equations into equation (4.191), we have

$$C = \frac{1}{\gamma_G p} + \frac{1 - \gamma_G}{\gamma_G \alpha_R p} + \frac{(1 - \gamma_G)(1 - \alpha_R)}{\gamma_G \alpha_R p} = \frac{3}{p} + \frac{2}{p} \cdot \frac{\theta_2 k_2 - 1}{k_2 - 1}. \tag{4.192}$$

By this equation, we can observe that a smaller value of θ_2 leads to a lower cost. Note that the parameter θ_2 of BIG-$(\theta_2, r_2, k_2, \lambda_2)$ is greater than or equal to one. Therefore, the lowest cost is achieved when $\theta_2 = 1$, leading to

$$C = \frac{3}{p} + \frac{2}{p} \cdot \frac{\theta_2 k_2 - 1}{k_2 - 1} \geq \frac{5}{p}. \tag{4.193}$$

In other words, the sufficient (also necessary) condition Q_2 on the lowest cost is $\theta_2 = 1$. Based on such a condition, the equation set E_2 contains one equation $\theta_2 = 1$, while the inequation set U_2 is empty. By adding the equation $\theta_2 = 1$ to set K, the set K now includes the following constraints.

1. The values of the free parameters should guarantee the existence of TDG-(n_1, μ_1, υ_1).
2. The values of the free parameters should guarantee that the TD-(n_1, μ_1, υ_1) is resolvable.
3. The values of the free parameters should guarantee the existence of TD-(n, μ, x) interconnection.
4. The values of the free parameters should guarantee the existence of BIG-$(\theta_2, r_2, k_2, \lambda_2)$.
5. The values of $\alpha_R p$ and $\gamma_G p$ are both positive integers with $0 < \alpha_R, \gamma_G < 1$.
6. The value of n_1 should be a factor of $\alpha_R p$.

10 The process of checking whether there exist feasible values under constraints in set K is omitted, as the devised topology in the termination stage is an example with the feasible values.

7. The value of υ_1 should be a factor of $(1-\gamma_G)p$ and x.
8. The value of x should be a common factor of $\upsilon_1 \cdot (1-\gamma_G)p$ and $(1-\alpha_R)p$.
9. The value of t should be a common factor of p and $\frac{\upsilon_1 \cdot (1-\gamma_G)p \cdot (1-\alpha_R)p}{x^2}$ with $t \leq \frac{\upsilon_1 \cdot (1-\gamma_G)p}{x}$.
10. The value of μ should be a common factor of $\frac{\upsilon_1 \cdot (1-\gamma_G)p \cdot (1-\alpha_R)p}{tx^2}$ and $\frac{(1-\alpha_R)p}{x}$.
11. The value of k_2 should be a factor of $\frac{p}{t}$.
12. $\gamma_G = 0.5$ and $(1-\alpha_R)/\alpha_R = (\theta_2 k_2 - 1)/(k_2 - 1)$ to set K.
13. $\theta_2 = 1$.

Since there exist feasible values under constraints in set K and $i < 5$, we add one to i (hence $i = 3$) to enter the next iteration.

- Iteration 3 ($i = 3$). For metric I_3 (i.e., scalability), we set target G_3 as achieving the highest scalability. According to analysis results on scalability (see Table 4.5), it follows that

$$S = \gamma_G p \cdot \upsilon_1 \alpha_R p \cdot (\theta_2 p/t) = \frac{\theta_2 \upsilon_1 \gamma_G \alpha_R}{t} \cdot p^3. \tag{4.194}$$

Note that we have $\gamma_G = 0.5, \theta_2 = 1$, and $\alpha_R = (\theta_2 k_2 - 1)/(\theta_2 k_2 + k_2 - 2) = 0.5$ based on the discussions on metrics I_1 and I_2. By substituting these three equations into equation (4.194), we have

$$S = \frac{\theta_2 \upsilon_1 \gamma_G \alpha_R}{t} \cdot p^3 = \frac{\upsilon_1}{4t} \cdot p^3. \tag{4.195}$$

By this equation, we can observe that a larger value of υ_1 leads to a higher scalability, while a smaller value of t leads to a higher scalability. Under the constraints in set K, υ_1 should be a common factor of $p/2$ and x. Hence, we have $\upsilon_1 \leq p/2$, where the equality holds only if $x \geq p/2$. Meanwhile, t should be a common factor of p and $\frac{\upsilon_1 p^2}{4x^2}$ with $t \leq \frac{\upsilon_1 p}{2x}$. This implies t is a positive integer (hence $t \geq 1$). Therefore, the highest scalability is achieved when $t = 1$, $\upsilon_1 = p/2$, and $x \geq p/2$, leading to

$$S = \frac{\upsilon_1}{4t} \cdot p^3 \geq \frac{1}{8} \cdot p^4. \tag{4.196}$$

In other words, the sufficient condition Q_3 on the highest scalability is $t = 1$, $\upsilon_1 = p/2$, and $x \geq p/2$. Based on this condition, the equation set E_3 contains two equations $t = 1$ and $\upsilon_1 = p/2$, while the inequation set U_3 contains one inequation $x \geq p/2$. By setting $K = K \cup E_3 \cup U_3$, the set K now includes the following constraints.

1. The values of the free parameters should guarantee the existence of TDG-(n_1, μ_1, υ_1).
2. The values of the free parameters should guarantee that the TD-(n_1, μ_1, υ_1) is resolvable.
3. The values of the free parameters should guarantee the existence of TD-(n, μ, x) interconnection.
4. The values of the free parameters should guarantee the existence of BIG-$(\theta_2, r_2, k_2, \lambda_2)$.
5. The values of $\alpha_R p$ and $\gamma_G p$ are both positive integers with $0 < \alpha_R, \gamma_G < 1$.
6. The value of n_1 should be a factor of $\alpha_R p$.
7. The value of υ_1 should be a factor of $(1 - \gamma_G)p$ and x.
8. The value of x should be a common factor of $\upsilon_1 \cdot (1 - \gamma_G)p$ and $(1 - \alpha_R)p$.
9. The value of t should be a common factor of p and $\frac{\upsilon_1 \cdot (1-\gamma_G)p \cdot (1-\alpha_R)p}{x^2}$ with $t \le \frac{\upsilon_1 \cdot (1-\gamma_G)p}{x}$.
10. The value of μ should be a common factor of $\frac{\upsilon_1 \cdot (1-\gamma_G)p \cdot (1-\alpha_R)p}{tx^2}$ and $\frac{(1-\alpha_R)p}{x}$.
11. The value of k_2 should be a factor of $\frac{p}{t}$.
12. $\gamma_G = 0.5$ and $(1 - \alpha_R)/\alpha_R = (\theta_2 k_2 - 1)/(k_2 - 1)$ to set K.
13. $\theta_2 = 1$.
14. $t = 1, \upsilon_1 = p/2$, and $x \ge p/2$.

Since there exist feasible values under constraints in set K and $i < 5$, we add one to i (hence $i = 4$) to enter the next iteration.

- Iteration 4 ($i = 4$). For metrics I_4 (i.e., fault-tolerance), we set target G_4 as no requirements. Hence, the set K remains unchanged. Since there exist feasible values under constraints in set K and $i < 5$, we add one to i (hence $i = 5$) to enter the next iteration.
- Iteration 5 ($i = 5$). For metrics I_5 (i.e., latency), we set target G_5 as no requirements. Hence, the set K remains unchanged. Since there exist feasible values under constraints in set K and $i = 5$, we jump out of these iterations and enter the next stage.

Stage 3: Termination. First, we need to search for a feasible value for each free parameter in W under constraints in K. The constraints in set K are listed as follows for reference.

1. The values of the free parameters should guarantee the existence of TDG-(n_1, μ_1, υ_1).
2. The values of the free parameters should guarantee that the TD-(n_1, μ_1, υ_1) is resolvable.

3. The values of the free parameters should guarantee the existence of TD-(n, μ, x) interconnection.
4. The values of the free parameters should guarantee the existence of BIG-$(\theta_2, r_2, k_2, \lambda_2)$.
5. The values of $\alpha_R p$ and $\gamma_G p$ are both positive integers with $0 < \alpha_R, \gamma_G < 1$.
6. The value of n_1 should be a factor of $\alpha_R p$.
7. The value of υ_1 should be a factor of $(1 - \gamma_G)p$ and x.
8. The value of x should be a common factor of $\upsilon_1 \cdot (1 - \gamma_G)p$ and $(1 - \alpha_R)p$.
9. The value of t should be a common factor of p and $\dfrac{\upsilon_1 \cdot (1 - \gamma_G)p \cdot (1 - \alpha_R)p}{x^2}$ with $t \leq \dfrac{\upsilon_1 \cdot (1 - \gamma_G)p}{x}$.
10. The value of μ should be a common factor of $\dfrac{\upsilon_1 \cdot (1 - \gamma_G)p \cdot (1 - \alpha_R)p}{tx^2}$ and $\dfrac{(1 - \alpha_R)p}{x}$.
11. The value of k_2 should be a factor of $\dfrac{p}{t}$.
12. $\gamma_G = 0.5$ and $(1 - \alpha_R)/\alpha_R = (\theta_2 k_2 - 1)/(k_2 - 1)$ to set K.
13. $\theta_2 = 1$.
14. $t = 1$, $\upsilon_1 = p/2$, and $x \geq p/2$.

Next, we search for feasible values of free parameters (i.e., $\alpha_R, \gamma_G, \upsilon_1, n_1, t, x, \mu, \theta_2$, and k_2) under these constraints.

- Parameters α_R and γ_G. The constraints in set K fix the values of α_R and γ_G as $\alpha_R = \gamma_G = 0.5$. Note that the values of $\alpha_R p$ and $\gamma_G p$ should be both positive integers with $0 < \alpha_R, \gamma_G < 1$. As a result, the value of $p/2$ should be a positive integer. This indicates that parameter p should be an even number.
- Parameters υ_1 and x. The constraints in set K not only fix the value of υ_1 as $\upsilon_1 = p/2$, but also require that υ_1 should be a factor of x. This indicates that the value of x should be a multiple of $p/2$. Besides, the value of x should also be a common factor of $\upsilon_1 \cdot (1 - \gamma_G)p$ (also equals $p^2/4$ as we set $\upsilon_1 = p/2$ and $\gamma_G = 0.5$) and $(1 - \alpha_R)p$ (also equals $p/2$ as we set $\alpha_R = 0.5$). In summary, the value of x should be a multiple of $p/2$, as well as a factor of $p/2$, which implies that the only feasible value of x is $x = p/2$. Hence, we set $x = p/2$.
- Parameter n_1. Note that the value of n_1 should guarantee the existence of TDG-$(n_1, \mu_1, p/2)$. According to Lemma 1.11, TDG-$(n_1, \mu_1, p/2)$ exists for any value of p if $n_1 = 2$ or $n_1 = 3$. Meanwhile, the value of n_1 should also be a factor of $p/2$. By assuming p as a multiple of 4 (i.e., $p = 4q$ with q as a positive integer), we set $n_1 = 2$, such that n_1 must be a factor of $p/2$.[11] By

11 The value of n_1 can be adjusted under different assumptions on the value of p. Suppose p is a multiple of 6, we should set $n_1 = 3$, such that n_1 is a factor of $p/2$. If $p/2$ is a prime power, we can set $n_1 = p/2$, such that n_1 is a factor of $p/2$ and TDG-$(n_1, \mu_1, p/2)$ exists by Lemma 1.11.

Theorem 1.22, TD-$(2, \mu_1, p/2)$ is resolvable. Therefore, $n_1 = 2$ is a feasible value if $p = 4q$, where q is a positive integer.
- Parameter t. The constraints in set K fix the value of t as $t = 1$.
- Parameter μ. Note that the value of μ should be a common factor of $p/2$ and 1. This implies that the only feasible value of μ is $\mu = 1$. Hence, we set $\mu = 1$.
- Parameters θ_2 and k_2. Based on the constraints in set K, the value of parameter θ_2 is fixed as $\theta_2 = 1$. By Lemma 1.7, BIG-$(\theta_2, r_2, k_2, \lambda_2)$ with $\theta_2 = 1$ exists if and only if $k_2 \geq 2$. Meanwhile, since $n = k_2$, the value of k_2 should guarantee the existence of TD-$(k_2, 1, p/2)$ interconnection. According to Lemma 2.1, TD-$(k_2, 1, p/2)$ interconnection exists for any value of p if $k_2 = 2$ or $k_2 = 3$. Besides, the value of k_2 should be a factor of p/t (also equals p as we set $t = 1$). Recall that we assume p to be a multiple of 4. To satisfy the aforementioned constraints, we set $k_2 = 2$.

Based on the aforementioned discussions, given $p = 4q$ with q as a positive integer, the values of free parameters are set as $\alpha_R = \gamma_G = 0.5$, $v_1 = 2q$, $n_1 = 2$, $t = 1$, $x = 2q$, $\mu = \theta_2 = 1$, and $k_2 = 2$. Given the fixed values of free parameters, the values of bound parameters can be derived by equations (4.83)–(4.93). Particularly, the values of parameters β_G and l are not determined by the values of free parameters, but fixed as $\beta_G = 0$ and $l = 1$. The values of other bound parameters are derived as follows:

$$\alpha_G = 1 - \gamma_G = 0.5, \tag{4.197}$$

$$\beta_R = 1 - \alpha_R = 0.5, \tag{4.198}$$

$$\mu_1 = \frac{(1-\gamma_G)p_G}{v_1} = 1, \tag{4.199}$$

$$c_1' = \frac{\alpha_R p_R}{n_1} - 1 = \frac{p}{4} - 1 = q - 1, \tag{4.200}$$

$$n = k_2 = 2, \tag{4.201}$$

$$w = \frac{x}{v_1} = 1, \tag{4.202}$$

$$r_2 = \frac{v_1 \cdot (1-\gamma_G)p_G \cdot (1-\alpha_R)p_R}{t\mu x^2} = \frac{p}{2} = 2q, \tag{4.203}$$

$$\lambda_2 = \frac{v_1 \cdot (1-\gamma_G)p_G \cdot (1-\alpha_R)p_R}{t\mu x^2} \cdot \frac{k_2 - 1}{\theta_2 k_2 - 1} = \frac{p}{2} = 2q, \tag{4.204}$$

$$c_2 = \frac{p_R}{tk_2} - 1 = \frac{p}{2} - 1 = 2q - 1. \tag{4.205}$$

Under the preceding parameter settings, given the port number of switches as $p = 4q$, where q is a positive integer, the complete design of devised topologies can be summarized as follows:

- *Intra-module design.* A TDG-(2, 1, $2q$) with $q - 1$ copies is employed as the internal graph for each module, where element nodes are GNs and block nodes are RNs.
- *Connection design.* In each module, all RNs are selected as dock nodes. Such dock nodes are grouped as ISNs-($2q$, 1, 1). Dock nodes in distinct modules are all connected under the bipartite mode. Specifically, two ISNs-($2q$, 1, 1) are interconnected by a composite edge which is a TD-(2, 1, $2q$) interconnection.
- *Inter-module design.* A BIG-(1, $2q$, 2, $2q$) with $2q - 1$ copies is employed as the skeleton graph to interconnect different modules. For each RN cluster, it is incident to $4q$ super-edges. These super-edges are partitioned into $2q$ groups, each of which is incident to a copy of modules (i.e., two modules). Each group of two super-edges are instantiated as a 1-parallelized TD-(2, 1, $2q$) interconnection.
- *Finalization.* Each GN in the modules is directly connected to $2q$ CNs.

An example of the devised topology with $p = 4$ and $q = 1$ is provided in Figure 4.8.

4.6.2.5 Design Pattern $< G$, TDG(element), $B >$

Under design pattern $< G$, TDG(element), $B >$, since there are no existing topologies, we exploit the similar target behind Fat-Tree (the existing topology of design

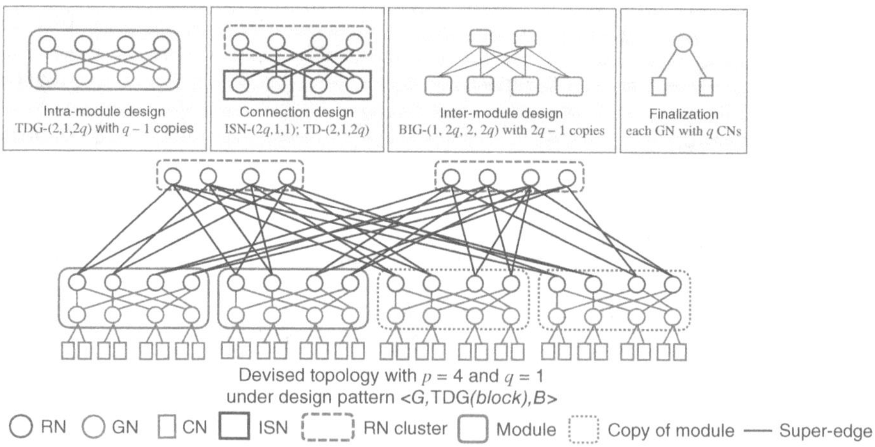

Figure 4.8 The complete design of a devised topology under design pattern $< G$, TDG(block), $B >$ with $p = 4$ and $q = 1$.

pattern < G, BIG(block), B >) to generate a topology of this design pattern. Such a generation is conducted under the guidance of the three-stage scheme.

Stage 1: Initialization. Given design pattern < G, TDG(element), B > as input, based on results in Section 4.3.2.12, we can study its free parameters and parameter constraints. Particularly, under this design pattern, the set W of free parameters contains nine free parameters, including α_G, γ_G, n_1, v_1, t, x, μ, θ_2, and k_2. Meanwhile, the set K contains the following constraints.

1. The values of the free parameters should guarantee the existence of TDG-(n_1, μ_1, v_1).
2. The values of the free parameters should guarantee the existence of TD-(n, μ, x) interconnection.
3. The values of the free parameters should guarantee the existence of BIG-$(\theta_2, r_2, k_2, \lambda_2)$.
4. The values of $\alpha_G p$, $\gamma_G p$, $(1-\alpha_G-\gamma_G)p$ are all positive integers such that $\alpha_G, \gamma_G > 0$ and $\alpha_G + \gamma_G < 1$.
5. The value of n_1 should be a common factor of p and x.
6. The value of v_1 should be a factor of $\alpha_G p$.
7. The value of x should be a common factor of $v_1 p$ and $(1-\alpha_G-\gamma_G)p$.
8. The value of t should be a common factor of p and $\dfrac{v_1 \cdot (1-\alpha_G-\gamma_G)p^2}{x^2}$ such that $t \le \dfrac{v_1 p}{x}$.
9. The value of μ should be a common factor of $\dfrac{v_1 \cdot (1-\alpha_G-\gamma_G)p^2}{tx^2}$ and $\dfrac{(1-\alpha_G-\gamma_G)p}{x}$.
10. The value of k_2 should be a factor of $\dfrac{p}{t}$.

Then we sort the five performance metrics by priority as I_1 (throughput) $\geq I_2$ (cost) $\geq I_3$ (scalability) $\geq I_4$ (fault-tolerance) $\geq I_5$ (latency). After initializing global counter i to one (i.e., $i = 1$), we enter the next stage.

Stage 2: Operation. This stage consists of five iterations controlled by the global counter i. In the following, we conduct each iteration.

- Iteration 1 ($i = 1$). For metric I_1 (i.e., throughput), we set target G_1 as $T_1 = T_2 = 1$, where T_1 and T_2 denote the values of intra-module throughput and inter-module throughput, respectively. By substituting $\beta_G = 1 - \alpha_G - \gamma_G$, and $\lambda_2/r_2 = (k_2-1)/(\theta_2 k_2 - 1)$ (see equations 4.104, 4.110, and 4.111) into analysis results on throughput in Table 4.11, it follows that

$$T_1 = \frac{\alpha_G}{\gamma_G}, \tag{4.206}$$

$$T_2 = \frac{\beta_G}{\gamma_G} \cdot \frac{\lambda_2}{r_2} = \frac{1-\alpha_G-\gamma_G}{\gamma_G} \cdot \frac{k_2-1}{\theta_2 k_2 - 1}. \tag{4.207}$$

Hence, the sufficient (also necessary) condition Q_1 on $T_1 = T_2 = 1$ is $\gamma_G = \alpha_G$ and $(1-\alpha_G-\gamma_G)/\gamma_G = (\theta_2 k_2 - 1)/(k_2 - 1)$. Based on this condition, the equation set E_1 contains two equations $\gamma_G = \alpha_G$ and $(1-\alpha_G-\gamma_G)/\gamma_G = (\theta_2 k_2 - 1)/(k_2 - 1)$, while the inequation set U_1 is empty. By adding these two equations to set K, the set K now includes the following constraints.

1. The values of the free parameters should guarantee the existence of TDG-(n_1, μ_1, υ_1).
2. The values of the free parameters should guarantee the existence of TD-(n, μ, x) interconnection.
3. The values of the free parameters should guarantee the existence of BIG-$(\theta_2, r_2, k_2, \lambda_2)$.
4. The values of $\alpha_G p, \gamma_G p, (1-\alpha_G-\gamma_G)p$ are all positive integers such that $\alpha_G, \gamma_G > 0$ and $\alpha_G + \gamma_G < 1$.
5. The value of n_1 should be a common factor of p and x.
6. The value of υ_1 should be a factor of $\alpha_G p$.
7. The value of x should be a common factor of $\upsilon_1 p$ and $(1-\alpha_G-\gamma_G)p$.
8. The value of t should be a common factor of p and $\frac{\upsilon_1 \cdot (1-\alpha_G-\gamma_G)p^2}{x^2}$ such that $t \leq \frac{\upsilon_1 p}{x}$.
9. The value of μ should be a common factor of $\frac{\upsilon_1 \cdot (1-\alpha_G-\gamma_G)p^2}{tx^2}$ and $\frac{(1-\alpha_G-\gamma_G)p}{x}$.
10. The value of k_2 should be a factor of $\frac{p}{t}$.
11. $\gamma_G = \alpha_G$ and $(1-\alpha_G-\gamma_G)/\gamma_G = (\theta_2 k_2 - 1)/(k_2 - 1)$.

Since there exist feasible values under constraints in set K and $i < 5$, we add one to i (hence $i = 2$) to enter the next iteration.[12]

- Iteration 2 ($i = 2$). For metric I_2 (i.e., cost), we set target G_2 as achieving the lowest cost. By substituting $\beta_G = 1-\alpha_G-\gamma_G$ (see equation (4.104)) into analysis results on cost in Table 4.7, it follows that

$$C = \frac{1}{\gamma_G p} + \frac{\alpha_G}{\gamma_G p} + \frac{\beta_G}{\gamma_G p} = \frac{2-\gamma_G}{\gamma_G p}. \tag{4.208}$$

12 The process of checking whether there exist feasible values under constraints in set K is omitted, as the devised topology in the termination stage is an example with the feasible values.

Note that we have $\gamma_G = \alpha_G$ and $\alpha_G = (k_2 - 1)/(\theta_2 k_2 + 2k_2 - 3)$ based on the discussions on metric I_1. By substituting these two equations into equation (4.208), we have

$$C = \frac{2 - \gamma_G}{\gamma_G p} = \frac{2(\theta_2 k_2 + 2k_2 - 3)}{(k_2 - 1)p} - \frac{1}{p}. \tag{4.209}$$

By this equation, we can observe that a smaller value of θ_2 leads to a lower cost. Note that the parameter θ_2 of BIG-$(\theta_2, r_2, k_2, \lambda_2)$ is greater than or equal to one. Therefore, the lowest cost is achieved when $\theta_2 = 1$, leading to

$$C = \frac{2(\theta_2 k_2 + 2k_2 - 3)}{(k_2 - 1)p} - \frac{1}{p} \geq \frac{5}{p}. \tag{4.210}$$

In other words, the sufficient (also necessary) condition Q_2 on the lowest cost is $\theta_2 = 1$. Based on this condition, the equation set E_2 contains one equation $\theta_2 = 1$, while the inequation set U_2 is empty. By adding equation $\theta_2 = 1$ to set K, the set K now includes the following constraints.

1. The values of the free parameters should guarantee the existence of TDG-(n_1, μ_1, υ_1).
2. The values of the free parameters should guarantee the existence of TD-(n, μ, x) interconnection.
3. The values of the free parameters should guarantee the existence of BIG-$(\theta_2, r_2, k_2, \lambda_2)$.
4. The values of $\alpha_G p, \gamma_G p, (1 - \alpha_G - \gamma_G)p$ are all positive integers such that $\alpha_G, \gamma_G > 0$ and $\alpha_G + \gamma_G < 1$.
5. The value of n_1 should be a common factor of p and x.
6. The value of υ_1 should be a factor of $\alpha_G p$.
7. The value of x should be a common factor of $\upsilon_1 p$ and $(1 - \alpha_G - \gamma_G)p$.
8. The value of t should be a common factor of p and $\frac{\upsilon_1 \cdot (1 - \alpha_G - \gamma_G)p^2}{x^2}$ such that $t \leq \frac{\upsilon_1 p}{x}$.
9. The value of μ should be a common factor of $\frac{\upsilon_1 \cdot (1 - \alpha_G - \gamma_G)p^2}{tx^2}$ and $\frac{(1 - \alpha_G - \gamma_G)p}{x}$.
10. The value of k_2 should be a factor of $\frac{p}{t}$.
11. $\gamma_G = \alpha_G$ and $(1 - \alpha_G - \gamma_G)/\gamma_G = (\theta_2 k_2 - 1)/(k_2 - 1)$.
12. $\theta_2 = 1$.

Since there exist feasible values under constraints in set K and $i < 5$, we add one to i (hence $i = 3$) to enter the next iteration.

- Iteration 3 ($i = 3$). For metric I_3 (i.e., scalability), we set target G_3 as achieving the highest scalability. According to analysis results on scalability (see Table 4.5), it follows that

$$S = \gamma_G p \cdot \upsilon_1 p \cdot (\theta_2 p/t) = \frac{\theta_2 \upsilon_1 \gamma_G}{t} \cdot p^3. \tag{4.211}$$

Note that we have $\theta_2 = 1$, $\alpha_G = (\theta_2 k_2 - 1)/(\theta_2 k_2 + k_2 - 2) = 1/3$, and $\gamma_G = \alpha_G = 1/3$ based on the discussions on metrics I_1 and I_2. By substituting these three equations into equation (4.211), the value of S is equal to

$$S = \frac{\theta_2 \upsilon_1 \gamma_G}{t} \cdot p^3 = \frac{\upsilon_1}{3t} \cdot p^3. \tag{4.212}$$

By this equation, we can observe that a larger value of υ_1 leads to a higher scalability, while a smaller value of t leads to a higher scalability. Under the constraints in set K, υ_1 should be a factor of $p/3$. Hence, the largest feasible value of υ_1 is $p/3$. Meanwhile, t should be a common factor of p and $\frac{\upsilon_1 \cdot p^2}{3x^2}$. This implies that t is a positive integer (hence $t \geq 1$). Therefore, the highest scalability is achieved when $t = 1$ and $\upsilon_1 = p/3$, leading to

$$S = \frac{\upsilon_1}{3t} \cdot p^3 \geq \frac{1}{9} \cdot p^4. \tag{4.213}$$

In other words, the sufficient condition Q_3 on the highest scalability is $t = 1$ and $\upsilon_1 = p/3$. Based on this condition, the equation set E_3 contains two equations $t = 1$ and $\upsilon_1 = p/3$, while the inequation set U_3 is empty. By adding these two equations to set K, the set K now includes the following constraints.

1. The values of the free parameters should guarantee the existence of TDG-(n_1, μ_1, υ_1).
2. The values of the free parameters should guarantee the existence of TD-(n, μ, x) interconnection.
3. The values of the free parameters should guarantee the existence of BIG-$(\theta_2, r_2, k_2, \lambda_2)$.
4. The values of $\alpha_G p, \gamma_G p, (1 - \alpha_G - \gamma_G)p$ are all positive integers such that $\alpha_G, \gamma_G > 0$ and $\alpha_G + \gamma_G < 1$.
5. The value of n_1 should be a common factor of p and x.
6. The value of υ_1 should be a factor of $\alpha_G p$.
7. The value of x should be a common factor of $\upsilon_1 p$ and $(1 - \alpha_G - \gamma_G)p$.
8. The value of t should be a common factor of p and $\frac{\upsilon_1 \cdot (1 - \alpha_G - \gamma_G)p^2}{x^2}$ such that $t \leq \frac{\upsilon_1 p}{x}$.
9. The value of μ should be a common factor of $\frac{\upsilon_1 \cdot (1 - \alpha_G - \gamma_G)p^2}{tx^2}$ and $\frac{(1 - \alpha_G - \gamma_G)p}{x}$.
10. The value of k_2 should be a factor of $\frac{p}{t}$.
11. $\gamma_G = \alpha_G$ and $(1 - \alpha_G - \gamma_G)/\gamma_G = (\theta_2 k_2 - 1)/(k_2 - 1)$.

12. $\theta_2 = 1$.

13. $t = 1$ and $\upsilon_1 = p/3$.

Since there exist feasible values under constraints in set K and $i < 5$, we add one to i (hence $i = 4$) to enter the next iteration.

- Iteration 4 ($i = 4$). For metrics I_4 (i.e., fault-tolerance), we set target G_4 as no requirements. Hence, the set K remains unchanged. Since there exist feasible values under constraints in set K and $i < 5$, we add one to i (hence $i = 5$) to enter the next iteration.
- Iteration 5 ($i = 5$). For metrics I_5 (i.e., latency), we set target G_5 as no requirements. Hence, the set K remains unchanged. Since there exist feasible values under constraints in set K and $i = 5$, we jump out of these iterations and enter the next stage.

Stage 3: Termination. First, we need to search for a feasible value for each free parameter in W under constraints in K. The constraints in set K are listed as follows for reference.

1. The values of the free parameters should guarantee the existence of TDG-(n_1, μ_1, υ_1).
2. The values of the free parameters should guarantee the existence of TD-(n, μ, x) interconnection.
3. The values of the free parameters should guarantee the existence of BIG-$(\theta_2, r_2, k_2, \lambda_2)$.
4. The values of $\alpha_G p$, $\gamma_G p$, $(1 - \alpha_G - \gamma_G)p$ are all positive integers such that $\alpha_G, \gamma_G > 0$ and $\alpha_G + \gamma_G < 1$.
5. The value of n_1 should be a common factor of p and x.
6. The value of υ_1 should be a factor of $\alpha_G p$.
7. The value of x should be a common factor of $\upsilon_1 p$ and $(1 - \alpha_G - \gamma_G)p$.
8. The value of t should be a common factor of p and $\dfrac{\upsilon_1 \cdot (1 - \alpha_G - \gamma_G)p^2}{x^2}$ such that $t \leq \dfrac{\upsilon_1 p}{x}$.
9. The value of μ should be a common factor of $\dfrac{\upsilon_1 \cdot (1 - \alpha_G - \gamma_G)p^2}{tx^2}$ and $\dfrac{(1 - \alpha_G - \gamma_G)p}{x}$.
10. The value of k_2 should be a factor of $\dfrac{p}{t}$.
11. $\gamma_G = \alpha_G$ and $(1 - \alpha_G - \gamma_G)/\gamma_G = (\theta_2 k_2 - 1)/(k_2 - 1)$.
12. $\theta_2 = 1$.
13. $t = 1$ and $\upsilon_1 = p/3$.

Next, we search for feasible values of free parameters (i.e., α_G, γ_G, n_1, υ_1, t, x, μ, θ_2, and k_2) under these constraints.

- Parameters α_G and γ_G. The constraints in set K fix the values of α_G and γ_G as $\alpha_G = \gamma_G = 1/3$. Note that the values of $\alpha_G p, \gamma_G p, (1 - \alpha_G - \gamma_G)p$ should be positive integers such that $\alpha_G, \gamma_G > 0$ and $\alpha_G + \gamma_G < 1$. As a result, the value of $p/3$ should be a positive integer. This indicates that parameter p should be a multiple of 3.
- Parameters υ_1 and t. The constraints in set K fix the values of υ_1 and t as $\upsilon_1 = p/3$ and $t = 1$.
- Parameter x. Note that the value of x should be a common factor of $p^2/3$ and $p/3$. To satisfy this constraint, we set $x = p/3$.
- Parameter n_1. Note that the value of n_1 should guarantee the existence of TDG-$(n_1, \mu_1, p/3)$. According to Lemma 1.11, TDG-$(n_1, \mu_1, p/2)$ exists for any value of p if $n_1 = 2$ or $n_1 = 3$. Meanwhile, the value of n_1 should also be a factor of $p/3$. By assuming p as a multiple of 6 (i.e., $p = 6q$ with q as a positive integer), we set $n_1 = 2$, such that n_1 must be a factor of $p/3$.[13]
- Parameter μ. Note that the value of μ should be a common factor of p and 1. This implies that the only feasible value of μ is $\mu = 1$. Hence, we set $\mu = 1$.
- Parameters θ_2 and k_2. Based on the constraints in set K, the value of parameter θ_2 is fixed as $\theta_2 = 1$. By Lemma 1.7, BIG-$(\theta_2, r_2, k_2, \lambda_2)$ with $\theta_2 = 1$ exists if and only if $k_2 \geq 2$. Meanwhile, since $n = k_2$, the value of k_2 should guarantee the existence of TD-$(k_2, 1, p/2)$ interconnection. According to Lemma 2.1, TD-$(k_2, 1, p/2)$ interconnection exists for any value of p if $k_2 = 2$ or $k_2 = 3$. Besides, the value of k_2 should be a factor of p/t (also equals p as we set $t = 1$). Recall that we assume p to be a multiple of 6. To satisfy the preceding constraints, we set $k_2 = 2$.

Based on the aforementioned discussions, given $p = 6q$ with q as a positive integer, the values of free parameters are set as $\alpha_G = \gamma_G = 0.5$, $\upsilon_1 = 2q$, $n_1 = 2$, $t = 1$, $x = 2q$, $\mu = \theta_2 = 1$, and $k_2 = 2$. Given the fixed values of free parameters, the values of bound parameters can be derived based on equations (4.102)–(4.112). Particularly, the values of parameters α_R, β_R, l are not determined by the values of free parameters, but fixed as $\alpha_R = 1$, $\beta_R = 0$, and $l = 2$. The values of other bound parameters are derived as follows:

$$\beta_G = 1 - \alpha_G - \gamma_G = 1/3, \tag{4.214}$$

$$\mu_1 = \frac{\alpha_G p}{\upsilon_1} = 1, \tag{4.215}$$

13 The value of n_1 can be adjusted under different assumptions on the value of p. Suppose p is a multiple of 9, we should set $n_1 = 3$, such that n_1 is a factor of $p/3$. If $p/3$ is a prime power, we can set $n_1 = p/3$, such that n_1 is a factor of $p/3$ and TDG-$(n_1, \mu_1, p/3)$ exists by Lemma 1.11.

$$c_1' = \frac{p}{n_1} - 1 = 3q - 1, \tag{4.216}$$

$$n = k_2 = 2, \tag{4.217}$$

$$w = x - 1 = 2q - 1, \tag{4.218}$$

$$r_2 = \frac{v_1 \cdot (1 - \alpha_G - \gamma_G)p^2}{t\mu x^2} = p = 6q, \tag{4.219}$$

$$\lambda_2 = \frac{v_1 \cdot (1 - \alpha_G - \gamma_G)p_G p_R}{t\mu x^2} \cdot \frac{k_2 - 1}{\theta_2 k_2 - 1} = p = 6q, \tag{4.220}$$

$$c_2 = \frac{p_R}{tk_2} - 1 = \frac{p}{2} - 1 = 3q - 1. \tag{4.221}$$

Under the preceding parameter settings, given the port number of switches as $p = 6q$, where q is a positive integer, the complete design of devised topologies can be summarized as follows:

- *Intra-module design.* A TDG-$(2, 1, 2q)$ with $3q - 1$ copies is employed as the internal graph for each module, where element nodes are GNs and block nodes are RNs.
- *Connection design.* In each module, all GNs are selected as dock nodes. Such dock nodes are grouped as ISNs-$(2q, 2q - 1, 1)$. Dock nodes in distinct modules are all connected under the bipartite mode. Specifically, two ISNs-$(2q, 2q - 1, 1)$ are interconnected by a composite edge which is a TD-$(2, 1, 2q)$ interconnection.

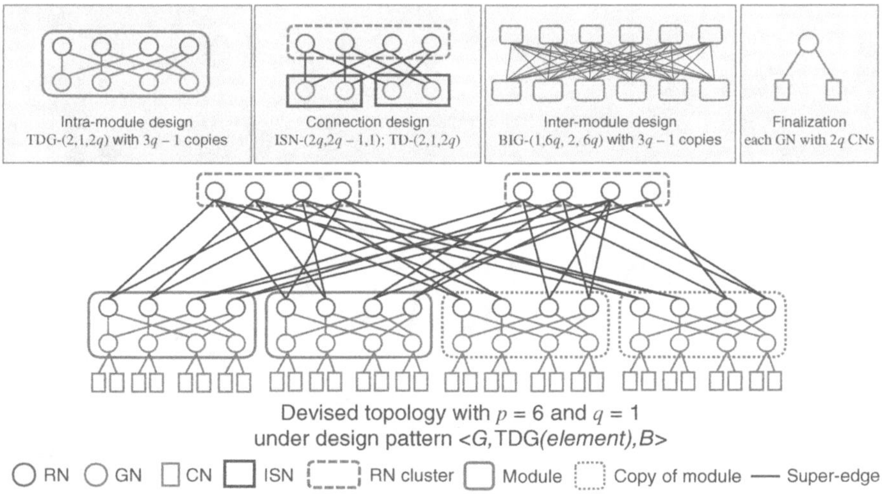

Figure 4.9 The complete design of a devised topology under design pattern $< G$, TDG(element), $B >$ with $p = 6$ and $q = 1$.

- *Inter-module design*. A BIG-$(1, 6q, 2, 6q)$ with $3q - 1$ copies is employed as the skeleton graph to interconnect different modules. For each RN cluster, it is incident to $6q$ super-edges. These super-edges are partitioned into $3q$ groups, each of which is incident to a copy of modules (i.e., two modules). Each group of two super-edges are instantiated as a 1-parallelized TD-$(2, 1, 2q)$ interconnection.
- *Finalization*. Each GN in the modules is directly connected to $2q$ CNs.

An example of the devised topology with $p = 6$ and $q = 1$ is provided in Figure 4.9.

5

Applications and Future Work

Through the first four chapters, we have introduced a generic topology design framework in a top-down manner. Besides the applications to classical data center networks, such framework can be applied to many emerging task-oriented network scenarios, including:

- AI clusters for the training and inference of large language model (LLM) (Gershon et al. 2024)
- Data centers with optical reconfigurability (Patronas et al. 2024)
- Space data centers and orbit AI (European Space Agency 2025; Beals et al. 2025)
- Low earth orbit (LEO) satellite networks (Pan et al. 2023)
- Artificial intelligence of things (AIoT) (Siam et al. 2025)
- Multi-agent networks (Oroojlooy and Hajinezhad 2023)

In this chapter, we focus on topology design of AI clusters (Gershon et al. 2024). In fact, with the rapid advancement of AI, the emergence of modern AI clusters has driven large-scale interconnection networks to evolve from single-domain systems into dual-domain architectures (NVIDIA 2017) that integrate scale-up domains with the scale-out domain. By examining a collection of representative and widely adopted AI cluster instances, we systematically reverse-engineer their existing topology designs into a unified design scheme, revealing the underlying design patterns and design steps that govern their constructions.

At the end of this chapter, we will discuss prospective directions for future work, ranging from theoretical exploration to practical extensions.

5.1 Topology Design for AI Clusters

As the complexity of large-scale AI model training and inference continues to increase, the demand for computational resources has grown rapidly (Stanford Institute for Human-Centered Artificial Intelligence [HAI] 2025). During training, models must process massive parameter spaces and datasets, with model sizes expanding from hundreds of millions to trillions of parameters (Shen et al. 2024). This growth has driven the expansion of compute clusters from thousands to hundreds of thousands of GPUs in order to enable efficient parallelism and accelerate training iterations (Epoch AI 2025). Although inference is generally less computationally intensive than training, the need to support real-time and high-concurrency applications has similarly increased resource demands (NVIDIA 2025g, l), motivating further scaling of compute clusters to ensure low latency and high throughput (NVIDIA 2025j). Together, these trends impose stringent scalability requirements on AI clusters (Sharma 2025).

5.1.1 Single Domain System

In traditional switch-centric data center networks (Chen et al. 2016), each server is equipped with only one NIC, which means the degree of each server in network topology graphs is only one; thus, each server is directly attached to a switch. In certain AI clusters where hardware resources are constrained or cost is tightly controlled (An et al. 2024; Gangidi et al. 2024), servers equipped with only a single NIC are still employed. According to the definition in Section 2.1, such servers are classified as CNs and satisfy the assumption of having a degree of one by being attached to one and only one node with routing function. Consequently, the topologies of these AI clusters can be directly interpreted and generated through our design framework presented in Chapter 2. A representative example of such AI clusters is the cost-performance–efficient DeepSeek Fire-Flyer 2 AI-HPC (An et al. 2024), a 10,000-GPU cluster for deep learning training. It employs servers equipped with eight NVIDIA A100 PCIe GPUs (NVIDIA 2020a) and a single 200 Gbps NVIDIA Mellanox ConnectX-6 (CX6) InfiniBand (IB) NIC (NVIDIA 2022) so as to significantly reduce hardware and energy expenditures. Through a two-zone, two-layer Fat-Tree network topology together with software–hardware co-design (e.g., HFReduce for efficient all-reduce communication, HaiScale for optimized parallelism, 3FS for high-throughput storage, and NVLink bridge expansion for LLM needs), Fire-Flyer 2 AI-HPC achieves 80% of the NVIDIA DGX-A100 cluster performance at merely 60% of the cost (An et al. 2024).

As shown in Figure 5.1, under the design framework, the topology of Fire-Flyer 2 AI-HPC is a switch-centric topology and its design pattern is $< G$, BIG(element), $B >$, where servers are represented by CNs, switches are represented by GNs and RNs, and the port number of switches is 40. Its design in each step is illustrated as follows:

- **Intra-module design**. The internal graph is a BIG-(1, 20, 40, 20), where element nodes and block nodes are viewed as GNs and RNs, respectively. In fact, each module in the design framework corresponds precisely to a zone in DeepSeek's topology design.

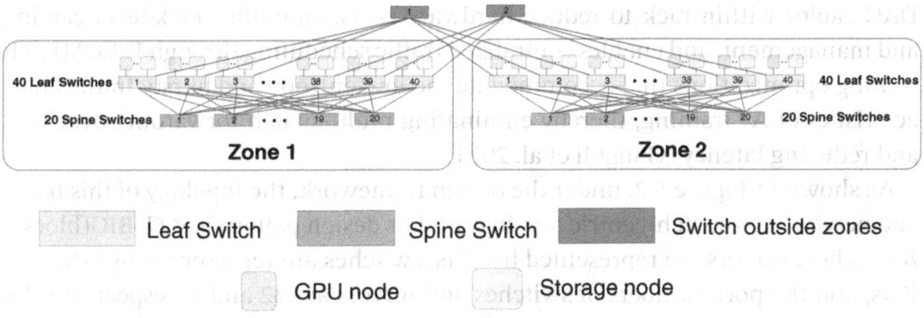

Figure 5.1 Topology of Fire-Flyer 2 AI-HPC.

- **Connection design**. In each module, all element nodes (GNs) are selected as dock nodes. Each dock node serves as an intermediate super-node, i.e., an ISN-(1, 1, 2). Then the dock node interconnection is constructed under the bipartite mode. The composite-edge between any two intermediate super-nodes (ISNs) is designed as a TD-(2, 1, 1) interconnection through an RN cluster.
- **Inter-module design**. The skeleton graph is a BIG-(1, 2, 2, 2). Such a graph contains 2 element nodes and 2 block nodes that correspond to modules and RN clusters, respectively. For any two modules, they are connected with a common RN cluster via two super-edges. Such super-edges form a 20-parallelized TD-(2, 1, 1) interconnection.
- **Finalization**. Each GN is directly connected with 16 CNs.

With the advancement of hardware technologies, it is now feasible for each server to contain multiple GPUs, each GPU to be equipped with multiple NICs, and each NIC to have multiple ports (Ren et al. 2025). However, in practice, a more commonly used configuration is that each server contains multiple GPUs, each GPU is equipped with a single NIC, and each NIC identically has one or two ports (Ren et al. 2025; Unat et al. 2024).

With such server configurations, in certain AI clusters, all NIC ports of a server are connected to a single switch (Meta Engineering Team 2024). Consequently, these servers can still be classified as CNs and satisfy the assumption of being attached to one and only one node with routing function; the only distinction is that each attachment consumes more than one degree of that node with routing function, which can be readily accommodated in the finalization step of our design framework. Therefore, the topologies of such AI clusters can likewise be directly interpreted and generated through the design framework presented in Chapter 2.

A representative example of such AI clusters is Meta's backend network proposed by Gangidi et al. (2024), where each server is attached to the sole switch in its rack using all eight of its 400G NIC ports via copper-based digital-to-analog converter (DAC) cables. This design enables exclusive utilization of copper-based

DAC cables within rack to reduce hardware costs, simplifies rack-level cabling and management, and enables centralized traffic scheduling through E-ECMP/TE routing optimizations that accommodate the low-entropy and bursty traffic characteristics of AI training, thereby eliminating multi-switch forwarding overhead and reducing latency (Gangidi et al. 2024).

As shown in Figure 5.2, under the design framework, the topology of this backend network is a switch-centric topology and its design pattern is $< G$, BIG(block), $B >$, where servers are represented by CNs, switches are represented by GNs and RNs, and the port numbers of switches and servers are 32 and 8, respectively. Its design in each step is illustrated as follows:

- **Intra-module design**. The internal graph is a BIG-(1, 16, 16, 16), where element nodes and block nodes are viewed as GNs and RNs, respectively. In fact, each module in the design framework corresponds precisely to an AI zone in Meta's topology design.
- **Connection design**. In each module, all block nodes (RNs) are selected as dock nodes. Each dock node serves as an intermediate super-node, i.e., an ISN-(1, 1). Then the dock node interconnection is constructed under the bipartite mode. The composite-edge between any two ISNs is designed as a TD-(2, 16, 1) interconnection through an RN cluster.
- **Inter-module design**. The skeleton graph is a BIG-(1, 1, 2, 1). Such a graph contains 2 element nodes and 1 block node that correspond to modules and an RN cluster, respectively. For any two modules, they are connected with a common RN cluster via two super-edges. Such super-edges form a 16-parallelized TD-(2, 16, 1) interconnection.
- **Finalization**. Each GN is directly attached to two CNs, with each attachment consuming 8 (the port number of servers) degrees of the GN.

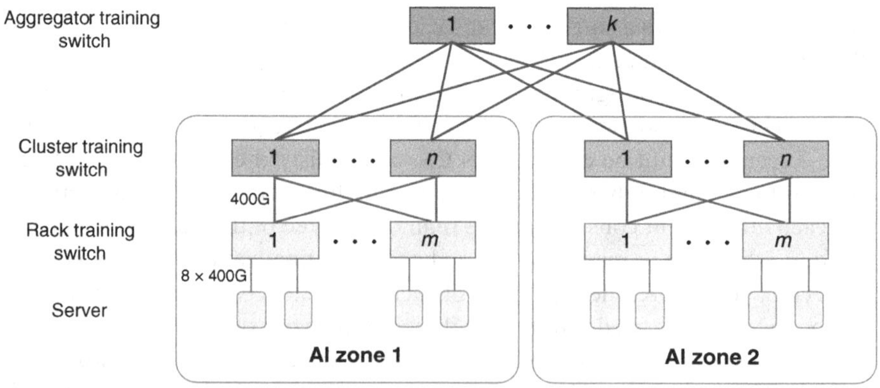

Figure 5.2 Topology of Meta's backend network.

5.1.2 Dual-domain System: Scale-up and Scale-out

The presence of multiple GPUs within each server necessitates intra-server interconnection (Sojoodi et al. 2025). Moreover, the resulting internal topology inevitably influences the design of the inter-server network (De Sensi et al. 2024). In AI clusters, as the number of GPUs housed within individual servers (high-bandwidth domains) continues to grow (Li et al. 2019; Ye et al. 2024), the intra-server (intra-high-bandwidth-domain) network has been conceptualized as the scale-up domain (NVIDIA 2017), while the inter-server (inter-high-bandwidth-domain) network is correspondingly conceptualized as the scale-out domain (NVIDIA 2017; Gangidi et al. 2024), featuring their respective roles in node-local and system-wide scalability (Ye et al. 2024).

Therefore, we can interpret existing AI clusters from the perspectives of the scale-up and scale-out domains.

Figure 5.3 illustrates the architecture of Nvidia's Rail-optimized network (NVIDIA 2023a, 2024a; Gangidi et al. 2024), with guidance on how to interpret this AI cluster from its scale-up and scale-out domains provided in Figure 5.4.

In Nvidia's Rail-optimized network (Gangidi et al. 2024), each high-bandwidth domain forms a scale-up domain, within which *K* GPUs are interconnected via high-bandwidth links. These scale-up domains are then interconnected through the NICs equipped on their constituent GPUs. Thereby, the interconnection among GPUs in different scale-up domains, i.e., the interconnection among the *M* scale-up domains, constitutes the scale-out domain.

We provide a clearer structural interpretation of AI clusters from the perspective of scale-up and scale-out domains in Figure 5.5, in which the AI cluster resembles a three-dimensional (3D) book spine. In the 3D book spine, the GPUs constitute

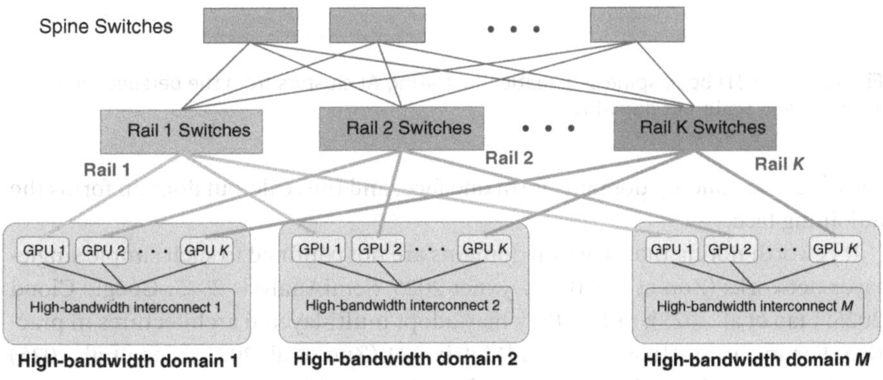

Figure 5.3 An illustration of Nvidia's Rail-optimized network.

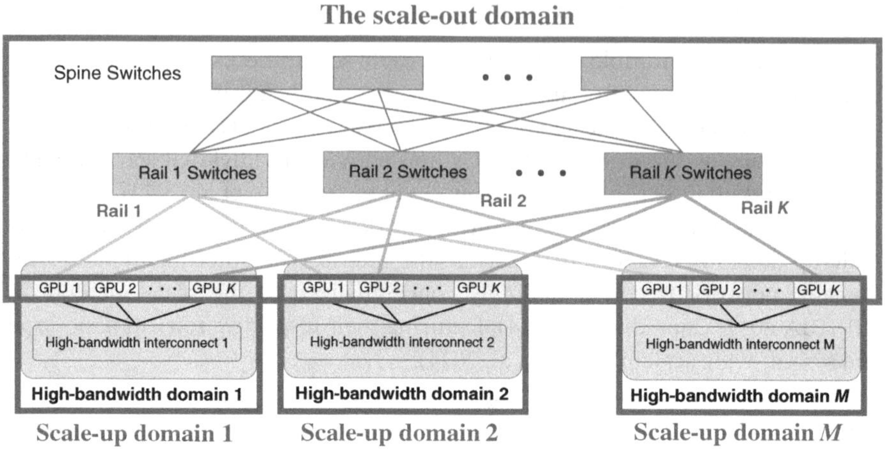

Figure 5.4 An illustration of Nvidia's Rail-optimized network's scale-up domains and scale-out domain.

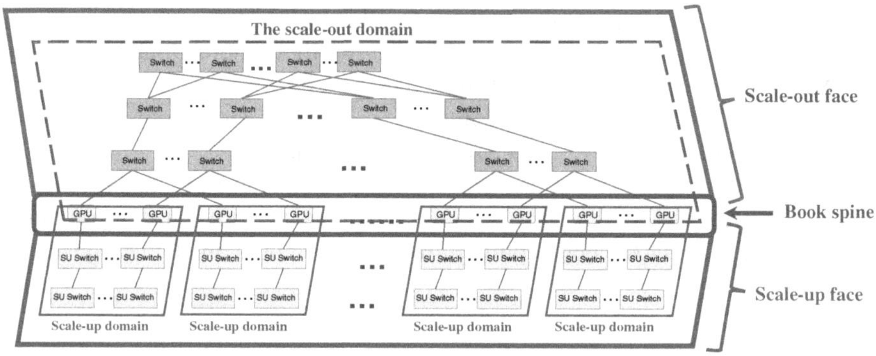

Figure 5.5 A 3D book-spine schematic illustrating AI clusters from the perspective of scale-up and scale-out domains.

the edge, the scale-up domains form one face, and the scale-out domain forms the adjoining face.

It is worth noting that scale-up domains are not confined to such simple single-layer networks (Zuo et al. 2025; Dyxnet 2025; SemiAnalysis 2025; Google Cloud 2025; Liao et al. 2025); rather, they may adopt multi-layered architectures in practice. For instance, Huawei's CloudMatrix 384 (Zuo et al. 2025; Xiao et al. 2025) constitutes a scale-up domain with a three-layer architecture.

As shown in Figure 5.6, we notice that the topology of CloudMatrix 384 (Zuo et al. 2025) can be directly interpreted and generated by the design framework.

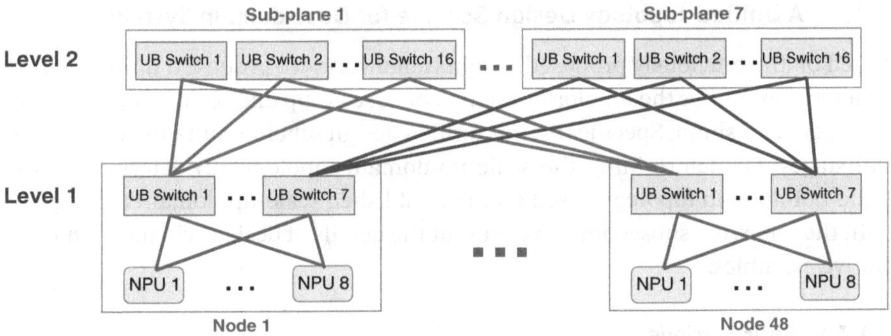

Figure 5.6 An illustration of a scale-up domain with a three-layer architecture—Huawei's CloudMatrix 384.

Specifically, according to the definition in Section 2.1, since neural processing units (NPUs) have both computing function and routing function within scale-up domains, NPUs can be classified as HNs within scale-up domains; since UB switch chips have only routing function and no direct access to CNs within scale-up domains, UB switch chips can be classified as RNs within scale-up domains. Consequently, under the design framework, the design pattern of the topology of CloudMatrix 384 is $< H$, BIG(block), $B >$, where NPUs serve as HNs and UB switch chips serve as RNs. Its design in each step is illustrated as follows:

- **Intra-module design**. The internal graph is a BIG-(1, 7, 8, 7), where element nodes and block nodes are viewed as HNs and RNs, respectively. In fact, each module under the design framework corresponds precisely to a node in Huawei's topology design (Zuo et al. 2025).
- **Connection design**. In each module, all block nodes (RNs) are selected as dock nodes. Each dock node serves as an intermediate super-node, i.e., an ISN-(1, 1). Then the dock node interconnection is constructed under the bipartite mode. Each composite-edge is designed as a TD-(48, 16, 1) interconnection among 48 ISNs through an RN cluster.
- **Inter-module design**. The skeleton graph is a BIG-(1, 7, 48, 7). Such a graph contains 48 element nodes and 7 block nodes that correspond to modules and RN clusters, respectively. For any 48 modules, they are connected with a common RN cluster via 48 super-edges. Such super-edges form a 1-parallelized TD-(48, 16, 1) interconnection. In fact, each RN cluster under the design framework corresponds precisely to a sub-plane in Huawei's topology design (Zuo et al. 2025).
- **Finalization**. This step is omitted as the topology contains no CNs.

5.1.3 A Unified Topology Design Scheme for Dual-domain Systems

Based on the previously proposed design framework, we propose a unified design scheme that unifies the topology design of both scale-up and scale-out domains in a sequential fashion. Specifically, our unified design scheme contains two sequential stages: (1) determining the scale-up-domain topology, (2) determining the scale-out-domain topology based on the established scale-up-domain topology.

In the following subsections, we present the details of design scheme with illustrative examples.

5.1.3.1 Basic Settings

AI clusters usually involve vast interconnected devices, such as GPUs and diverse switches (Ye et al. 2024). In general, devices in AI clusters execute two types of functions: computing and routing (Ye et al. 2024; Tang 2025). For clarity of presentation, the categorization of nodes corresponding to network devices is specified separately for the scale-up domain and the scale-out domain, as summarized in Table 5.1.

For network devices within the scale-up domains, AI accelerators (e.g., GPUs, TPUs, and NPUs) perform both computing and intra–scale-up-domain routing functions (NVIDIA 2025m, o), whereas scale-up-domain switches are responsible solely for intra–scale-up-domain routing (NVIDIA 2021). Likewise, switches within the scale-out domain are solely responsible for scale-out-domain routing (NVIDIA 2023a). Consequently, for AI clusters, in scale-up domains, we refer to nodes corresponding to devices that perform both computing and routing functions within their scale-up domains as scale-up hybrid nodes (SUHNs), and to nodes corresponding to devices that perform only routing function within their scale-up domains as scale-up routing nodes (SURNs); in the scale-out domain, we refer to nodes corresponding to devices that perform routing function within the scale-out domain as routing nodes (RNs).

For instance, as presented in Figure 5.4, in the topology of Nvidia's Rail-optimized network, within the scale-up domains, the GPUs are SUHNs, since they perform both computing and routing functions, and the switches within the high-bandwidth domains are SURNs, since they perform only routing function; within

Table 5.1 The categorization of nodes in AI clusters.

Associated domain	Node name	Function in network
Scale-up domains	Scale-up hybrid node (SUHN)	Computing and routing
	Scale-up routing node (SURN)	Routing only
Scale-out domain	Routing node (RN)	Routing

the scale-out domain, the rail switches and spine switches are RNs since they perform routing function.

Based on the aforementioned setup, the topology of an AI cluster is modeled as an undirected graph in which the scale-up domains and the scale-out domain are distinguished explicitly. In the undirected graph, within the scale-up domains, nodes correspond to SUHNs and SURNs; within the scale-out domain, apart from the SUHNs, nodes correspond to RNs; edges correspond to communication links.

To address the difficulty of joint topology design across scale-up and scale-out domains, we decouple the topology design of the entire AI cluster into two sequential stages: (1) topology design for the scale-up domains, followed by (2) topology design for the scale-out domain based on the established scale-up topology. The coordination and integration between these two domains are captured in the latter stage, as the topology design for the scale-out domain explicitly factors in the established topology of the scale-up domains. In this manner, the complete topology design is achieved through a sequential process in which the later stage accounts for the outcomes of the preceding one.

This sequential manner enables us to reduce the substantial design complexity inherent in a monolithic dual-domain joint design, and to more readily adopt individual concrete tools for each part of the design.

5.1.3.2 Scale-up Domain

1. **Design Patterns**

   According to Section 5.1.3.1, the scale-up domains consist of only two types of nodes: SUHNs and SURNs, whose functions within the scale-up domains correspond exactly to the functions of HNs and RNs in large-scale networks as defined in Chapter 2, respectively. Further, aside from the constraint that SUHNs cannot be connected to CNs, the settings of SUHNs and SURNs within the scale-up domains are entirely identical to those of HNs and RNs within large-scale networks in Chapter 2, respectively. Accordingly, based on Chapter 2 and Chapter 4, the design procedure, performance metrics, guiding principles, and concrete tools of topology design for scale-up domains in AI cluster fully aligns with those of topology design for large-scale networks in Chapter 2. Consequently, the single-iteration design patterns of topology design for scale-up domains, as summarized in Table 5.2, are exactly the summary of single-iteration design patterns with HNs in the internal graph for large-scale networks in Section 4.2.

   Owing to constraints imposed by high-bandwidth interconnect technologies, the scale of scale-up domains is inherently limited (Sojoodi et al. 2025; Ding et al. 2025). Consequently, most practical scale-up-domain topologies can be fully specified within a single iteration under the design scheme

Table 5.2 A summary of design patterns of scale-up domains.

Numbering of design pattern	Intra-module design	Dock node selection	Dock node interconnection	Inter-module design
1	CG	Any nodes as dock nodes	Plain mode	CG
2			Bipartite mode	BIG
3	CBG	Any nodes as dock nodes	Plain mode	CG
4			Bipartite mode	BIG
5	BIG	Element nodes as dock nodes	Plain mode	CG
6			Bipartite mode	BIG
7		Block nodes as dock nodes	Plain mode	CG
8			Bipartite mode	BIG
9	TDG	Element nodes as dock nodes	Plain mode	CG
10			Bipartite mode	BIG
11		Block nodes as dock nodes	Plain mode	CG
12			Bipartite mode	BIG

for scale-up domains (Ye et al. 2024). In light of this observation, the following discussion concentrates on design patterns arising from a single iteration. Extensions to multi-iteration design patterns arise naturally from the design framework and will be illustrated by concrete examples in the following section.

For clarity of presentation, the design patterns of each iteration are described separately. In particular, according to the design pattern naming scheme in Section 4.2.2, the first term in the naming *triple* is fixed to H for design patterns of each iteration pertaining to the topology design of scale-up domains. Thus, we omit the first term and abbreviate the naming of design patterns of each iteration pertaining to the topology design of scale-up domains to the remaining two-tuple. In naming the design pattern of iteration i ($i \geq 2$), since its intra-module design adopts the topology generated in iteration $i - 1$, the left-hand component of the first term in the two-tuple, which stands for intra-module design, is denoted as iteration $i - 1$.

2. **Revisit of Existing Scale-up Domain Topologies**
By following the preceding design scheme for scale-up domains, a great number of existing scale-up-domain topologies can be reverse-engineered, as summarized in Table 5.3. In the following paragraphs, we illustrate the corresponding designs in each step of the scheme of 17 representative real-world scale-up–domain topologies, respectively. Note that the design step of

finalization is omitted as the scale-up-domain topology contains no CNs. Moreover, if the scale-up-domain topology consists of only a single module, then design steps of both connection design and inter-module design are omitted, and only the design step of intra-module design is presented.

Table 5.3 A summary of scale-up-domain topologies in existing AI clusters.

AI cluster	Intra-module design			Connection Design		Inter-module design	
	CG	CBG	BIG	PM	BM	CG	BIG
AMD Instinct MI300X Platform (AMD 2024)	•						
Intel Gaudi3 AI Accelerator HLB-325 Baseboard (Intel 2025)	•						
NVIDIA DGX H100 System (NVIDIA 2025m)			•				
NVIDIA DGX H200 System (NVIDIA 2025m)			•				
NVIDIA DGX B200 System (NVIDIA 2025a)			•				
NVIDIA DGX B300 System (NVIDIA 2025b)			•				
NVIDIA GB200 NVL72 (NVIDIA 2025h)			•				
Baidu Kunlun 32-XPU SuperNode (Baidu Cloud 2025)			•				
China Mobile Open-architecture 64-AI Accelerator Supernode (China Mobile 2025)			•				
Alibaba Panjiu AL64 Supernode (Alibaba Cloud 2025)			•				
Alibaba Panjiu AL128 Supernode (Alibaba Cloud 2025)			•				
Tencent ETH-X Ultra Supernode (Open Data Center Committee [ODCC] 2025)			•				

(Continued)

Table 5.3 (Continued)

Huawei CloudMatrix 384 (Zuo et al. 2025)		•		•		•
China Mobile Cascaded Open-architecture 1024-AI Accelerator Supernode (Dyxnet 2025)		•		•		•
Amazon EC2 Trn3 UltraServers (Gen1) (SemiAnalysis 2025)	•		▲	•	▲	•
Amazon EC2 Trn3 UltraServers (Gen2) (SemiAnalysis 2025)		•	▲	•	▲	•
Google Ironwood Rack (Google Cloud 2025)	•		■		■	
Huawei UB-Mesh-Rack (Liao et al. 2025)	•		•		•	
Huawei UB-Mesh-Pod (Liao et al. 2025)	•		•		•	

Note: • for iteration 1; ▲ for iteration 2; ■ for iterations 1, 2, 3, 4, 5.
Abbreviations: BIG, bipartite incidence graph; BM, bipartite mode; CBG, complete bipartite graph; CG, complete graph; PM, plain mode.

AMD Instinct MI300X Platform: The AMD Instinct MI300X platform (AMD 2024) is the world's first single-node AI accelerator delivering 1.5 TB of HBM3 memory and 5.3 TB/s bandwidth. It is built on the advanced CDNA3 architecture and 4th-generation Infinity Fabric, specifically engineered for trillion-parameter LLMs. Its core strengths lie in full 8-GPU all-to-all connectivity with 896 GB/s bottleneck-free P2P bandwidth, native 2 : 4 structured sparsity acceleration, 41.8 PFLOPS peak FP8/INT8 performance (AMD 2023), and 1.8× higher Llama-70B inference throughput than an 8×H100 system (AMD 2025b), leading in memory capacity, bandwidth, and cost-performance ratio. Already deployed at scale by leading customers including Oracle, Vultr, and Meta, the MI300X surpassed $5 billion in 2024 sales, becoming AMD's fastest product ever to reach the $1 billion milestone and establishing dominance in cloud-based large-model training and inference markets (AMD 2025a). Under the design scheme for scale-up domains, the design pattern of the topology of AMD Instinct MI300X platform (AMD 2024) is <CG(-), ->, where MI300X GPUs serve as SUHNs with a degree of 7. Its design in each step is illustrated as follows:

- **Intra-module design**. The internal graph is a CG with 8 SUHNs.

Intel Gaudi3 AI Accelerator HLB-325 Baseboard: The Intel Gaudi3 AI Accelerator HLB-325 Baseboard Intel (2024a, 2025) integrates eight Gaudi3 OAM cards with a switchless, direct all-to-all interconnect topology, delivering 4.2 TB/s of bidirectional bandwidth, purpose-built for efficient training of trillion-parameter LLMs. Its core technical strengths include eight PCIe Gen5 ×16 host interfaces with on-board retimers for pristine signal integrity, six OSFP ports providing 1.2 TB/s RoCE-based scale-out bandwidth, and a streamlined, high-density 7.6 kW design that achieves exceptional system simplicity and cost-performance ratio. Offered as the fully integrated HLFB-325L subsystem, it has been widely adopted by leading manufacturers such as Supermicro and HPE, powering large-scale cloud AI training and inference clusters while accelerating the rapid deployment and volume production of the Intel Gaudi ecosystem (Intel 2024b).

Under the design scheme for scale-up domains, the design pattern of the topology of Intel Gaudi3 AI accelerator HLB-325 baseboard (Intel 2025) is <CG(-), ->, where Gaudi3 AI accelerators serve as SUHNs with a degree of 7. Its design in each step is illustrated as follows:

- **Intra-module design**. The internal graph is a CG with 8 SUHNs.

NVIDIA DGX H100/H200 System: The NVIDIA DGX H100 and DGX H200 systems (NVIDIA 2025m) represent NVIDIA's flagship enterprise-grade AI supercomputing platforms. Both systems integrate eight GPUs connected through a fully non-blocking NVSwitch fabric, while DGX H200 system further enhances memory capacity and bandwidth through HBM3e, offering 1.76× the capacity and 1.43× the bandwidth of DGX H100 system—an upgrade that substantially improves training efficiency for trillion-parameter models.

DGX H100 system combines eight Hopper-architecture GPUs providing 640 GB of HBM3, 3.35 TB/s of memory bandwidth, a 7.2 TB/s NVSwitch interconnect, and 1 TB/s of network bandwidth, delivering up to 32 PFLOPS of FP8 performance for large-scale LLM training and high-performance computing workloads.

DGX H200 system retains the same NVSwitch and networking architecture, therefore sharing the same scale-up-domain topology with DGX H100 system (NVIDIA 2025m), but expands to 1,128 GB of HBM3e with 4.8 TB/s of bandwidth, enabling 2–4× acceleration in large-model inference (NVIDIA 2024c).

Designed for advanced AI training, inference, and HPC applications, DGX H100/H200 systems have become the backbone of modern AI cluster worldwide. Supported by NVIDIA's mature software ecosystem and end-to-end platform integration, they have been widely deployed, exceeding 10,000 units in 2024, and serve as foundational systems for large-scale AI clusters at giant enterprises such as Meta and Oracle (NVIDIA 2023c; Oracle 2024b).

As NVIDIA DGX H100 system and DGX H200 system share the same scale-up domain topology, the following exposition concentrates on the NVIDIA DGX H200 system. Under the design scheme for scale-up domains, the design pattern of the topology of NVIDIA DGX H200 system (NVIDIA 2025m) is <BIG(-), ->, where H200 Tensor Core GPUs are represented by SUHNs with a degree of 4, and

NVSwitches are represented by SURNs with a degree of 8. Its design in each step is illustrated as follows:

- **Intra-module design**. The internal graph is a BIG-(1, 4, 8, 4), where element nodes and block nodes are viewed as SUHNs and SURNs, respectively.

NVIDIA DGX B200/B300 System: As NVIDIA DGX B200 system (NVIDIA 2025a) and NVIDIA DGX B300 system (NVIDIA 2025b) share the same NVSwitch and networking architecture, their scale-up-domain topologies are identical. Therefore, the following exposition concentrates on the NVIDIA DGX B200 system.

Under the design scheme for scale-up domains, the design pattern of the topology of NVIDIA DGX B200 system (NVIDIA 2025a) is <BIG(-), ->, where B200 Tensor Core GPUs are represented by SUHNs with a degree of 2, and NVSwitches are represented by SURNs with a degree of 8. Its design in each step is illustrated as follows:

- **Intra-module design**. The internal graph is a BIG-(1, 2, 8, 2), where element nodes and block nodes are viewed as SUHNs and SURNs, respectively.

NVIDIA GB200 NVL72: The NVIDIA GB200 NVL72 (NVIDIA 2025h) is the world's first rack-scale, liquid-cooled AI supercomputer integrating 72 Blackwell GPUs and 36 Grace CPUs within a single system. Leveraging 5th-generation NVLink, it enables 72 GPUs to operate in a fully connected NVLink domain, marking a significant shift in exascale-class AI system design (NVIDIA 2025i).

Key capabilities include 1,440 NVFP4 petaFLOPS of computational performance, 13.4 TB of HBM3E memory with 576 TB/s of aggregate bandwidth, and 5th-generation NVLink providing 1.8 TB/s of inter-GPU bandwidth. Compared with an equivalent H100-based configuration, the NVL72 delivers 30× faster real-time LLM inference, 4× faster training, alongside 25× improved energy efficiency (NVIDIA 2024b).

Designed for large-scale AI inference facilities, trillion-parameter LLM training, and emerging workloads such as quantum error correction (NVIDIA 2025h), the NVIDIA GB200 NVL72 has rapidly become a cornerstone of next-generation AI and HPC infrastructure (NVIDIA 2025k). With rack-level pricing around $3 million and tight integration into the NVLink Fusion ecosystem, shipments are expected to exceed 10,000 racks in 2025, supporting deployments by partners including Oracle, Meta, Microsoft Azure, and AWS (Oracle 2024a; Microsoft 2025; Amazon Web Services 2025).

Under the design scheme for scale-up domains, the design pattern of the topology of NVIDIA GB200 NVL72 (NVIDIA 2025h) is <BIG(-), ->, where NVIDIA

B200 Tensor Core GPU are represented by SUHNs with a degree of 18, and 5th-generation NVSwitch chips are represented by SURNs with a degree of 72. Its design in each step is illustrated as follows:

- **Intra-module design**. The internal graph is a BIG-(1, 18, 72, 18), where element nodes and block nodes are viewed as SUHNs and SURNs, respectively.

Baidu Kunlun 32-XPU SuperNode: The Baidu Kunlun 32-XPU SuperNode (Baidu Cloud 2025) integrates 32 Baidu P800 Kunlun XPUs in a fully connected XPU Link domain. Its key features include a 1U four-card liquid-cooled design with high compute density and low system weight that supports single-person maintenance. The supernode supports demanding AI workloads such as ERNIE 5.0 multimodal training, large-scale MoE model training and inference, intelligent customer service, and financial risk modeling (China Daily 2025). Validated through Baidu Cloud's UltraServer public testing, the supernode has already been scaled to cluster deployment exceeding 30,000 XPUs, strengthening Baidu's capabilities for AI cluster self-sufficiency (Baidu 2025).

Under the design scheme for scale-up domains, the design pattern of the topology of Baidu Kunlun 32-XPU SuperNode (Baidu Cloud 2025) is <BIG(-), ->, where Baidu P800 Kunlun XPU are represented by SUHNs with a degree of 4, and switch trays are represented by SURNs with a degree of 32. Its design in each step is illustrated as follows:

- **Intra-module design**. The internal graph is a BIG-(1, 4, 32, 4), where element nodes and block nodes are viewed as SUHNs and SURNs, respectively.

China Mobile Open-architecture 64-AI Accelerator Supernode: Under the design scheme for scale-up domains, the design pattern of the topology of China Mobile open-architecture 64-AI accelerator supernode (China Mobile 2025) is <BIG(-), ->, where AI accelerators are represented by SUHNs with a degree of 8, and switches are represented by SURNs with a degree of 64. Its design in each step is illustrated as follows:

- **Intra-module design**. The internal graph is a BIG-(1, 8, 64, 8), where element nodes and block nodes are viewed as SUHNs and SURNs, respectively.

Alibaba Panjiu AL64 Supernode: Under the design scheme for scale-up domains, the design pattern of the topology of Alibaba Panjiu AL64 Supernode (Alibaba Cloud 2025) is <BIG(-), ->, where Alibaba CIPU 2.0 AI accelerators are represented by SUHNs with a degree of 16, and ALink Switch chips are represented by SURNs with a degree of 64. Its design in each step is illustrated as follows:

- **Intra-module design**. The internal graph is a BIG-(1, 16, 64, 16), where element nodes and block nodes are viewed as SUHNs and SURNs, respectively.

Alibaba Panjiu AL128 Supernode: Under the design scheme for scale-up domains, the design pattern of the topology of Alibaba Panjiu AL128 Supernode (Alibaba Cloud 2025) is <BIG(-), ->, where Alibaba CIPU 2.0 AI accelerators are represented by SUHNs with a degree of 32, and ALink Switch chips are represented by SURNs with a degree of 128. Its design in each step is illustrated as follows:

- **Intra-module design**. The internal graph is a BIG-(1, 32, 128, 32), where element nodes and block nodes are viewed as SUHNs and SURNs, respectively.

Tencent ETH-X Ultra Supernode: Under the design scheme for scale-up domains, the design pattern of the topology of Tencent ETH-X Ultra supernode (Open Data Center Committee [ODCC] 2025) is <BIG(-), ->, where GPUs are represented by SUHNs with a degree of 16, and scale-up switches are represented by SURNs with a degree of 512. Its design in each step is illustrated as follows:

- **Intra-module design**. The internal graph is a BIG-(1, 16, 512, 16), where element nodes and block nodes are viewed as SUHNs and SURNs, respectively.

Huawei CloudMatrix 384: Under the design scheme for scale-up domains, the design pattern of the topology of Huawei CloudMatrix 384 (Zuo et al. 2025) is <BIG(block), B >, where Ascend 910 NPUs serve as SUHNs and UB switch chips serve as SURNs. Its design in each step is illustrated as follows:

- **Intra-module design**. The internal graph is a BIG-(1, 7, 8, 7), where element nodes and block nodes are viewed as SUHNs and SURNs, respectively. In fact, each module under the design scheme for scale-up domains corresponds precisely to an Ascend 910 node in Huawei's topology design.
- **Connection design**. In each module, all block nodes (SURNs) are selected as dock nodes. Each dock node serves as an intermediate super-node, i.e., an ISN-(1, 1). Then the dock node interconnection is constructed under the bipartite mode. Each composite-edge is designed as a TD-(48, 16, 1) interconnection among 48 ISNs through an SURN cluster.
- **Inter-module design**. The skeleton graph is a BIG-(1, 7, 48, 7). Such a graph contains 48 element nodes and 7 block nodes that correspond to modules and SURN clusters, respectively. For any 48 modules, they are connected with a common SURN cluster via 48 super-edges. Such super-edges form a 1-parallelized TD-(48, 16, 1) interconnection. In fact, each SURN cluster under the design scheme for scale-up domains corresponds precisely to a sub-plane in Huawei's topology design.

China Mobile Cascaded Open-architecture 1024-AI Accelerator Supernode: Under the design scheme for scale-up domains, the design pattern of the topology of China Mobile Cascaded Open-architecture 1024-AI accelerator supernode (Dyxnet 2025) is <BIG(block), B >, where AI accelerators serve as SUHNs and switches serve as SURNs. Its design in each step is illustrated as follows:

- **Intra-module design**. The internal graph is a BIG-(1, 8, 64, 8), where element nodes and block nodes are viewed as SUHNs and SURNs, respectively. In fact, each module under the design scheme for scale-up domains corresponds precisely to a SuperPoD in China Mobile's cascaded open-architecture supernode topology design.
- **Connection design**. In each module, all block nodes (SURNs) are selected as dock nodes. Each dock node serves as an intermediate super-node, i.e., an ISN-(1, 1). Then the dock node interconnection is constructed under the bipartite mode. Each composite-edge is designed as a TD-(16, 8, 1) interconnection among 16 ISNs through an SURN cluster.
- **Inter-module design**. The skeleton graph is a BIG-(1, 8, 16, 8). Such a graph contains 16 element nodes and 8 block nodes that correspond to modules and SURN clusters, respectively. For any 16 modules, they are connected with a common SURN cluster via 16 super-edges. Such super-edges form a 1-parallelized TD-(16, 8, 1) interconnection. In fact, each SURN cluster under the design scheme for scale-up domains corresponds precisely to a Plane in China Mobile's cascaded open-architecture supernode topology design.

Amazon EC2 Trn3 UltraServers (Gen1): Under the design scheme for scale-up domains, the topology of Amazon EC2 Trn3 UltraServers (Gen1) (SemiAnalysis 2025) is constructed by two rounds of iteration, where SUHNs and SURNs correspond to AWS Trainium3 chips and AWS NeuronSwitch-v1, respectively. In iteration 1, the design pattern for scale-up domains is <CG(any), B >. In iteration 2, the topologies generated in iteration 1 are taken as modules, and the design pattern for scale-up domains evolves into <iteration 1(SURN), P >. The design of the topology of Amazon EC2 Trn3 UltraServers (Gen1) at each step of the two iterations is illustrated as follows:

- **Iteration 1**:
  - **Intra-module design**. The internal graph is a CG with 2 SUHNs.
  - **Connection design**. In each module, all SURNs are selected as dock nodes. Each dock node serves as an intermediate super-node, i.e., an ISN-(1, 1). Then the dock node interconnection is constructed under the bipartite mode. Each composite-edge is designed as a TD-(16, 8, 1) interconnection among 16 ISNs through an SURN cluster.

  - **Inter-module design**. The skeleton graph is a BIG-(1, 2, 16, 2). Such a graph contains 16 element nodes and 2 block nodes that correspond to modules and SURN clusters, respectively. For any 16 modules, they are connected with a common SURN cluster via 16 super-edges. Such super-edges form a 1-parallelized TD-(16, 8, 1) interconnection.
- **Iteration 2**:
  - **Intra-module design**. The topologies generated in iteration 1 are treated as modules.
  - **Connection design**. In each module, all SUHNs are selected as dock nodes. Each dock node serves as an intermediate super-node, i.e., an ISN-(1, 1, 3). Then the dock node interconnection is constructed under the plain mode. Each composite-edge between two ISNs is designed as an all-to-all interconnection.
  - **Inter-module design**. The skeleton graph is a CG with 2 nodes. Each node corresponds to a module. Each super-edge between two modules is instantiated as a 32-parallelized all-to-all interconnection.

Amazon EC2 Trn3 UltraServers (Gen2): Under the design scheme for scale-up domains, the topology of Amazon EC2 Trn3 UltraServers (Gen2) (SemiAnalysis 2025) is constructed by two rounds of iteration, where SUHNs and SURNs correspond to AWS Trainium3 chips and AWS NeuronSwitch-v1, respectively. In iteration 1, the design pattern for scale-up domains is <BIG(element), B >. In iteration 2, the topologies generated in iteration 1 are taken as modules, and the design pattern for scale-up domains turns into <iteration 1(SURN), P >. The design of the topology of Amazon EC2 Trn3 UltraServers (Gen2) at each step of the two iterations is illustrated as follows:

- **Iteration 1**:
  - **Intra-module design**. The internal graph is a BIG-(1, 8, 4, 8), where element nodes and block nodes are viewed as SUHNs and SURNs, respectively.
  - **Connection design**. In each module, all SUHNs are selected as dock nodes. Each dock node serves as an intermediate super-node, i.e., an ISN-(1, 1). Then the dock node interconnection is constructed under the bipartite mode. Each composite-edge is designed as a TD-(18, 10, 1) interconnection among 18 ISNs through an SURN cluster.
  - **Inter-module design**. The skeleton graph is a BIG-(1, 4, 18, 4). Such a graph contains 18 element nodes and 4 block nodes that correspond to modules and SURN clusters, respectively. For any 18 modules, they are connected with a common SURN cluster via 18 super-edges. Such super-edges form a 1-parallelized TD-(18, 10, 1) interconnection.

- **Iteration 2**:
  - **Intra-module design**. The topologies generated in iteration 1 are treated as modules.
  - **Connection design**. In each module, all SUHNs are selected as dock nodes. Each dock node serves as an intermediate super-node, i.e., an ISN-(1, 1, 4). Then the dock node interconnection is constructed under the plain mode. Each composite-edge between two ISNs is designed as an all-to-all interconnection.
  - **Inter-module design**. The skeleton graph is a CG with 2 nodes. Each node corresponds to a module. Each super-edge between two modules is instantiated as a 72-parallelized all-to-all interconnection.

Google Ironwood Rack: Under the design scheme for scale-up domains, the topology of Google Ironwood Rack (Google Cloud 2025) can be constructed by five rounds of iteration, where Google Ironwood TPUs serve as SUHNs. In iteration 1, the design pattern for scale-up domains is <CG(any), P>. In iteration i ($i = 2, 3, 4, 5$), the topologies generated in iteration ($i - 1$) are taken as modules, and the design pattern for scale-up domains evolves into <iteration ($i - 1$)(SURN), P>. The design of the topology of Google Ironwood Rack at each step of the five iterations is illustrated as follows:

- **Iteration 1**:
  - **Intra-module design**. The internal graph is a CG with 2 SUHNs.
  - **Connection design**. In each module, all SUHNs are selected as dock nodes. Each dock node serves as an intermediate super-node, i.e., an ISN-(1, 1). Then the dock node interconnection is constructed under the plain mode. Each composite-edge between two ISNs is designed as an all-to-all interconnection.
  - **Inter-module design**. The skeleton graph is a CG with 2 nodes. Each node corresponds to a module. Each super-edge between two modules is instantiated as a 2-parallelized all-to-all interconnection.
- **Iteration** i ($i = 2,3,4,5$):
  - **Intra-module design**. The topologies generated in iteration ($i - 1$) are treated as modules.
  - **Connection design**. In each module, all SURNs are selected as dock nodes. Each dock node serves as an intermediate super-node, i.e., an ISN-(1, 1, i). Then the dock node interconnection is constructed under the plain mode. Each composite-edge between two ISNs is designed as an all-to-all interconnection.
  - **Inter-module design**. The skeleton graph is a CG with 2 nodes. Each node corresponds to a module. Each super-edge between two modules is instantiated as a 2^i-parallelized all-to-all interconnection.

Huawei UB-Mesh-Rack: Under the design scheme for scale-up domains, the design pattern of the topology of Huawei UB-Mesh-Rack (Liao et al. 2025) is <CG(any), *P* >, where NPUs serve as SUHNs. Its design in each step is illustrated as follows:

- **Intra-module design**. The internal graph is a CG with 8 SUHNs. In fact, each module under the design scheme for scale-up domains corresponds precisely to a one-dimensional (1D) full-mesh NPU board in Huawei's topology design.
- **Connection design**. In each module, all SUHNs are selected as dock nodes. Each dock node serves as an intermediate super-node, i.e., an ISN-(1, 1). Then the dock node interconnection is constructed under the plain mode. Each composite-edge between two ISNs is designed as an all-to-all interconnection.
- **Inter-module design**. The skeleton graph is a CG with 8 nodes. Each node corresponds to a module. Each super-edge between two modules is instantiated as a 8-parallelized all-to-all interconnection. This topology is described as a two-dimensional (2D) full-mesh topology within rack by Huawei.

Huawei UB-Mesh-Pod: Under the design scheme for scale-up domains, regarding Huawei UB-Mesh-Racks as SUHNs, the design pattern of Huawei UB-Mesh-Pod's topology (Liao et al. 2025) is <CG(any), *P* >. Its design in each step is illustrated as follows:

- **Intra-module design**. The internal graph is a CG with 4 SUHNs. In fact, each module under the design scheme corresponds precisely to a 1D full-mesh rack row in Huawei's topology design.
- **Connection design**. In each module, all SUHNs are selected as dock nodes. Each dock node serves as an intermediate super-node, i.e., an ISN-(1, 1). Then the dock node interconnection is constructed under the plain mode. Each composite-edge between two ISNs is designed as an all-to-all interconnection.
- **Inter-module design**. The skeleton graph is a CG with 4 nodes. Each node corresponds to a module. Each super-edge between two modules is instantiated as a 4-parallelized all-to-all interconnection. Huawei refers to this topology figuratively as a 16-rack 2D full-mesh topology.

Advances in high–compute-density rack designs, long-reach high-speed copper interconnects, active electrical cable (AEC) technologies, high-bandwidth low-latency optical interconnect solutions, and OCS systems have collectively facilitated a substantial increase in the scale of scale-up domains (Ding et al. 2025). As demonstrated by the examples earlier, the capacity of a single scale-up domain has expanded from the initial 8 AI accelerators to 32, 64, 72, 128, 144, 384, 512, 1024, and even Google Ironwood TPU Superpod's 9,216 AI accelerators (Google Cloud 2025).

For a range of inference workloads, such as real-time dialogue generation with Llama-70B (Meta 2024), real-time multimodal queries with Llama-405B (Meta AI 2024), complex mathematical reasoning with DeepSeek-R1 (DeepSeek-AI 2025), and agent-style code generation with Claude Opus 4.5 (Anthropic 2025), a single, sufficiently large scale-up domain (e.g., configurations with 384, 512, or 9,216 chips) is already capable of delivering efficient inference performance (Xu et al. 2024).

However, the training of trillion-parameter frontier models, such as GPT-4, OpenAI Orion, Llama 4, and xAI Grok 3, demands clusters comprising tens of thousands of AI accelerators (Stanford Institute for Human-Centered Artificial Intelligence [HAI] 2025). In these scenarios, the capacity of a single scale-up domain remains far from adequate, thereby necessitating the use of the scale-out domain to interconnect multiple scale-up domains (Ye et al. 2024; Gangidi et al. 2024).

5.1.3.3 Scale-out Domain

In Section 5.1.3.2, we designed the topologies of scale-up domains under our unified design scheme for scale-up domains through one or multiple iterations. As discussed earlier, the scale of a single scale-up domain remains insufficient to satisfy the computational demands of training trillion-parameter models. Therefore, it is necessary to employ high-performance interconnect technologies such as IB (InfiniBand Trade Association 2023) and Ethernet (Wikipedia 2025; IEEE Standards Association 2022), which provide scalability far beyond that of the high-bandwidth networks within scale-up domains while offering significantly lower cost (Aramide 2024; Pichetti 2024), to construct the scale-out domain that interconnects multiple scale-up domains, thereby forming an AI accelerator cluster of sufficient scale (Ye et al. 2024).

In scale-out domain that contains multiple scale-up domains, topology designs for distinct scale-up domains may be different. However, due to the requirement for synchronization in parallel strategies, which necessitates symmetry in network topologies of AI clusters (Lizier et al. 2023), as well as for ease of analysis, we assume that all scale-up domains in an AI cluster share the same topology. The subsequent discussion of our unified design scheme for the scale-out domain is conducted under this assumption.

In our unified design scheme for AI clusters, the topology of the scale-out domain is designed on the basis of the previously established scale-up-domain topology. Specifically, the topology of the scale-out domain is designed by treating the established scale-up-domain topology as a module and initiating new iterations under our design framework. More explicitly, the topology design of the scale-out domain follows immediately after the topology design of the scale-up domains, together forming a standard design procedure under our design framework that consists of multiple iterations. In this procedure, the topology design of the scale-out domain corresponds to the design steps starting from the iteration that treats the scale-up-domain topology, which was just generated in the proceeding iteration, as a module.

In Figure 5.7, we present the complete workflow of the unified design scheme for AI clusters.

1. **Design Patterns**
   Since the scale-out-domain network relies on a communication stack that is fundamentally different from that of the high-bandwidth networks within scale-up domains (Aramide 2024), GPUs (i.e., SUHNs under the design scheme for scale-up domains) must be equipped with network interface cards (NICs) to access scale-out-domain networks such as IB and Ethernet (AFL 2024; Peter Jones 2025; Ye et al. 2024; Gangidi et al. 2024).

   This implies that, for AI clusters with scale-out-domain networks such as IB and Ethernet, under our unified topology design scheme, only SUHNs (i.e., GPUs) in its scale-up-domain topology can have direct connections to nodes outside the scale-up domain they are in.

   Consequently, in the design procedure of our unified topology design scheme for AI clusters, for the iterations pertaining to the topology design of the scale-out domain, i.e., the iterations starting from the one that treats the scale-up-domain topology as a module, in their connection design, dock nodes cannot be selected from any nodes in the scale-up-domain topology other than SUHNs (i.e., GPUs), i.e., SURNs.

   In particular, for the iteration at which the topology design of the scale-out domain initiates, i.e., the iteration that treats the scale-up-domain topology as

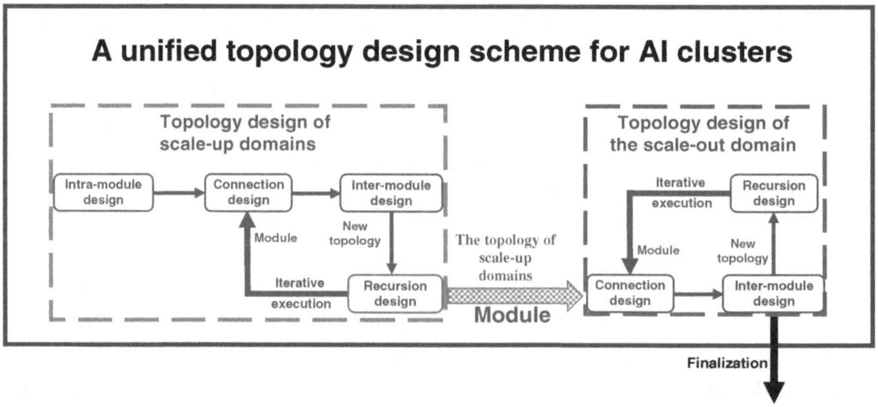

Figure 5.7 A unified topology design scheme for AI clusters.

a module, in its connection design, dock nodes can only be selected from the SUHNs in the scale-up-domain topology.

It is worth noting that, in a small number of AI cluster architectures, the switches they employed for scale-out-domain interconnections can be directly connected with the switches they leveraged in scale-up-domain interconnections. For example, in Huawei's UB-Mesh (Liao et al. 2025), the low-radix switches (LRS) utilized in scale-up-domain interconnections—the intra-Rack and intra-Pod interconnects, can be directly connected with the high-radix switches (HRS) deployed for scale-out-domain interconnections—the inter-Pod and inter-SuperPod interconnections; furthermore, the high-radix switches (HRS) can be further directly connected with the DCN switches that Huawei adopts for even larger-scale interconnections—large-scale data-center network scaling to 100K NPUs or more.

For such AI cluster architectures, the topology design of scale-up domains and the scale-out domain can be fused into a single, standard multi-iteration design procedure under our unified design framework, without imposing any additional domain-specific constraints. Consequently, the topology design of such AI clusters can be carried out directly by following our unified design framework, without explicitly distinguishing between the topology design of scale-up and scale-out domains, and therefore without the need to consider the aforementioned dock node selection constraints in the topology design of the scale-out domain.

Moreover, although GPUs (i.e., SUHNs) in distinct scale-up domains can be directly connected to each other through the NICs with which they are equipped, such direct GPU-to-GPU connections are generally not adopted in scale-out domains (De Sensi et al. 2024; Ren et al. 2025). This is because scale-out networking inherently requires switches to construct multi-hierarchical topologies that provide scalability, full connectivity, non-blocking bandwidth, and fault tolerance (Ye et al. 2024), whereas direct GPU-to-GPU connections are only suitable for two-node configurations and do not scale efficiently to large-scale AI and HPC clusters (NADDOD 2025; NVIDIA 2023b).

Consequently, for large-scale AI accelerator clusters, under the design procedure of our unified topology design scheme for AI clusters, in the iteration at which the topology design of the scale-out domain initiates, i.e., the iteration that treats the scale-up-domain topology as a module, the connection design generally selects GPUs (i.e., SUHNs) as dock nodes (Ye et al. 2024; Gangidi et al. 2024; AFL 2024; Peter Jones 2025), constructs the dock node interconnection under the bipartite mode, designs each composite-edge as a transversal design (TD), and accordingly, the inter-module design designs the skeleton graph as a block incidence graph (BIG) in which element nodes correspond to modules (i.e., scale-up domains) and block nodes correspond to RN clusters.

As a result, the design pattern of this iteration, which bridges the topology design of scale-up domains and that of the scale-out domain, is generally fixed as <scale-up-domain topology(SUHN), B >.

For iterations following the iteration at which the topology design of the scale-out domain initiates, that is, iterations pertaining to the topology design of the scale-out domain beyond this initiating iteration, their design patterns are consistent with the summary of design patterns for iteration i ($i \geq 2$) under the unified design framework, with the only restriction that SURNs cannot be selected as dock nodes in their connection design. Therefore, we do not restate the details here.

2. **Revisit of Existing AI Cluster Topologies**
   By following a unified design scheme for AI clusters, a great number of existing AI cluster topologies can be reverse-engineered, as summarized in Table 5.4. In the following paragraphs, we illustrate the corresponding scale-out-domain topologies for 12 representative real-world AI clusters, respectively. Note that the design step of finalization is omitted since there are no CNs.

NVIDIA DGX H100 SuperPOD: The scale-up domains of NVIDIA DGX H100 SuperPOD (NVIDIA 2023a) are NVIDIA DGX H100 Systems (NVIDIA 2025m), which have been reverse-engineered under the design scheme for scale-up domains. In particular, the design pattern of NVIDIA DGX H100 System is <BIG (-), -> where H100 Tensor Core GPUs serve as SUHNs and NVSwitches serve as SURNs, and its specific topology is a BIG-(1, 4, 8, 4) where element nodes and block nodes are viewed as SUHNs and SURNs, respectively.

On the basis of this scale-up-domain topology, under the design scheme for the scale-out domain, the topology of the scale-out domain of NVIDIA DGX H100 SuperPOD (NVIDIA 2023a) can be constructed by two rounds of iteration, where Leaf and Spine switches serve as RNs. In iteration 1, the established scale-up-domain topology is taken as modules, and the design pattern for the scale-out domain is <scale-up-domain topology(SUHN), B >. In iteration 2, the topologies generated in iteration 1 are taken as modules, and the design pattern for the scale-out domain evolves into <iteration 1(BIG-block), B >.[1] The design of the topology of the scale-out domain of NVIDIA DGX H100 SuperPOD at each step of the two iterations is illustrated as follows.

1 As the right-hand component of the term d_2 in the naming of the design pattern of an iteration under the design scheme for the scale-out domain, BIG-block means that dock nodes are selected from the block nodes of the skeleton graph in the inter-module design of the previous iteration, which is a BIG.

Table 5.4 A summary of scale-out-domain topologies in existing AI clusters.

AI cluster	Scale-out-domain topology design				
	Iteration count	Connection Design		Inter-module Design	
		PM	BM	CG	BIG
NVIDIA DGX H100 SuperPOD (NVIDIA 2023a)	2		•		•
NVIDIA DGX B200 SuperPOD (NVIDIA 2025d)	2		•		•
NVIDIA DGX B300 SuperPOD (NVIDIA 2025e)	2		•		•
NVIDIA DGX GB300 SuperPOD (NVIDIA 2025f)	2		•		•
Meta Rail-only Network (Wang et al. 2024)	1		•		•
Intel Gaudi 3 AI Accelerator Cluster (Intel 2025)	2		•		•
DeepSeek Multi-Plane Fat-Tree Network (Zhao et al. 2025)	2		•		•
DeepSeek Fire-Flyer AI-HPC (An et al. 2024)	3		•		•
Alibaba HPN's Backend Network (Qian et al. 2024)	3		•		•
Tencent Astral Network Architecture (Meng et al. 2025)	3		•		•
China Mobile HPN1.0 (Dyxnet 2025)	3		•		•
ZCube (Yan et al. 2025)	$k-1$		•		•

Abbreviations: BIG, bipartite incidence graph; BM, bipartite mode; CG, complete graph; PM, plain mode.

- **Iteration 1**:
  - **Intra-module design**. The established scale-up-domain topology is treated as modules.
  - **Connection design**. In each module, all SUHNs (element nodes of the scale-up-domain topology BIG-$(1, 4, 8, 4)$ under the design scheme for scale-up domains) are selected as dock nodes. Each dock node serves as an

intermediate super-node, i.e., an ISN-(1, 1, 2). Then the dock node interconnection is constructed under the bipartite mode. Each composite-edge is designed as a TD-(32, 1, 1) interconnection among 32 ISNs through an RN cluster.

- **Inter-module design**. The skeleton graph is a BIG-(1, 8, 32, 8). Such a graph contains 32 element nodes and 8 block nodes that correspond to modules and RN clusters, respectively. For any 32 modules, they are connected with a common RN cluster via 32 super-edges. Such super-edges form a 1-parallelized TD-(32, 1, 1) interconnection.

In fact, each topology generated by iteration 1 under the design scheme for the scale-out domain corresponds precisely to a scalable unit (SU) in NVIDIA's topology design.

- **Iteration 2**:
  - **Intra-module design**. The topologies generated in iteration 1 are treated as modules.
  - **Connection design**. In each module, all RNs in the block nodes (RN clusters) of the skeleton graph of iteration 1's inter-module design are selected as dock nodes. Each dock node serves as an intermediate super-node, i.e., an ISN-(1, 1, 3). Then the dock node interconnection is constructed under the bipartite mode. Each composite-edge is designed as a TD-(4, 16, 1) interconnection among 4 ISNs through an RN cluster.
  - **Inter-module design**. The skeleton graph is a BIG-(1, 1, 4, 1). Such a graph contains 4 element nodes and one block node that correspond to modules and an RN cluster, respectively. For any 4 modules, they are connected with a common RN cluster via 4 super-edges. Such super-edges form a 8-parallelized TD-(4, 16, 1) interconnection.

The topology of the scale-out domain of NVIDIA DGX H100 SuperPOD is described as a Rail-optimized full Fat-Tree topology in NVIDIA's official documentation. Under the design scheme for the scale-out domain, this is reflected by the fact that the 32 scale-up domains, i.e., NVIDIA DGX H100 Systems, in each topology generated by iteration 1 are rail-aligned, and the layer of switches introduced in iteration 1 (the Leaf switch layer) is fully meshed with the layer of switches introduced in iteration 2 (the Spine switch layer).

NVIDIA DGX B200 SuperPOD: The scale-up domains of NVIDIA DGX B200 SuperPOD (NVIDIA 2025d) are NVIDIA DGX B200 Systems (NVIDIA 2025a), which have been reverse-engineered under the design scheme for scale-up domains. In particular, the design pattern of NVIDIA DGX B200 System is <BIG (-), -> where B200 Tensor Core GPUs serve as SUHNs and NVSwitches serve as SURNs, and its specific topology is a BIG-(1, 2, 8, 2) where element nodes and block nodes are viewed as SUHNs and SURNs, respectively.

On the basis of this scale-up-domain topology, under the design scheme for the scale-out domain, the topology of the scale-out domain of NVIDIA DGX B200 SuperPOD (NVIDIA 2025d) can be constructed by two rounds of iteration, where Leaf and Spine switches serve as RNs. In iteration 1, the established scale-up-domain topology is taken as modules, and the design pattern for the scale-out domain is <scale-up-domain topology(SUHN), B >. In iteration 2, the topologies generated in iteration 1 are taken as modules, and the design pattern for the scale-out domain evolves into <iteration 1(BIG-block), B >. The design of the topology of the scale-out domain of NVIDIA DGX B200 SuperPOD at each step of the two iterations is illustrated as follows.

- **Iteration 1**:
  - **Intra-module design**. The established scale-up-domain topology is treated as modules.
  - **Connection design**. In each module, all SUHNs (element nodes of the scale-up-domain topology BIG-(1, 2, 8, 2) under the design scheme for scale-up domains) are selected as dock nodes. Each dock node serves as an intermediate super-node, i.e., an ISN-(1, 1, 2). Then the dock node interconnection is constructed under the bipartite mode. Each composite-edge is designed as a TD-(32, 1, 1) interconnection among 32 ISNs through an RN cluster.
  - **Inter-module design**. The skeleton graph is a BIG-(1, 8, 32, 8). Such a graph contains 32 element nodes and 8 block nodes that correspond to modules and RN clusters, respectively. For any 32 modules, they are connected with a common RN cluster via 32 super-edges. Such super-edges form a 1-parallelized TD-(32, 1, 1) interconnection.

  In fact, each topology generated by iteration 1 under the design scheme for the scale-out domain corresponds precisely to a SU in NVIDIA's topology design.

- **Iteration 2**:
  - **Intra-module design**. The topologies generated in iteration 1 are treated as modules.
  - **Connection design**. In each module, all RNs in the block nodes (RN clusters) of the skeleton graph of iteration 1's inter-module design are selected as dock nodes. Each dock node serves as an intermediate super-node, i.e., an ISN-(1, 1, 3). Then the dock node interconnection is constructed under the bipartite mode. Each composite-edge is designed as a TD-(4, 16, 1) interconnection among 4 ISNs through an RN cluster.
  - **Inter-module design**. The skeleton graph is a BIG-(1, 1, 4, 1). Such a graph contains 4 element nodes and one block node that correspond to modules and an RN cluster, respectively. For any 4 modules, they are connected with a common RN cluster via 4 super-edges. Such super-edges form a 8-parallelized TD-(4, 16, 1) interconnection.

In fact, as the 32 scale-up domains, i.e., NVIDIA DGX B200 Systems, in each topology generated by iteration 1 are rail-aligned, and the layer of switches introduced in iteration 1 (the Leaf switch layer) is fully meshed with the layer of switches introduced in iteration 2 (the Spine switch layer), the topology of the scale-out domain of NVIDIA DGX B200 SuperPOD is also a Rail-optimized Fat-Tree topology.

NVIDIA DGX B300 SuperPOD: The scale-up domains of NVIDIA DGX B300 SuperPOD (NVIDIA 2025e) are NVIDIA DGX B300 Systems (NVIDIA 2025b), which have been reverse-engineered under the design scheme for scale-up domains. In particular, the design pattern of NVIDIA DGX B300 System is <BIG (-), -> where B300 Tensor Core GPUs serve as SUHNs and NVSwitches serve as SURNs, and its specific topology is a BIG-(1, 2, 8, 2) where element nodes and block nodes are viewed as SUHNs and SURNs, respectively.

On the basis of this scale-up-domain topology, under the design scheme for the scale-out domain, the topology of the scale-out domain of NVIDIA DGX B300 SuperPOD (NVIDIA 2025e) can be constructed by two rounds of iteration, where Leaf and Spine switches serve as RNs. In iteration 1, the established scale-up-domain topology is taken as modules, and the design pattern for the scale-out domain is <scale-up-domain topology(SUHN), B >. In iteration 2, the topologies generated in iteration 1 are taken as modules, and the design pattern for the scale-out domain evolves into <iteration 1(BIG-block), B >. The design of the topology of the scale-out domain of NVIDIA DGX B300 SuperPOD at each step of the two iterations is illustrated as follows.

- **Iteration 1**:
  - **Intra-module design**. The established scale-up-domain topology is treated as modules.
  - **Connection design**. In each module, all SUHNs (element nodes of the scale-up-domain topology BIG-(1, 2, 8, 2) under the design scheme for scale-up domains) are selected as dock nodes. Each dock node serves as an intermediate super-node, i.e., an ISN-(1, 1, 2). Then the dock node interconnection is constructed under the bipartite mode. Each composite-edge is designed as a TD-(72, 1, 1) interconnection among 72 ISNs through an RN cluster.
  - **Inter-module design**. The skeleton graph is a BIG-(1, 8, 72, 8). Such a graph contains 72 element nodes and 8 block nodes that correspond to modules and RN clusters, respectively. For any 72 modules, they are connected with a common RN cluster via 72 super-edges. Such super-edges form a 1-parallelized TD-(72, 1, 1) interconnection.

In fact, each topology generated by iteration 1 under the design scheme for the scale-out domain corresponds precisely to a SU in NVIDIA's topology design.

- **Iteration 2**:
  - **Intra-module design**. The topologies generated in iteration 1 are treated as modules.
  - **Connection design**. In each module, all RNs in the block nodes (RN clusters) of the skeleton graph of iteration 1's inter-module design are selected as dock nodes. Each dock node serves as an intermediate super-node, i.e., an ISN-(1, 1, 3). Then the dock node interconnection is constructed under the bipartite mode. Each composite-edge is designed as a TD-(8, 36, 1) interconnection among 8 ISNs through an RN cluster.
  - **Inter-module design**. The skeleton graph is a BIG-(1, 1, 8, 1). Such a graph contains 8 element nodes and one block node that correspond to modules and an RN cluster, respectively. For any 8 modules, they are connected with a common RN cluster via 8 super-edges. Such super-edges form a 8-parallelized TD-(8, 36, 1) interconnection.

In fact, as the 72 scale-up domains, i.e., NVIDIA DGX B300 Systems, in each topology generated by iteration 1 are rail-aligned, and the layer of switches introduced in iteration 1 (the Leaf switch layer) is fully meshed with the layer of switches introduced in iteration 2 (the Spine switch layer), the topology of the scale-out domain of NVIDIA DGX B300 SuperPOD is also a Rail-optimized Fat-Tree topology.

NVIDIA DGX GB300 SuperPOD: The scale-up domains of NVIDIA DGX GB300 SuperPOD (NVIDIA 2025f) are NVIDIA DGX GB300 systems (NVIDIA 2025c). NVIDIA DGX GB300 system, namely NVIDIA GB300 NVL72, retains the NVSwitch and networking architecture of NVIDIA GB200 NVL72 (NVIDIA 2025h), while replacing the NVIDIA B200 Tensor Core GPUs with NVIDIA B300 Tensor Core GPUs, thereby increasing the per-GPU HBM3e capacity from 192 to 288 GB and achieving a 1.5× FP4 compute throughput improvement. Hence, NVIDIA DGX GB300 system shares the same scale-up-domain topology with NVIDIA GB200 NVL72, which has been reverse-engineered under the design scheme for scale-up domains. In particular, the design pattern of NVIDIA DGX GB300 system is <BIG(-), -> where NVIDIA B300 Tensor Core GPUs serve as SUHNs and NVSwitches serve as SURNs, and its specific topology is a BIG-(1, 18, 72, 18) where element nodes and block nodes are viewed as SUHNs and SURNs, respectively.

On the basis of this scale-up-domain topology, under the design scheme for the scale-out domain, the topology of the scale-out domain of NVIDIA DGX GB300 SuperPOD (NVIDIA 2025f) can be constructed by two rounds of iteration, where Q3400-RD Quantum-X800 switches serve as RNs. In iteration 1, the established scale-up-domain topology is taken as modules, and the design pattern for the scale-out domain is <scale-up-domain topology(SUHN), B>. In iteration 2, the topologies generated in iteration 1 are taken as modules, and the design pattern for the scale-out domain evolves into <iteration 1(BIG-block), B >. The design of the topology of the scale-out domain of NVIDIA DGX GB300 SuperPOD at each step of the two iterations is illustrated as follows.

- **Iteration 1**:
  - **Intra-module design**. The established scale-up-domain topology is treated as modules.
  - **Connection design**. In each module, all SUHNs (element nodes of the scale-up-domain topology BIG-(1, 18, 72, 18) under the design scheme for scale-up domains) are selected as dock nodes. Each dock node serves as an intermediate super-node, i.e., an ISN-(1, 1, 2). Then the dock node interconnection is constructed under the bipartite mode. Each composite-edge is designed as a TD-(4, 1, 1) interconnection among 4 ISNs through an RN cluster.
  - **Inter-module design**. The skeleton graph is a BIG-(1, 4, 4, 4). Such a graph contains 4 element nodes and 4 block nodes that correspond to modules and RN clusters, respectively. For any 4 modules, they are connected with a common RN cluster via 4 super-edges. Such super-edges form a 18-parallelized TD-(4, 1, 1) interconnection.

  In fact, each topology generated by iteration 1 under the design scheme for the scale-out domain corresponds precisely to half of a SU in NVIDIA's topology design. Moreover, the 4 block nodes (RN clusters) in each topology generated by iteration 1 under the design scheme for the scale-out domain belong respectively to the 4 rails in each SU in NVIDIA's topology design, with each of the 4 block nodes (RN clusters) respectively constituting half of each of the 4 rails.

- **Iteration 2**:
  - **Intra-module design**. The topologies generated in iteration 1 are treated as modules.
  - **Connection design**. In each module, all RNs in the block nodes (RN clusters) of the skeleton graph of iteration 1's inter-module design are selected as dock nodes. Each dock node serves as an intermediate super-node, i.e., an ISN-(1, 1, 3). Then the dock node interconnection is constructed under the bipartite mode. Each composite-edge is designed as a TD-(32, 72, 1) interconnection among 32 ISNs through an RN cluster.
  - **Inter-module design**. The skeleton graph is a BIG-(1, 1, 32, 1). Such a graph contains 32 element nodes and one block node that correspond to modules and an RN cluster, respectively. For any 32 modules, they are connected with a common RN cluster via 32 super-edges. Such super-edges form a 4-parallelized TD-(32, 72, 1) interconnection.

Under the design scheme for the scale-out domain, the two iterations respectively introduce the first and second layers of switches for the scale-out domain of NVIDIA DGX GB300 SuperPOD, corresponding to NVIDIA's two-layer design.

Furthermore, as the 4 scale-up domains, i.e., NVIDIA DGX GB300 Systems, in each topology generated by iteration 1 are rail-aligned, and the layer of switches introduced in iteration 1 (the leaf layer) is fully meshed with the layer of switches introduced in iteration 2 (the spine layer), the topology of the scale-out domain of NVIDIA DGX GB300 SuperPOD is also a Rail-optimized Fat-Tree topology.

Meta Rail-only Network: In Meta's rail-only network, each high-bandwidth domain corresponds to a scale-up domain under the unified topology design scheme for AI clusters, where GPUs serve as SUHNs; however, its topology has not been publicly disclosed.

The design pattern of the topology of the scale-out domain of Meta Rail-only Network is <scale-up-domain topology(SUHN), B >, where rail switches serve as RNs, and its design in each step is illustrated as follows:

- **Intra-module design**. The established scale-up-domain topology is treated as modules.
- **Connection design**. In each module, all SUHNs are selected as dock nodes. Each dock node serves as an intermediate super-node, i.e., an ISN-$(1, 1, d_{HBD})$. Then the dock node interconnection is constructed under the bipartite mode. Each composite-edge is designed as a TD-$(M, 1, 1)$ interconnection among M ISNs through an RN cluster.
- **Inter-module design**. The skeleton graph is a BIG-$(1, K, M, K)$. Such a graph contains M element nodes and K block nodes that correspond to modules and RN clusters, respectively. For any M modules, they are connected with a common RN cluster via M super-edges. Such super-edges form a 1-parallelized TD-$(M, 1, 1)$ interconnection. In fact, each composite-edge TD-$(M, 1, 1)$ under the design scheme for the scale-out domain corresponds precisely to a rail in Meta's topology design.

Intel Gaudi 3 AI Accelerator Cluster

The scale-up domains of Intel Gaudi 3 AI Accelerator Cluster (Intel 2025) are Intel Gaudi3 AI Accelerator HLB-325 Baseboards (Intel 2024a), which have been reverse-engineered under the design scheme for scale-up domains. In particular, the design pattern of Intel Gaudi3 AI Accelerator HLB-325 Baseboard is <CG(-), -> where Gaudi3 AI accelerators serve as SUHNs, and its specific topology is a CG with 8 SUHNs.

On the basis of this scale-up-domain topology, under the design scheme for the scale-out domain, the topology of the scale-out domain of Intel Gaudi 3 AI Accelerator Cluster (Intel 2025) can be constructed by two rounds of iteration, where Ethernet switches serve as RNs. In iteration 1, the established scale-up-domain topology is taken as modules, and the design pattern for the scale-out domain is <scale-up-domain topology(SUHN), B >. In iteration 2, the topologies generated in iteration 1 are taken as modules, and the design pattern for the scale-out domain evolves into <iteration 1(BIG-block), B >. The design of the topology of the scale-out domain of Intel Gaudi 3 AI Accelerator Cluster at each step of the two iterations is illustrated as follows.

- **Iteration 1**:
  - **Intra-module design**. The established scale-up-domain topology is treated as modules. In fact, each module of iteration 1 under the design scheme for the scale-out domain corresponds precisely to a node in Intel's topology design.
  - **Connection design**. In each module, all six OSFPs with which the SUHNs equipped are selected as dock nodes. Each dock node serves as an intermediate super-node, i.e., an ISN-(1, 1). Then the dock node interconnection is constructed under the bipartite mode. Each composite-edge is designed as a TD-(16, 1, 1) interconnection among 16 ISNs through an RN cluster.
  - **Inter-module design**. The skeleton graph is a BIG-(1, 3, 16, 3). Such a graph contains 16 element nodes and 3 block nodes that correspond to modules and RN clusters, respectively. For any 16 modules, they are connected with a common RN cluster via 16 super-edges. Such super-edges form a 2-parallelized TD-(16, 1, 1) interconnection.

  In fact, each topology generated by iteration 1 under the design scheme for the scale-out domain corresponds precisely to a sub-cluster in Intel's topology design.
- **Iteration 2**:
  - **Intra-module design**. The topologies generated in iteration 1 are treated as modules.
  - **Connection design**. In each module, all RNs in the block nodes (RN clusters) of the skeleton graph of iteration 1's inter-module design are selected as dock nodes. Each dock node serves as an intermediate super-node, i.e., an ISN-(1, 1, 2). Then the dock node interconnection is constructed under the bipartite mode. Each composite-edge is designed as a TD-(32, 16, 1) interconnection among 32 ISNs through an RN cluster.
  - **Inter-module design**. The skeleton graph is a BIG-(1, 3, 32, 3). Such a graph contains 32 element nodes and 3 block nodes that correspond to modules and RN clusters, respectively. For any 32 modules, they are connected with a common RN cluster via 32 super-edges. Such super-edges form a 1-parallelized TD-(32, 16, 1) interconnection.

DeepSeek Multi-Plane Fat-Tree Network: The scale-up domains of DeepSeek Multi-Plane Fat-Tree network (Zhao et al. 2025) are NVIDIA H800 GPU SXM systems (NVIDIA 2025n). NVIDIA H800 SXM GPU system is a weakened variant of NVIDIA DGX H100 system. Specifically, the NVLink bandwidth is reduced from 900 to 400 GB/s, whereas the NVSwitch and networking architecture are preserved. Hence, NVIDIA H800 SXM GPU system shares the same scale-up-domain topology with DGX H100 system (NVIDIA 2025m), which has been reverse-engineered under the design scheme for scale-up domains.

In particular, the design pattern of NVIDIA H800 SXM GPU system is <BIG (-), -> where H800 GPUs serve as SUHNs and NVSwitches serve as SURNs, and its specific topology is a BIG-(1, 4, 8, 4) where element nodes and block nodes are viewed as SUHNs and SURNs, respectively.

On the basis of this scale-up-domain topology, under the design scheme for the scale-out domain, the topology of the scale-out domain of DeepSeek Multi-Plane Fat-Tree Network (Zhao et al. 2025) can be constructed by two rounds of iteration, where IB switches serve as RNs. In iteration 1, the established scale-up-domain topology is taken as modules, and the design pattern for the scale-out domain is <scale-up-domain topology(SUHN), B >. In iteration 2, the topologies generated in iteration 1 are taken as modules, and the design pattern for the scale-out domain evolves into <iteration 1(BIG-block), B >. Steps of the topology design for scale-out domain of DeepSeek Multi-Plane Fat-Tree Network are illustrated as follows.

- **Iteration 1**:
  - **Intra-module design**. The established scale-up-domain topology is treated as modules.
  - **Connection design**. In each module, all SUHNs (element nodes of the scale-up-domain topology BIG-(1, 4, 8, 4) under the design scheme for scale-up domains) are selected as dock nodes. Each dock node serves as an intermediate super-node, i.e., an ISN-(1, 1, 2). Then the dock node interconnection is constructed under the bipartite mode. Each composite-edge is designed as a TD-(32, 1, 1) interconnection among 32 ISNs through an RN cluster.
  - **Inter-module design**. The skeleton graph is a BIG-(1, 8, 32, 8). Such a graph contains 32 element nodes and 8 block nodes that correspond to modules and RN clusters, respectively. For any 32 modules, they are connected with a common RN cluster via 32 super-edges. Such super-edges form a 1-parallelized TD-(32, 1, 1) interconnection.

In fact, the 8 block nodes (RN clusters) in each topology generated by iteration 1 under the design scheme for the scale-out domain belong respectively to the eight planes in DeepSeek's topology design.

- **Iteration 2**:
  - **Intra-module design**. The topologies generated in iteration 1 are treated as modules.
  - **Connection design**. In each module, all RNs in the block nodes (RN clusters) of the skeleton graph of iteration 1's inter-module design are selected as dock nodes. Each dock node serves as an intermediate super-node, i.e., an ISN-(1, 1, 3). Then the dock node interconnection is constructed under the bipartite mode. Each composite-edge is designed as a TD-(64, 32, 1) interconnection among 64 ISNs through an RN cluster.
  - **Inter-module design**. The skeleton graph is a BIG-(1, 8, 64, 8). Such a graph contains 64 element nodes and 8 block nodes that correspond to modules and RN clusters, respectively. For any 64 modules, they are connected with a common RN cluster via 64 super-edges. Such super-edges form a 1-parallelized TD-(64, 32, 1) interconnection.

  In fact, each composite-edge TD-(64, 32, 1) in iteration 2 under the design scheme for the scale-out domain corresponds precisely to a plane in DeepSeek's topology design. Collectively, the eight composite edges constitute DeepSeek's eight-plane design.

Under the design scheme for the scale-out domain, iterations 1 and 2 respectively introduce the first and second layers of switches, i.e., the layer of Leaf switches and the layer of Spine switches, together constituting DeepSeek's two-layer scale-out network design.

DeepSeek Fire-Flyer AI-HPC: In DeepSeek Fire-Flyer AI-HPC (An et al. 2024), each GPU compute node containing 8 NVIDIA PCIe A100 GPUs (NVIDIA 2020a) and a single 200 Gbps NVIDIA Mellanox Connect-6 (CX6) IB NIC (NVIDIA 2022) corresponds to a scale-up domain with a degree of one to the scale-out domain under the unified topology design scheme for AI clusters. Therefore, one can either view each GPU compute node as a CN and directly apply the unified design framework, as demonstrated in Section 5.1.1, or carry out the design scheme for the scale-out domain on top of these scale-up domains, which is detailed here.

Under the design scheme for the scale-out domain, the topology of the scale-out domain of DeepSeek Fire-Flyer AI-HPC (An et al. 2024) can be constructed by three rounds of iteration, where Mellanox QM8700 IB Switches (NVIDIA 2020b), each offering 40 ports at 200 Gbps, serve as RNs.

In iteration 1, the established scale-up domains are taken as modules, and the design pattern for the scale-out domain is <scale-up-domain topology, B>. In iteration 2, the topologies generated in iteration 1 are taken as modules, and the design pattern for the scale-out domain evolves into <iteration 1(BIG-block), B>. In iteration 3, the topologies generated in iteration 2 are taken as modules, and the design pattern for the scale-out domain evolves into <iteration 2(BIG-element),

$B >$.[2] The design of the topology of the scale-out domain of DeepSeek Fire-Flyer AI-HPC at each step of the three iterations is illustrated as follows.

- **Iteration 1**:
  - **Intra-module design**. The established scale-up domains are treated as modules.
  - **Connection design**. Each module itself serves as a dock node and correspondingly an intermediate super-node, i.e., an ISN-(1, 1, d_{cn}). Then the dock node interconnection is constructed under the bipartite mode. Each composite-edge is designed as a TD-(15, 1, 1) interconnection among 15 ISNs through an RN cluster.
  - **Inter-module design**. The skeleton graph is a BIG-(1, 1, 15, 1). Such a graph contains 15 element nodes and one block node that correspond to modules and an RN cluster, respectively. For any 15 modules, they are connected with a common RN cluster via 15 super-edges. Such super-edges form a 1-parallelized TD-(15, 1, 1) interconnection.

  In fact, each RN cluster of iteration 1 under the design scheme for the scale-out domain corresponds precisely to a leaf switch in DeepSeek's topology design. Moreover, iteration 1 under the design scheme for the scale-out domain corresponds precisely to the finalization step under the unified design framework.
- **Iteration 2**:
  - **Intra-module design**. The topologies generated in iteration 1 are treated as modules.
  - **Connection design**. In each module, all RNs in the block nodes (RN clusters) of the skeleton graph of iteration 1's inter-module design are selected as dock nodes. Each dock node serves as an intermediate super-node, i.e., an ISN-(1, 1, 1 + d_{cn}). Then the dock node interconnection is constructed under the bipartite mode. Each composite-edge is designed as a TD-(40, 20, 1) interconnection among 40 ISNs through an RN cluster.
  - **Inter-module design**. The skeleton graph is a BIG-(1, 1, 40, 1). Such a graph contains 40 element nodes and one block node that correspond to modules and an RN cluster, respectively. For any 40 modules, they are connected with a common RN cluster via 40 super-edges. Such super-edges form a 1-parallelized TD-(40, 20, 1) interconnection.

2 As the right-hand component of the term d_2 in the naming of the design pattern of an iteration under the design scheme for the scale-out domain, BIG-element means that dock nodes are selected from the element nodes of the skeleton graph in the inter-module design of the previous iteration, which is a BIG.

In fact, each topology generated by iteration 2 under the design scheme for the scale-out domain corresponds precisely to a zone in DeepSeek's topology design.

- **Iteration 3**:
  - **Intra-module design**. The topologies generated in iteration 2 are treated as modules.
  - **Connection design**. In each module, all RNs in the element nodes of the skeleton graph of iteration 2's inter-module design are selected as dock nodes. Each dock node serves as an intermediate super-node, i.e., an ISN-$(1, 1, 3 + d_{cn})$. Then the dock node interconnection is constructed under the bipartite mode. Each composite-edge is designed as a TD-$(2, 1, 1)$ interconnection among 2 ISNs through an RN cluster.
  - **Inter-module design**. The skeleton graph is a BIG-$(1, 2, 2, 2)$. Such a graph contains 2 element nodes and 2 block nodes that correspond to modules and RN clusters, respectively. For any 2 modules, they are connected with a common RN cluster via 2 super-edges. Such super-edges form a 20-parallelized TD-$(2, 1, 1)$ interconnection.

Alibaba HPN's Backend Network: In Alibaba HPN's backend network (Qian et al. 2024), each host containing 8 GPUs corresponds to a scale-up domain under the unified topology design scheme for AI clusters, where GPUs serve as SUHNs; however, its topology has not been publicly disclosed.

Under the design scheme for the scale-out domain, the topology of the scale-out domain of Alibaba HPN's backend network (Qian et al. 2024) can be constructed by three rounds of iteration, where the ToR, Aggregation, and Core switches serve as RNs.

In iteration 1, the established scale-up-domain topology is taken as modules, and the design pattern for the scale-out domain is <scale-up-domain topology (SUHN), B >. In iteration 2, the topologies generated in iteration 1 are taken as modules, and the design pattern for the scale-out domain evolves into <iteration 1(BIG-block), B >. In iteration 3, the topologies generated in iteration 2 are taken as modules, and the design pattern for the scale-out domain evolves into <iteration 2(BIG-block), B >. Steps of topology design for scale-out domain of Alibaba HPN's backend network are illustrated as follows.

- **Iteration 1**:
  - **Intra-module design**. The established scale-up-domain topology is treated as modules.
  - **Connection design**. In each module, all SUHNs are selected as dock nodes. Each dock node serves as an intermediate super-node, i.e., an ISN-$(1, 1, d_{\text{host}})$. We show the ISNs-$(1, 1, d_{\text{host}})$ in a module of iteration 1 in Figure 5.8. Then the dock node interconnection is constructed under the

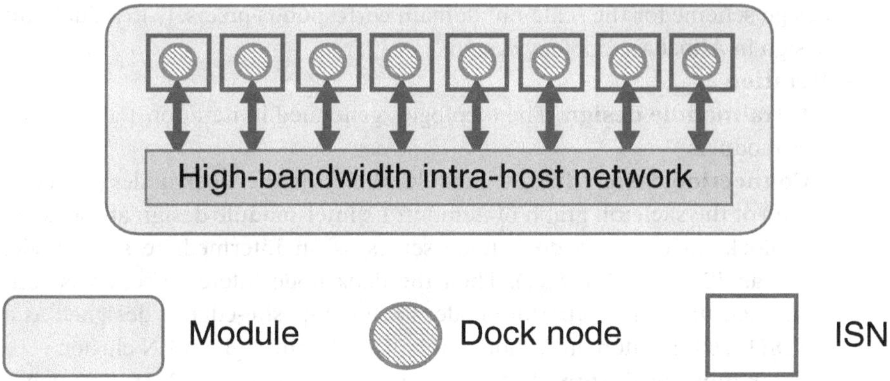

Figure 5.8 ISNs-$(1, 1, d_{\text{host}})$ in a module of iteration 1 under the design scheme for the scale-out domain of Alibaba HPN's backend network.

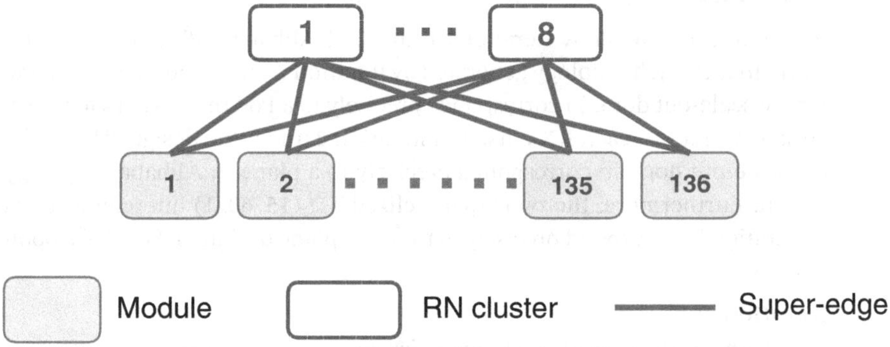

Figure 5.9 The skeleton graph BIG-(1, 8, 136, 8) of inter-module design in iteration 2 under the design scheme for the scale-out domain of Alibaba HPN's backend network.

bipartite mode. Each composite-edge is designed as a TD-(136, 2, 1) interconnection among 136 ISNs through an RN cluster.

- **Inter-module design**. The skeleton graph is a BIG-(1, 8, 136, 8). Such a graph contains 136 element nodes and 8 block nodes that correspond to modules and RN clusters, respectively, as illustrated in Figure 5.9. For any 136 modules, they are connected with a common RN cluster via 136 super-edges. Such super-edges form a 1-parallelized TD-(136, 2, 1) interconnection.

In fact, each topology generated by iteration 1 under the design scheme for the scale-out domain corresponds precisely to a segment in Alibaba's topology design. Moreover, each composite-edge TD-(136, 2, 1) in iteration 1 under the design scheme for the scale-out domain corresponds precisely to a dual-ToR design in Alibaba's topology design.

- **Iteration 2**:
  - **Intra-module design**. The topologies generated in iteration 1 are treated as modules.
  - **Connection design**. In each module, all RNs in the block nodes (RN clusters) of the skeleton graph of iteration 1's inter-module design are selected as dock nodes. Each dock node serves as an intermediate super-node, i.e., an ISN-$(1, 1, 1 + d_{\text{host}})$. Then the dock node interconnection is constructed under the bipartite mode. Each composite-edge is designed as a TD-(15, 60, 1) interconnection among 15 ISNs through an RN cluster.
  - **Inter-module design**. The skeleton graph is a BIG-(1, 2, 15, 2). Such a graph contains 15 element nodes and 2 block nodes that correspond to modules and RN clusters, respectively. For any 15 modules, they are connected with a common RN cluster via 15 super-edges. Such super-edges form a 8-parallelized TD-(15, 60, 1) interconnection.

In Figure 5.10, we show iterations 1 and 2 of Alibaba HPN's backend network. In fact, each topology generated by iteration 2 under the design scheme for the scale-out domain corresponds precisely to a Pod in Alibaba's topology design. Moreover, each RN cluster of iteration 2 under the design scheme for the scale-out domain corresponds precisely to a plane in Alibaba's topology design. Furthermore, the two 8-parallelized TD-(15, 60, 1) interconnections in iteration 2 correspond precisely to the dual-plane design in Alibaba's topology design.

- **Iteration 3**:
  - **Intra-module design**. The topologies generated in iteration 2 are treated as modules.

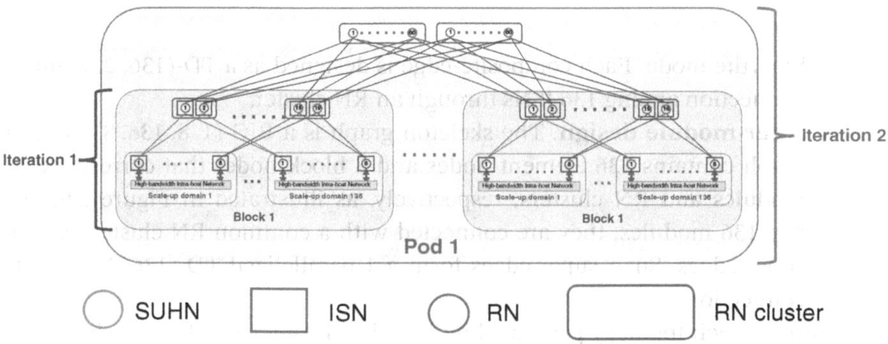

Figure 5.10 Iterations 1 and 2 under the design scheme for the scale-out domain of Alibaba HPN's backend network.

- **Connection design**. In each module, all RNs in the block nodes (RN clusters) of the skeleton graph of iteration 2's inter-module design are selected as dock nodes. Each dock node serves as an intermediate super-node, i.e., an ISN-(1, 1, 2). Then the dock node interconnection is constructed under the bipartite mode. Each composite-edge is designed as a TD-(N, 8, 1) interconnection among N ISNs through an RN cluster.
- **Inter-module design**. The skeleton graph is a BIG-(1, 8, N, 8). Such a graph contains N element nodes and 8 block nodes that correspond to modules and RN clusters, respectively. For any N modules, they are connected with a common RN cluster via N super-edges. Such super-edges form a 15-parallelized TD-(N, 8, 1) interconnection.

In Figure 5.11, we show iteration 3 of Alibaba HPN's backend network.

Tencent Astral Network Architecture: For the scale-up domains of Tencent Astral network architecture (Meng et al. 2025), under the design scheme for scale-up domains, their design pattern is <BIG(-), ->, where GPUs serve as SUHNs and NVSwitches serve as SURNs (NVIDIA 2025o). And the specific topology of the scale-up domains under the design scheme for scale-up domains is a BIG-(1, r_{SU}, 8, r_{SU}), where element nodes and block nodes are viewed as SUHNs and SURNs, respectively.

On the basis of this scale-up-domain topology, under the design scheme for the scale-out domain, the topology of the scale-out domain of Tencent Astral network architecture (Meng et al. 2025) can be constructed by three rounds of iteration, where the ToR, Aggregation, and Core switches serve as RNs. In iteration 1, the established scale-up-domain topology is taken as modules, and the design pattern for the scale-out domain is <scale-up-domain topology(SUHN), B>. In iteration 2, the topologies generated in iteration 1 are taken as modules, and the design pattern for the scale-out domain evolves into <iteration 1(BIG-block), B>. In iteration 3,

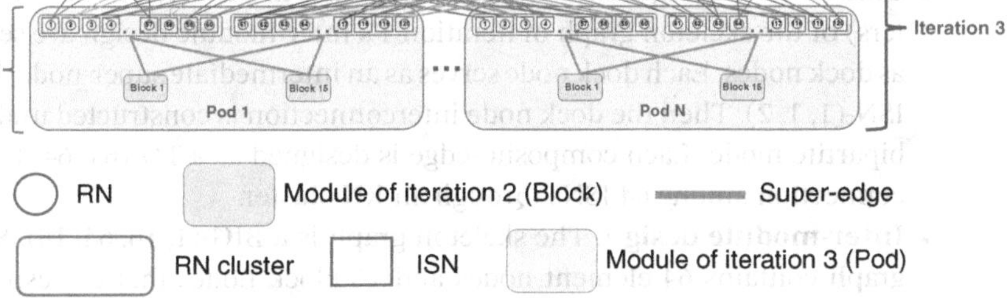

Figure 5.11 Iteration 3 under the design scheme for the scale-out domain of Alibaba HPN's backend network.

the topologies generated in iteration 2 are taken as modules, and the design pattern for the scale-out domain evolves into <iteration 2(BIG-block), B >. The design of the topology of the scale-out domain of Tencent Astral Network Architecture at each step of the three iterations is illustrated as follows.

- **Iteration 1**:
  - **Intra-module design**. The established scale-up-domain topology is treated as modules. In fact, each module of iteration 1 under the design scheme for the scale-out domain corresponds precisely to a GPU Server in Tencent's topology design.
  - **Connection design**. In each module, all SUHNs (element nodes of the scale-up-domain topology BIG-$(1, r_{SU_I}, 8, r_{SU_I})$ under the design scheme for scale-up domains) are selected as dock nodes. Each dock node serves as an intermediate super-node, i.e., an ISN-(1, 1, 2). Then the dock node interconnection is constructed under the bipartite mode. Each composite-edge is designed as a TD-(128, 2, 1) interconnection among 128 ISNs through an RN cluster.
  - **Inter-module design**. The skeleton graph is a BIG-(1, 8, 128, 8). Such a graph contains 128 element nodes and 8 block nodes that correspond to modules and RN clusters, respectively. For any 128 modules, they are connected with a common RN cluster via 128 super-edges. Such super-edges form a 1-parallelized TD-(128, 2, 1) interconnection.

  In fact, each topology generated by iteration 1 under the design scheme for the scale-out domain corresponds precisely to a Block in Tencent's topology design. Moreover, each RN cluster of iteration 1 under the design scheme for the scale-out domain corresponds precisely to a ToR switch Group in tier 1 in Tencent's topology design.

- **Iteration 2**:
  - **Intra-module design**. The topologies generated in iteration 1 are treated as modules.
  - **Connection design**. In each module, all RNs in the block nodes (RN clusters) of the skeleton graph of iteration 1's inter-module design are selected as dock nodes. Each dock node serves as an intermediate super-node, i.e., an ISN-(1, 1, 2). Then the dock node interconnection is constructed under the bipartite mode. Each composite-edge is designed as a TD-(64, 64, 1) interconnection among 64 ISNs through an RN cluster.
  - **Inter-module design**. The skeleton graph is a BIG-(1, 16, 64, 16). Such a graph contains 64 element nodes and 16 block nodes that correspond to modules and RN clusters, respectively. For any 64 modules, they are connected with a common RN cluster via 64 super-edges. Such super-edges form a 1-parallelized TD-(64, 64, 1) interconnection.

In fact, each topology generated by iteration 2 under the design scheme for the scale-out domain corresponds precisely to a Pod in Tencent's topology design. Moreover, each RN cluster of iteration 2 under the design scheme for the scale-out domain corresponds precisely to half of an Aggregation switch Group in tier 2 in Tencent's topology design.

- **Iteration 3**:
  - **Intra-module design**. The topologies generated in iteration 2 are treated as modules.
  - **Connection design**. In each module, all RNs in the block nodes (RN clusters) of the skeleton graph of iteration 2's inter-module design are selected as dock nodes. Each dock node serves as an intermediate super-node, i.e., an ISN-(1, 1, 3). Then the dock node interconnection is constructed under the bipartite mode. Each composite-edge is designed as a TD-(8, 64, 1) interconnection among 8 ISNs through an RN cluster.
  - **Inter-module design**. The skeleton graph is a BIG-(1, 64, 8, 64). Such a graph contains 8 element nodes and 64 block nodes that correspond to modules and RN clusters, respectively. For any 8 modules, they are connected with a common RN cluster via 8 super-edges. Such super-edges form a 16-parallelized TD-(8, 64, 1) interconnection.

  In fact, each RN cluster of iteration 3 under the design scheme for the scale-out domain corresponds precisely to a Core switch Group in tier 3 in Tencent's topology design.

China Mobile HPN1.0: In China Mobile HPN1.0 (Dyxnet 2025), each server containing 8 GPUs corresponds to a scale-up domain under the unified topology design scheme for AI clusters, where GPUs serve as SUHNs; however, its topology has not been publicly disclosed.

Under the design scheme for the scale-out domain, the topology of the scale-out domain of China Mobile HPN1.0 (Dyxnet 2025) can be constructed by three rounds of iteration, where the Leaf, Spine, and SuperSpine switches serve as RNs.

In iteration 1, the established scale-up-domain topology is taken as modules, and the design pattern for the scale-out domain is <scale-up-domain topology(SUHN), *B* >. In iteration 2, the topologies generated in iteration 1 are taken as modules, and the design pattern for the scale-out domain evolves into <iteration 1(BIG-block), *B* >. In iteration 3, the topologies generated in iteration 2 are taken as modules, and the design pattern for the scale-out domain evolves into <iteration 2(BIG-block), *B* >. The design of the topology of the scale-out domain of China Mobile HPN1.0 at each step of the three iterations is illustrated as follows.

- **Iteration 1**:
  - **Intra-module design**. The established scale-up-domain topology is treated as modules.

- **Connection design**. In each module, all SUHNs are selected as dock nodes. Each dock node serves as an intermediate super-node, i.e., an ISN-$(1, 1, d_{\text{server}})$. Then the dock node interconnection is constructed under the bipartite mode. Each composite-edge is designed as a TD-(128, 2, 1) interconnection among 128 ISNs through an RN cluster.
- **Inter-module design**. The skeleton graph is a BIG-(1, 8, 128, 8). Such a graph contains 128 element nodes and 8 block nodes that correspond to modules and RN clusters, respectively. For any 128 modules, they are connected with a common RN cluster via 128 super-edges. Such super-edges form a 1-parallelized TD-(128, 2, 1) interconnection.

In fact, each topology generated by iteration 1 under the design scheme for the scale-out domain corresponds precisely to a segment in China Mobile's topology design.

- **Iteration 2**:
  - **Intra-module design**. The topologies generated in iteration 1 are treated as modules.
  - **Connection design**. In each module, all RNs in the block nodes (RN clusters) of the skeleton graph of iteration 1's inter-module design are selected as dock nodes. Each dock node serves as an intermediate super-node, i.e., an ISN-$(1, 1, 1 + d_{\text{server}})$. Then the dock node interconnection is constructed under the bipartite mode. Each composite-edge is designed as a TD-(16, 32, 1) interconnection among 16 ISNs through an RN cluster.
  - **Inter-module design**. The skeleton graph is a BIG-(1, 8, 16, 8). Such a graph contains 16 element nodes and 8 block nodes that correspond to modules and RN clusters, respectively. For any 16 modules, they are connected with a common RN cluster via 16 super-edges. Such super-edges form a 2-parallelized TD-(16, 32, 1) interconnection.

  In fact, each topology generated by iteration 2 under the design scheme for the scale-out domain corresponds precisely to a PoD in China Mobile's topology design. Moreover, each RN cluster of iteration 2 under the design scheme for the scale-out domain corresponds precisely to a Rail in China Mobile's topology design.

- **Iteration 3**:
  - **Intra-module design**. The topologies generated in iteration 2 are treated as modules.
  - **Connection design**. In each module, all RNs in the block nodes (RN clusters) of the skeleton graph of iteration 2's inter-module design are selected as dock nodes. Each dock node serves as an intermediate super-node, i.e., an ISN-$(1, 1, 2 + d_{\text{server}})$. Then the dock node interconnection is constructed under the bipartite mode. Each composite-edge is designed as a TD-(8, 32, 1) interconnection among 8 ISNs through an RN cluster.

  - **Inter-module design**. The skeleton graph is a BIG-(1, 32, 8, 32). Such a graph contains 8 element nodes and 32 block nodes that correspond to modules and RN clusters, respectively. For any 8 modules, they are connected with a common RN cluster via 8 super-edges. Such super-edges form a 8-parallelized TD-(8, 32, 1) interconnection.

  In fact, each RN cluster of iteration 3 under the design scheme for the scale-out domain corresponds precisely to a Plane in China Mobile's topology design.

ZCube: In ZCube (Yan et al. 2025), intra–scale-up-domain networks are assumed to be optimized by manufacturers. Therefore, the scale-up-domain topology is excluded from ZCube's scope of consideration. Its focus lies solely on the scale-out-domain topology between GPUs and scale-out-domain switches.

Under the design scheme for the scale-out domain, by regarding GPUs and scale-out-domain switches as HNs and RNs, respectively, the scale-out-domain topology ZCube(n, k) ($n, k \geq 2$) by Yan et al. (2025) can be constructed by $k - 1$ rounds of iteration.

In iteration 1, the design pattern for the scale-out domain is < H, BIG(block and element), B >. In iteration i ($2 \leq i \leq k - 1$), the topologies generated in iteration $i - 1$ are taken as modules, and the design pattern for the scale-out domain evolves into <iteration $i - 1$(BIG-block and CN), B >. The design of the scale-out-domain topology ZCube(n, k) ($n, k \geq 2$) at each step of the $k - 1$ iterations is illustrated as follows.

- **Iteration 1**:
  - **Intra-module design**. The internal graph is a BIG-(1, 1, n, 1), where element nodes and block nodes are viewed as HNs and SURNs, respectively.
  - **Connection design**. In each module, all element nodes (HNs) and block nodes (RNs) are selected as dock nodes. Each dock node serves as an intermediate super-node, i.e., an ISN-(1, 2). Then the dock node interconnection is constructed under the bipartite mode. Each composite-edge is designed as a TD-(n, 1, 1) interconnection among n ISNs through an RN cluster.
  - **Inter-module design**. The skeleton graph is a BIG-(1, n, n, n). Such a graph contains n element nodes and n block nodes that correspond to modules and RN clusters, respectively. For any n modules, they are connected with a common RN cluster via n super-edges. Such super-edges form a 2-parallelized TD-(n, 1, 1) interconnection.
- **Iteration i ($2 \leq i \leq k - 1$)**:
  - **Intra-module design**. The topologies generated in iteration $i - 1$ are treated as modules.
  - **Connection design**. In each module, all RNs in the block nodes (RN clusters) of the skeleton graph of iteration ($i - 1$)'s inter-module design and all

HNs are selected as dock nodes. Each dock node serves as an intermediate super-node, i.e., an ISN-$(1, 1, 2i)$. Then the dock node interconnection is constructed under the bipartite mode. Each composite-edge is designed as a TD-$(n, 1, 1)$ interconnection among n ISNs through an RN cluster.

- **Inter-module design**. The skeleton graph is a BIG-$(1, n^i, n, n^i)$. Such a graph contains n element nodes and n^i block nodes that correspond to modules and an RN clusters, respectively. For any n modules, they are connected with a common RN cluster via n super-edges. Such super-edges form a 2-parallelized TD-$(n, 1, 1)$ interconnection.

Under the unified scheme, the $k-1$ iterations respectively introduce the level-$1, 2, \cdots, k-1$ switches for the scale-out-domain topology ZCube(n, k) $(n, k \geq 2)$, corresponding to ZCube's recursive construction method.

5.1.4 3D Torus

A 3D Torus is a grid of nodes (like TPUs) arranged in a 3D lattice (width, height, and depth), where the edges of each dimension "wrap around" to the opposite side.

3D torus is adopted as the topology of Google's AI cluster. It offers exceptional cost-efficiency for large-scale AI clusters by utilizing direct node-to-node connections instead of expensive switches. However, it suffers from a large network diameter, which increases latency for global communication and necessitates highly complex, topology-aware task scheduling. Additionally, performance can be significantly hindered by resource fragmentation and node failures, requiring sophisticated reconfiguration techniques like optical circuit switching (OCS) to maintain efficiency. In the context of Mixture of Experts (MoE) models (a standard paradigm of LLM), where tokens must be frequently reshuffled across all participating chips (all-to-all communication), the 3D Torus topology faces several significant bottlenecks.

In short, the engineering implementation of the 3D Torus remains more of an art than a science and is beyond the scope of our current topology design framework. A potential topic of future work is to integrate the 3D Torus into a more generalized topology design framework from the perspective of automorphism group.

5.2 Future Work

So far, we have provided a comprehensive introduction to the topology design framework for large-scale networks. This framework is not the final word on network topology design. Instead, it will continue to develop as application scenarios shift and change.

Potential topics of theoretical exploration include:

- A unified algebraic analysis framework through automorphism group and tensor decomposition
- Enrich subgraph options by adding random graphs such as random regular graph, random biregular bipartite graph and random geometric graph
- Enrich subgraph options by adding Cayley graphs such as cycle graph, Hypercube, and Torus
- Asymmetric network topology design through partial BIBD and partial transversal design
- Joint design of network topologies with routing algorithms
- Joint design of network topologies with traffic engineering
- Design of time-varying topologies

Potential topics of practical extensions include:

- Simulation and testbed verification of newly designed topologies
- Parameterized joint topology design for multi-domain, multi-plane AI cluster
- Dynamic topology design with optical reconfigurability
- Topology design for space data center, orbit AI and LEO satellite networks
- Topology design for AIoT and edge intelligence
- Topology design for robot networks, multi-robot systems, and swarm robotics
- Topology design for multi-agent networks
- Adaptive parameter selection algorithms through reinforcement learning and deep learning

5.3 Epilogue: The End of the Beginning

As we close the final chapter of this book, I look back on this 10-year scientific odyssey. It has been a process of seeking light in the darkness—traversing from heuristic engineering art to rigorous mathematical science. We have stripped away the complexity of large-scale networks to reveal the elegant, recursive skeletons underneath.

In doing so, we find that the "end" of our research—a general topology design framework—brings us back to the "beginning" : a first principle approach through combinatorial design and algebraic graph theory. This journey resonates deeply with the lines from T.S. Eliot's Four Quartets, which capture the essence of our recursive design philosophy:

"In my beginning is my end. In my end is my beginning."

Just as our topology framework uses the end of a micro-structure to form the beginning of a macro-structure, the conclusion of this book serves not as a termination, but as a foundation. The cycle of knowledge is complete, yet it turns anew.

However, as we stand on the precipice of the AI era—where AI will redefine the human world and human civilization—we must remain humble. We have solved the topology design problem for today, but tomorrow brings scale and constraints we have yet to dream of.

Therefore, regarding the work presented here and the vast unknown that lies ahead, I borrow the words of Winston Churchill to define where we stand in the grand timeline of discovery:

"Now this is not the end. It is not even the beginning of the end. But it is, perhaps, the end of the beginning."

The theory is established. The framework is ready. May this book serve as a solid stepping stone for those who will travel the road less traveled, venturing further into the unknown.

The real journey has just begun.

A

Proofs

A.1 Proofs for Chapter 1

A.1.1 Proof of Lemma 1.4

Proof: According to Høholdt and Janwa (2012), for all biregular bipartite graphs containing v nodes with degree r and b nodes with degree k, where $v \leq b$, the tightest lower bound of SLEs of their adjacency matrices is

$$\sqrt{\frac{r(v-k)}{v-1}}. \tag{A.1}$$

For such biregular bipartite graphs, the number of edges is equal to vr or bk, which implies that $vr = bk$. Let $\theta = v/k = b/r$. For all biregular bipartite graphs given θk nodes with degree r and θr nodes with degree k, the lower bound in equation A.1 can be rewritten as

$$\sqrt{\frac{r(v-k)}{v-1}} = \sqrt{\frac{rk(\theta-1)}{\theta k-1}}, \tag{A.2}$$

where by $v \leq b$, we have $r \geq k$.

Note that parameter θ is a positive number greater than or equal to one. In the following, we consider two cases.

In the case where $\theta = 1$, a BIG-(θ, r, k, λ) is a CBG. The SLE of the adjacency matrix of such a BIG is zero (Brouwer and Haemers 2011). By equation A.2, it follows that when $\theta = 1$, the lower bound is achieved with a value of zero.

In the case where $\theta > 1$, also, by Høholdt and Janwa (2012), a BIG-(θ, r, k, λ) with v element nodes ($v = \theta k$) and b block nodes ($b = \theta r$) achieves such a lower bound (A.2).[1] In other words, the SLE of the adjacency matrix of a BIG-(θ, r, k, λ) is

---

1 Note that when $\theta > 1$, the BIG-(θ, r, k, λ) has $r \geq k$. The reason is that by Lemma 1.3, the BIG-(θ, r, k, λ) is an incidence graph of a BIBD-(v, b, r, k, λ) with $r \geq k$.

$$\lambda_2^{eig}(BIG) = \sqrt{\frac{rk(\theta - 1)}{\theta k - 1}}. \tag{A.3}$$

The proof is completed. ■

A.1.2 Proof of Corollary 1.4

Proof: According to Lemma 1.4, the SLE of a BIG-(θ, r, k, λ) satisfies

$$\lambda_2^{eig}(BIG) = \sqrt{\frac{rk(\theta - 1)}{\theta k - 1}} = \sqrt{\frac{r[(\theta k - 1) - (k - 1)]}{\theta k - 1}} = \sqrt{r - \frac{r(k - 1)}{\theta k - 1}}. \tag{A.4}$$

By Lemma 1.6, it follows that

$$\frac{r(k - 1)}{\theta k - 1} = \lambda. \tag{A.5}$$

Since $\lambda \geq 1$ and $k \geq 2$, then we have

$$\lambda_2^{eig}(BIG) = \sqrt{r - \lambda} < \sqrt{r - 1} + \sqrt{k - 1}. \tag{A.6}$$

Moreover, recall that in a BIG-(θ, r, k, λ), the degree of each node is either r or k. Hence, BIGs satisfy the definition of bipartite Ramanujan graphs, and the proof is completed. ■

A.1.3 Proof of Lemma 1.5

Proof: Note that the distance between two nodes is the length of the shortest path between such two nodes. This proof can be conducted by considering the shortest path between distinct types of node pairs (element nodes and block nodes), respectively. Details are omitted due to space constraints. ■

A.1.4 Proof of Lemma 1.6

Proof: Considering that parameter θ is a positive integer, we proceed with the proof in the following cases.

The first case is when $\theta = 1$. In this case, according to Lemma 1.3, a BIG is a CBG where each element node is adjacent to all block nodes. Hence, each pair of block nodes are adjacent to all block nodes. By the definition of BIG-(θ, r, k, λ), the parameter λ describes the number of common block nodes that each pair of element nodes are adjacent to, while the parameter θr describes the number of block nodes. Therefore, we have $\lambda = \theta r = r$, which is equivalent to Equation A.5 with $\theta = 1$.

The second case is when $\theta > 1$. In this case, according to Lemma 1.3, a BIG-(θ, r, k, λ) corresponds to a BIBD-(v, b, r, k, λ) with $v = \theta k$ and $b = \theta r$. Since the parameters of the BIBD-(v, b, r, k, λ) satisfy the equation $r(k-1) = \lambda(v-1)$ (*cf.* Lemma 1.1 in Section 1.6), we have $r(k-1) = \lambda(\theta k - 1)$. By the definition of BIG, we have $k \geq 2$. Hence, by dividing $\theta k - 1$ at both sides of the equation, we obtain Equation A.5. The proof is completed. ■

A.1.5 Proof of Lemma 1.7

Proof: By Lemma 1.3, a BIG-(θ, r, k, λ) with $\theta > 1$ corresponds to a BIBD-(v, b, r, k, λ), where $v = \theta k$ and $b = \theta r$. Hence, the existence of a BIBD-(v, b, r, k, λ) with $v = \theta k$ and $b = \theta r$ implies the existence of a BIG-(θ, r, k, λ). Next, we turn to prove the existence of BIBDs under different sufficient conditions, respectively.

- Given an integer q such that $q \geq 2$, there exists a BIBD-$(q+1, q+1, q, q, q-1)$ (Colbourn and Dinitz 2006). Its incidence graph is a BIG-$\left(1 + \frac{1}{q}, q, q, q-1\right)$.
- Given an even number q such that $2 \leq q \leq 212$, there exists a BIBD-$\left(2q-1, 2q-1, q, q, \frac{q}{2}\right)$ (Colbourn and Dinitz 2006), which is known as a Hadamard BIBD. Its incidence graph is a BIG-$\left(2 - \frac{1}{q}, q, q, \frac{q}{2}\right)$.
- Given an odd number q such that $3 \leq q \leq 211$, there exists a BIBD-$\left(2q+1, 2q+1, q, q, \frac{q}{2}\right)$ (Colbourn and Dinitz 2006), which is known as a Hadamard BIBD. Its incidence graph is a BIG-$\left(2 + \frac{1}{q}, q, q, \frac{q-1}{2}\right)$.
- Given a positive integer q, there exists a BIBD-$\left(q+1, \frac{q(q+1)}{2}, q, 2, 1\right)$ (Colbourn and Dinitz 2006). Its incidence graph is a BIG-$\left(\frac{q+1}{2}, q, 2, 1\right)$.
- Given a prime power p, there exists a BIBD-$(p^2, p^2 + p, p+1, p, 1)$, which is known as an affine BIBD (Colbourn and Dinitz 2006). By setting $q = p + 1$, there exists a BIBD-$((q-1)^2, q(q-1), q, q-1, 1)$, where $q - 1$ is a prime power. Its incidence graph is a BIG-$(q-1, q, q-1, 1)$.
- Given an integer p such that $p = 1$ or p is a prime power, there exists a BIBD-$(p^2 + p + 1, p^2 + p + 1, p + 1, p + 1, 1)$, which is known as a projective BIBD (Colbourn and Dinitz 2006). By setting $q = p + 1$, there exists a BIBD-$(q^2 - q + 1, q^2 - q + 1, q, q, 1)$, where $q = 2$ or $q - 1$ is a prime power. Its incidence graph is a BIG-$\left(q - 1 + \frac{1}{q}, q, q, 1\right)$.

The proof is completed. ■

A.1.6 Proof of Lemma 1.8

Proof: We first show that any TDG-(n, μ, v) is an incidence graph of a TD-(n, μ, v). Given a TDG-(n, μ, v), we construct a set of elements and a multi-set of blocks as follows. For each element node in the TDG, an element corresponding to this element node is added to the element set. For each block node in the BIG, a block corresponding to this block node is added to the block set. Next, we define the following containing relations between elements and blocks. If an element node is adjacent to a block node in the BIG, then the corresponding element is contained in the block. Meanwhile, the elements are partitioned into groups in the same way as element nodes in the TDG. By the definition of incidence graphs (*cf.* Section 1.4.6.6), we turn to show that the elements, groups, and blocks constitute a TD-(n, μ, v).

Since there are nv element nodes with degree μv and μv^2 block nodes with degree n, the element set contains nv elements and μv^2 blocks, such that each block contains n elements. Meanwhile, there are n groups in the TDG, each containing v element nodes. Accordingly, the elements are also partitioned into n groups and each group contains v elements. Moreover, each pair of element nodes in distinct groups are adjacent to μ block nodes. Correspondingly, each pair of distinct elements are contained in either exactly μ common blocks or one group. Based on these properties, the elements, groups, and blocks form a TD-(n, μ, v).

Next, we show that for any TD-(n, μ, v), its incidence graph is a TDG-(n, μ, v). Recall that a TD-(n, μ, v) consists of nv elements and μv^2 blocks, such that each block contains n elements and each element is contained in μv blocks. Correspondingly, its incidence graph is a biregular bipartite graph that contains nv element nodes with degree μv and μv^2 block nodes with degree n. Meanwhile, each pair of elements are contained in either exactly μ common blocks or one group. Hence, each pair of element nodes in distinct groups are adjacent to a number of μ common blocks. Since the elements in TD-(n, μ, v) are partitioned into n groups, the element nodes are also partitioned into n groups. By the definition of TDG, the incidence graph of a TD-(n, μ, v) is a TDG-(n, μ, v).

The proof is completed. ∎

A.1.7 Proof of Lemma 1.9

Proof: First, we consider a more general design called partial geometric design. According to van Dam and Spence (2005), a *partial geometric design* with parameters (r, k, t, c) is jointly defined by a set of elements $\mathcal{X}$ and a multi-set of blocks $\mathcal{B}$ such that

- each block is a subset of elements;
- each element is contained in r blocks;

- each block is contained in k elements;
- for any element $x \in \mathcal{X}$ and any block $B \in \mathcal{B}$, the number of element-block pairs $\{(y, C) | C \in \mathcal{B}, x \in C, y \in \mathcal{X} \cap B \cap C, y \neq x, C \neq B\}$ equals c if $x \in B$ or t otherwise.

For such a design, the SLE of the adjacency matrix of its incidence graph (*cf.* Section 1.4.6.6) is $\sqrt{r + k - 1 + c - t}$ (van Dam and Spence 2005).

Based on the definition, a TD-(n, μ, υ) is a special case of partial geometric design. Particularly, by the definition of a TD-(n, μ, υ), each element is contained in $\mu\upsilon$ blocks and each block is contained in n elements, which implies $r = \mu\upsilon$ and $k = n$. For each element x and block B, depending on whether $x \in B$ or not, we consider the following two cases to discuss the values of c and t, respectively.

- If $x \in B$, since each block contains n elements, there are $n - 1$ elements such that each element y satisfies $y \in B$ and $y \neq x$. By the definition of TD, each pair of elements are contained in either the same group or μ blocks. Hence, such $n - 1$ elements belong to distinct groups with the element x. For each element y, there are $\mu - 1$ blocks such that each block C satisfies $x \in C$, $y \in C$, and $C \neq B$. Therefore, there are $(n - 1)(\mu - 1)$ element-block pairs such that each element-block pair (y, C) satisfies $C \in \mathcal{B}, x \in C, y \in \mathcal{X} \cap B \cap C$, $y \neq x, C \neq B$, which implies that $c = (\mu - 1)(n - 1)$.
- If $x \notin B$, since each block contains n elements such that each element y satisfies $y \in B$ and $y \neq x$. By the definition of TD, each pair of elements are contained in either the same group or μ blocks. Note that each block in TD contains precisely one element from each group (Wallis and George 2016). Hence, there are $n - 1$ elements such that each element y and its corresponding block C satisfy $y \in B$, $y \neq x$, and $x \in C$. Meanwhile, for each element y, there are μ blocks such that each block C satisfies $x \in C$, $y \in C$, and $C \neq B$. Therefore, there are $\mu(n - 1)$ element-block pairs such that each element-block pair (y, C) satisfies $C \in \mathcal{B}, x \in C, y \in \mathcal{X} \cap B \cap C, y \neq x, C \neq B$, which implies that $t = \mu(n - 1)$.

Therefore, a TD-(n, μ, υ) is a partial geometric design with parameters $r = \mu\upsilon$, $k = n$, $c = (\mu - 1)(n - 1)$, and $t = \mu(n - 1)$.

Finally, by Lemma 1.8, a TDG-(n, μ, υ) is the incidence graph of a TD-(n, μ, υ). Hence, the SLE of the adjacency matrix of a TDG-(n, μ, υ) is

$$\sqrt{r + k - 1 + c - t} = \sqrt{\mu\upsilon + n - 1 + (\mu - 1)(n - 1) - \mu(n - 1)} = \sqrt{\mu\upsilon}, \tag{A.7}$$

thus the proof is completed. ■

A.1.8 Proof of Corollary 1.5

Proof: From Lemma 1.9, the SLE of a TDG-(n, μ, v) satisfies

$$\lambda_2^{eig}(TDG) = \sqrt{\mu v}. \tag{A.8}$$

By the definition of TDG, we have $\mu v \geq 2$, $n \geq 2$, and following derivation:

$$\mu v < \mu v + 2\sqrt{\mu v - 1} = (\mu v - 1) + 2\sqrt{\mu v - 1} + 1 = (\sqrt{\mu v - 1} + 1)^2. \tag{A.9}$$

Therefore,

$$\lambda_2^{eig}(TDG) = \sqrt{\mu v} < \sqrt{\mu v - 1} + 1 \leq \sqrt{\mu v - 1} + \sqrt{n - 1}. \tag{A.10}$$

Furthermore, the degree of each node in the TDG-(n, μ, v) is either μv or n. Hence, TDGs satisfy the definition of bipartite Ramanujan graphs.

The proof is completed. ■

A.1.9 Proof of Lemma 1.11

Proof: By Lemma 1.8, for any TD-(n, μ, v), its incidence graph is a TDG-(n, μ, v). Hence, the existence of a TD-(n, μ, v) implies the existence of a TDG-(n, μ, v). Next, we turn to prove the existence of TDs in the two cases, respectively.

The first case is when v is a prime power; i.e., $v = p^e$, where p is a prime number and e is a positive integer. In this case, by the property of TD (*cf.* Lemma 1.2 in Section 1.6), there exists a TD-(n, μ, v) for any integer n such that $2 \leq n \leq p^e + 1$, or equivalently, $2 \leq n \leq v + 1$.

In the second case, we consider a TD-(n, μ, v) with $n = 2$ or $n = 3$. In this case, we consider the prime factorization of v as $p_1^{e_1} p_2^{e_2} ... p_l^{e_l}$, where integers p_i are distinct primes and integers e_i are positive numbers. By the property of TD (*cf.* Lemma 1.2 in Appendix 1.6), if $n \leq \min\{p_i^{e_i} + 1 : 1 \leq i \leq l\}$, there exists a TD-$(n, \mu, v)$. Hence, if $n = 2$ or $n = 3$, there exists a TD-(n, μ, v). Therefore, if $n = 2$ or $n = 3$, there exists a TD-(n, μ, v) for any integer $v \geq 2$.

The proof is completed. ■

A.2 Proofs for Chapter 2

A.2.1 Proof of Theorem 2.1

Proof: By the property of CG, the distance between any two nodes in the graph is one. Hence, any group of s nodes form an ISN-$(s, s, 1)$, i.e., $x = w = s$. ■

A.2.2 Proof of Theorem 2.2

Proof: By the property of CBG, the distance between each pair of nodes in the same node partition is two, while the distance between each pair of nodes in distinct node partitions is one. Following we prove this lemma by considering two cases with $x = 1$ and $x \geq 2$, respectively.

Consider an ISN-(x, w, l) I containing one node a (i.e., $x = 1$). Regarding another node in the same node partition with node a, there is only one dock node in ISN I within a distance of two. As for the node in different node partition with node a, there is only one dock node in ISN I within a distance of one. Therefore, any node forms an ISN-(x, w, l) with $x = w = 1$ and $l = 2$.

Consider an ISN-(x, w, l) with $x \geq 2$. Let symbols E_1 and E_2 denote two node partitions, containing m nodes and n nodes respectively. Suppose node partitions E_1 and E_2 contain q_1 and q_2 dock nodes in the ISN, respectively. Then $q_1 \leq m$, $q_2 \leq n$. Without loss of generality, we assume $q_1 \geq q_2$. Then $q_2 \leq \min\{m, n\}$. For each node in node partition E_1, the q_2 dock nodes in node partition E_2 are within a distance of one. For each node in node partition E_2, the q_1 dock nodes in node partition E_1 are within a distance of one. Hence, if $q_1 \geq q_2 > 0$, the parameters of ISN-(x, w, l) satisfy $l = 1$ and $w = q_2$. Note that we have $q_1 + q_2 = x$. This implies $2q_2 \leq q_1 + q_2 = x$, i.e., $q_2 \leq x/2$. Note that q_2 is an integer. Hence, we have $w = q_2 \leq \min\left\{m, n, \left\lfloor \frac{x}{2} \right\rfloor\right\}$.

The proof is completed. ■

A.2.3 Proof of Theorem 2.3

Proof: By the property of BIGs, the distance between each pair of element nodes (GNs/HNs) is two. Hence, we have $l \geq 2$ and any group of s element nodes form an ISN-$(s, s, 2)$, i.e., $x = w = s$. ■

A.2.4 Proof of Theorem 2.4

Proof: For each ISN-(x, w, l) with $l = 1$ (i.e., ISN-(x, w)), since its dock nodes are selected from block nodes in the BIG-(θ, r, k, λ) with c copies, then the degree of each dock node is $k(c + 1)$. Hence, all x dock nodes (i.e., block nodes) in each ISN-(x, w) are adjacent to $xk(c + 1)$ element nodes. Meanwhile, in the BIG-(θ, r, k, λ) with c copies, there are $\theta k(c + 1)$ element nodes. Each element node is adjacent to at least w block nodes in the ISN-(x, w). As a result, all block nodes in each ISN-(x, w) are adjacent to at least $w\theta k(c + 1)$ element nodes, thus $xk(c + 1) \geq w\theta k(c + 1)$. In other words, we have $x \geq \theta w$. The proof is completed. ■

A.2.5 Proof of Theorem 2.5

Proof: By the property of TDG, the distance between each pair of element nodes (GNs/HNs) is two or four. Hence, we have $l \geq 2$. By the definition of ISNs-$(x, w, 2)$, we have $w \leq x$, in which the equality holds if and only if all nodes in an ISN-$(x, w, 2)$ is within distance two from any GN or HN. We first consider the case where $x = w$. Specifically, given an element node, its distance from any other element node in the same group is four. As a result, it is infeasible to construct an ISN-$(x, w, 2)$ with $l = 2$ when $w = x$. Therefore, the equation $w < x$ holds.

Next, we consider the case where $w = x - 1$. By the property of TDG, given an element node, its distance from any other element node in distinct groups is two. Then given a number of x element nodes selected from distinct groups, for each GN or HN, at least $x - 1$ element nodes are not contained in its group. In other words, such x element nodes form an ISN-$(x, x-1, 2)$ with $w = x - 1$. Therefore, we have $w \leq x - 1$. The proof is completed. ■

A.2.6 Proof of Theorem 2.6

Proof: For each ISN-(x, w, l) with $l = 1$ (i.e., ISN-(x, w)), since its dock nodes are selected from block nodes in the TDG-(n, μ, v) with c copies, then the degree of each dock node is $n(c+1)$. Hence, all x dock nodes (i.e., block nodes) in each ISN-(x, w) are adjacent to $xn(c+1)$ element nodes. Meanwhile, in a TDG-(n, μ, v) with c copies, there are $nv(c+1)$ element nodes. Each element node is adjacent to at least w block nodes in each ISN-(x, w). As a result, all block nodes in each ISN-(x, w) are adjacent to at least $wnv(c+1)$ element nodes, thus $xn(c+1) \geq wnv(c+1)$. In other words, we have $x \geq vw$. The proof is completed. ■

A.2.7 Proof of Lemma 2.1

Proof: First, we consider the case where $x = 1$. In this case, the TD-(n, μ, x) interconnection is a CBG which exists for any integer $n \geq 2$.

The second case is when x is a prime power ($x \geq 2$); i.e., $x = p^e$, where p is a prime number and e is a positive integer. In this case, the TD-(n, μ, x) interconnection is constructed by a TD-(n, μ, x). Hence, the existence of a TD-(n, μ, x) implies the existence of a TD-(n, μ, x) interconnection. By the property of TD (*cf.* Lemma 1.2 in Section 1.6), there exists a TD-(n, μ, x) for any n such that $2 \leq n \leq p^e + 1$, or equivalently, $2 \leq n \leq x + 1$. As a result, in this case, there exists a TD-(n, μ, x) interconnection for any integer $x \geq 1$.

In the third case, we consider a TD-(n, μ, x) interconnection with $n = 2$ or $n = 3$. Since $x \geq 1$, we consider the following two sub-cases.

- The first sub-case is $x = 1$. In this sub-case, the existence of TD-(n, μ, x) interconnection is implied by the proof in the first case.
- The second sub-case is $x \geq 2$. In this sub-case, the TD-(n, μ, x) interconnection is constructed by a TD-(n, μ, x). Hence, the existence of a TD-(n, μ, x) implies the existence of a TD-(n, μ, x) interconnection. We consider the prime factorization of x as $p_1^{e_1} p_2^{e_2} ... p_l^{e_l}$, where integers p_i are distinct primes and integers e_i are positive numbers. By the property of TD (*cf.* Lemma 1.2 in Section 1.6), if $n \leq \min\{p_i^{e_i} + 1 : 1 \leq i \leq l\}$, there exists a TD-$(n, \mu, x)$. Hence, if $n = 2$ or $n = 3$, there exists a TD-(n, μ, x) interconnection.

Therefore, if $n = 2$ or $n = 3$, there exists a TD-(n, μ, x) for any integer $x \geq 1$.

The proof is completed. ■

A.3 Proofs for Chapter 3

A.3.1 Proof of Theorem 3.1

Proof: This proof is conducted by calculating the latency and the intra-module throughput of tree-like topologies, respectively.

Latency. By Definition 2.6, the latency is the maximum distance between any pair of nodes with computing function. In tree-like topologies, there are no HNs and the nodes with computing function are all CNs. To calculate the latency of tree-like topologies, we focus on the distance between any pair of CNs. Recall that in the tree-like topologies, any two CNs can be attached to the same module or be attached to distinct modules. Next, we consider the distance between two CNs in such two cases, respectively.

- *Case 1:* Two CNs are attached to the same module. In this case, such two CNs can be attached to the same GN or two distinct GNs in the same module. The distance between two CNs in the same module is considered in such two sub-cases, respectively.
  - *Sub-case 1:* Two CNs are attached to the same GN. In this sub-case, there is a length-2 path between them that goes through the GN. Note that such two CNs are not adjacent. Hence, the distance between such two CNs is two.
  - *Sub-case 2:* Two CNs are attached to two distinct GNs in the same module. In this sub-case, the shortest path between them must go through their two attached GNs. In the BIG-$(1, q, q, q)$ as the internal graph, the distance between any two GNs is two. Since the distance between each CN and its attached GN is one, the distance between two CNs attached to two distinct GNs in the same module is four.

- *Case 2:* Two CNs are attached to two distinct modules. For ease of description, we denote such two modules as symbols M_1 and M_2, respectively. Without loss of generality, let symbols a_1 and b_2 denote a GN in module M_1 and a GN in module M_2, respectively. Note that the distance between GN a_1 and GN b_2 is at least four. Since the inter-module design is a BIG *with copies*, modules M_1 and M_2 can be duplicated from a common module or not. Following we consider these two sub-cases, respectively.
  - *Sub-case 1:* Modules M_1 and M_2 are duplicated from a common module. In this sub-case, they have the same connection pattern with RN clusters. Note that the internal graph is the BIG-$(1, q, q, q)$ for all modules. For GN a_1 in module M_1, we denote its counterpart GN in module M_2 by a_2. Likewise, for GN b_2 in module M_2, we denote its counterpart GN in module M_1 by b_1. According to Definition 1.23, there exists an RN (denoted by c_1) in module M_1, such that GNs a_1 and b_1 are connected through RN c_1. Likewise, GNs a_2 and b_2 are also connected through RN c_2, the counterpart RN of RN c_1 in module M_2. Since modules M_1 and M_2 have the same connection pattern with RN clusters, RNs c_1 and c_2 must be adjacent to some common RN in the third layer, which is denoted by d. Then there exists a length-4 path between GN a_1 and GN b_2, i.e., $a_1 - c_1 - d - c_2 - b_2$. In addition, since the distance between two dock nodes in distinct modules is at least two and the distance between a dock node and a GN in the same module is at least one, the distance between two GNs in distinct modules is no smaller than four. Hence, the distance between GN a_1 and GN b_2 is four. Therefore, the distance between each CN attached to GN a_1 and any CN attached to GN b_2 is six.
  - *Sub-case 2:* Modules M_1 and M_2 are not duplicated from a common module. In this sub-case, by Definition 1.23, there exist at least one RN cluster between them that interconnects two ISNs (denoted by I_1 and I_2, respectively) in modules M_1 and M_2. By the definition of ISN (Definition 2.10), there exists at least one RN (denoted by e_1) in ISN I_1 that is adjacent to GN a_1. Likewise, there exists at least one RN (denoted by e_2) in ISN I_2 that is adjacent to GN b_2. According to the definition of TD-(n, μ, x) interconnection, there exist μ length-2 paths between RN e_1 and RN e_2 through μ distinct RNs in the RN cluster. Let f denote one of the μ RNs in the RN cluster. Then there exists a length-4 path between GN a_1 and GN b_2, i.e., $a_1 - e_1 - f - e_2 - b_2$. In addition, since the distance between two dock nodes in distinct modules is at least two and the distance between a dock node and a GN in the same module is at least one, the distance between two GNs in distinct modules is no smaller than four. Hence, the distance between GN a_1 and GN b_2 is four. Therefore, the distance between each CN attached to GN a_1 and any CN attached to GN b_2 is six.

In conclusion, the latency of tree-like topologies is six.

Intra-module throughput. Recall that tree-like topologies adopt BIG-$(1, q, q, q)$ as the internal graph for each module, where element nodes are GNs and block nodes are RNs. Since each GN is attached with q CNs, the attachment number of each GN is q. Meanwhile, for any two GNs in the same module, there exist at least q edge-disjoint paths with length two between them, each of which goes through an RN. By Definition 2.7, the intra-module throughput of tree-like topologies is

$$T_1 = \frac{q}{q} = 1. \tag{A.11}$$

The proof is completed. ■

A.3.2 Proof of Theorem 3.2

Proof: This proof is conducted by considering the intra-module throughput and the latency HyperX-like topologies, respectively.

Intra-module throughput. Recall that HyperX-like topologies adopt CG as the internal graph for each module, where nodes are GNs. Since each GN is attached with p_0 CNs, the attachment number of each GN is p_0. Meanwhile, for any two GNs in the same module, there exist p_1 edge-disjoint paths. One of these p_1 paths are the edge between two GNs with length one. For the other $p_1 - 1$ paths, each of them goes through an GN in the same module with length 2. By Definition 2.7, the intra-module throughput of HyperX-like topologies is

$$T_1 = \frac{p_1}{p_0}. \tag{A.12}$$

Latency. By Definition 2.6, the latency is the maximum distance between any pair of nodes with computing function. In HyperX-like topologies, there are no HNs and the nodes with computing function are all CNs. To calculate the latency of HyperX-like topologies, we focus on the distance between any pair of CNs. Recall that in the HyperX-like topologies, any two CNs can be attached to the same module or be attached to distinct modules. Next, we consider the distance between two CNs in such two cases, respectively.

- *Case 1:* Two CNs are attached to the same module. In this case, such two CNs can be attached to the same GN or two distinct GNs in the same module. The distance between two CNs in the same module is considered in such two sub-cases, respectively.
  - *Sub-case 1:* Two CNs are attached to the same GN. In this sub-case, there is a length-2 path between them that goes through the GN. Note that such two CNs are not adjacent. Hence, the distance between such two CNs is two.

- *Sub-case 2:* Two CNs are attached to two distinct GNs in the same module. In this sub-case, the shortest path between them must go through their attached two GNs. In the CG as the internal graph, the distance between any two GNs is one. Since the distance between each CN and its attached GN is one, the distance between two CNs attached to two distinct GNs in the same module is three.

Therefore, the maximum distance between two CNs attached to the same module is three, while implies that the latency of the topology is at least three.

- *Case 2:* Two CNs are attached to two distinct modules. This case is considered in the following three sub-cases, respectively. In the first sub-case, each GN is directly connected to all GNs in distinct modules. In the second sub-case, each GN is directly connected to part of (but not all) GNs in all distinct modules. In the third sub-case, each GN is not directly to any node in some distinct modules.
  - *Sub-case 1:* Each GN is directly connected to all GNs in distinct modules. In this sub-case, each GN is directly connected to all other GNs (in the same module or in distinct modules). Hence, the GNs (without attaching with CNs) in the topology prototype forms a complete graph. By the property of complete graph, the number of GNs is equal to the degree of each GN plus one. Note that there are m modules in the topology prototype, each of which contains $p_1 + 1$ GNs. Meanwhile, the degree of GNs without attaching with CNs is $p - p_0 = p_1 + p_2$. Hence, this sub-case is equivalent to $m(p_1 + 1) = p_1 + p_2 + 1$, i.e., $m = 1 + \frac{p_2}{p_1 + 1}$. Furthermore, in this sub-case, the distance between each pair of GNs is equal to one, while implies that the distance between each pair of CNs is equal to three. In other words, the latency of the topology is three if and only if $m = 1 + \frac{p_2}{p_1 + 1}$.
  - *Sub-case 2:* Each GN is directly connected to part of (not all) GNs in all distinct modules. In this sub-case, each GN in one module participates the super-edge between such a module and all other modules. Note that each super-edge between two modules (i.e., a t-parallelized all-to-all interconnection) involves tx GNs as dock nodes in each module. Since each module contains $p_1 + 1$ GNs, we have $tx = p_1 + 1$. Meanwhile, since each GN is not directly connected to all GNs in distinct modules, we have $m \neq 1 + \frac{p_2}{p_1 + 1}$ according to the discussions in the first sub-case. Hence, this sub-case corresponds to the parameter settings with $tx \geq p_1 + 1$ and $m \neq 1 + \frac{p_2}{p_1 + 1}$.
  Next, we consider the latency in this sub-case. For ease of description, we denote a pair of modules as symbols M_1 and M_2, respectively. Without loss of generality, let symbols a and b denote a GN in module M_1 and a GN in module M_2, respectively, such that GNs a and b are not directly connected.

In this sub-case, GN a is directly connected to some GN (denoted by symbol c) in module M_2. Hence, there exists a length-2 path between GNs a and b (i.e. $a - c - b$). This implies that the distance between each CN attached to GN a and any CN attached to GN b is four. Therefore, the latency of the topology is four in this sub-case (i.e., $m \neq 1 + \frac{p_2}{p_1+1}$ and $tx \geq p_1 + 1$).

- *Sub-case 3:* Each GN is not directly to any node in some distinct modules. By the discussion in these two sub-cases, this sub-case corresponds to the parameter settings with $tx < p_1 + 1$ and $m \neq 1 + \frac{p_2}{p_1+1}$. Next, we prove that the latency is less than or equal to five in this sub-case. For ease of description, we denote a pair of modules as symbols M_1 and M_2, respectively. Without loss of generality, let symbols a and b denote a GN in module M_1 and a GN in module M_2, respectively, such that GN a is not directly connected to any GN in module M_2 and b is not directly connected to any GN in module M_1. Note that there exists a super-edge between modules M_1 and M_2. Let symbols c and d denote the dock nodes in modules M_1 and M_2 that participates such a super-edge, respectively. Then GN a is directly connected to GN c and GN b is directly connected to GN d in module M_2. Hence, there exists a length-3 path between GNs a and d (i.e. $a - c - d - b$). This implies that the distance between each CN attached to GN a and any CN attached to GN b is less than or equal to five. Therefore, the latency of the topology is less than or equal to five in this sub-case (i.e., $m \neq 1 + \frac{p_2}{p_1+1}$ and $tx < p_1 + 1$).

In conclusion, for the HyperX-like topologies, their latency is equal to 3, 4, or 5. Particularly, the value of latency is three if and only if their parameters satisfy $m = 1 + \frac{p_2}{p_1+1}$. If $m \neq 1 + \frac{p_2}{p_1+1}$, the latency is four if $tx \geq p_1 + 1$. The proof is completed. ■

A.4 Proofs for Chapter 4

A.4.1 Proof of Lemma 4.1

Proof: Note that the number of nodes with computing function attached to each module (i.e., the value of S_1) is equal to the product of the number of nodes with computing function attached to each GN (or each HN) and the number of GNs (or HNs) in each module. Hence, we introduce the following two propositions regarding the number of nodes with computing function attached to each GN (or each HN) and the number of GNs (or HNs) in each module, respectively.

Proposition 1 Let term S_1^a denote the number of nodes with computing function attached to each GN (or each HN). The value of S_1^a is determined by the selected option of term d_1. Specifically,

- if $d_1 = G$, we have $S_1^a = \gamma_G p_G$;
- if $d_1 = H$, we have $S_1^a = \gamma_H p_H + 1$.

Proof of Proposition 1. We consider the number of nodes with computing function attached to each GN or each HN (i.e., the value of S_1^a) for design patterns with $d_1 = G$ or $d_1 = H$, respectively.

- For design patterns with $d_1 = G$, the value of term S_1^a is the number of nodes with computing function attached to each GN. For each GN, its attached nodes with computing function contain its directly connected CNs. Recall that parameter γ_G denotes the proportion of degrees for connections with CNs for each GN and parameter p_G denotes the degree of GNs. Hence, the value of S_1^a is equal to $\gamma_G p_G$ (i.e., $S_1^a = \gamma_G p_G$) if $d_1 = G$.
- For design patterns with $d_1 = H$, the value of term S_1^a is the number of nodes with computing function attached to each HN. For each HN, its attached nodes with computing function contain its directly connected CNs and itself. Recall that parameter γ_H denotes the proportion of degrees for connections with CNs for each HN and parameter p_H denotes the degree of HNs. Hence, the value of S_1^a is equal to $\gamma_H p_H + 1$ (i.e., $S_1^a = \gamma_H p_H + 1$) if $d_1 = H$.

In summary, the value of term S_1^a is equal to $\gamma_G p_G$ if $d_1 = G$ or $\gamma_H p_H + 1$ if $d_1 = H$. The proof of this proposition is completed.

Proposition 2 Let term S_1^b denote the number of GNs (or HNs) in each module. The value of S_1^b is determined by the selected options of term d_1 and term d_2. Specifically,

- for design patterns with $d_2 =$ CG(any), we have $S_1^b = \alpha_G p_G + 1$ if $d_1 = G$ or $S_1^b = \alpha_H p_H + 1$ if $d_1 = H$;
- for design patterns with $d_2 =$ CBG(any), we have $S_1^b = 2\alpha_G p_G$ if $d_1 = G$ or $S_1^b = 2\alpha_H p_H$ if $d_1 = H$;
- for design patterns with $d_2 =$ BIG(block), we have $S_1^b = \theta_1 \alpha_R p_R$;
- for design patterns with $d_2 =$ BIG(element), we have $S_1^b = \theta_1 p_R$;
- for design patterns with $d_2 =$ TDG(block), we have $S_1^b = \upsilon_1 \alpha_R p_R$;
- for design patterns with $d_2 =$ TDG(element), we have $S_1^b = \upsilon_1 p_R$.

Proof of Proposition 2. We consider the number of GNs or HNs in each module (i.e., the value of S_1^b) for design patterns with different options of term d_2, respectively.

- For design patterns with $d_2 = \text{CG(any)}$, CG is employed as the internal graph, where nodes are all GNs or all HNs. Note that the number of nodes in a CG is equal to the degree of each node plus one. Hence, we consider the following two cases where nodes are all GNs or all HNs, respectively.
  - If nodes are all GNs (i.e., $d_1 = G$), the inner-degree of each GN is $\alpha_G p_G$, where parameter α_G denotes the proportion of degrees of each GN for connections with other nodes in the same module. Hence, the value of S_1^b is equal to $\alpha_G p_G + 1$ (i.e., $S_1^b = \alpha_G p_G + 1$) if $d_1 = G$.
  - If nodes are all HNs (i.e., $d_1 = H$), the inner-degree of each HN is $\alpha_H p_H$, where parameter α_H denotes the proportion of degrees of each HN for connections with other nodes in the same module. Hence, the value of S_1^b is equal to $\alpha_H p_H + 1$ (i.e., $S_1^b = \alpha_H p_H + 1$) if $d_1 = H$.
- For design patterns with $d_2 = \text{CBG(any)}$, CBG is employed as the internal graph, where nodes are all GNs or all HNs. Note that the number of nodes in a CBG is equal to twice the degree of each node. Hence, we consider the following two cases where nodes are all GNs or all HNs, respectively.
  - If nodes are all GNs (i.e., $d_1 = G$), the inner-degree of each GN is $\alpha_G p_G$, where parameter α_G denotes the proportion of degrees of each GN for connections with other nodes in the same module. Hence, the value of S_1^b is equal to $2\alpha_G p_G$ (i.e., $S_1^b = 2\alpha_G p_G$) if $d_1 = G$.
  - If nodes are all HNs (i.e., $d_1 = H$), the inner-degree of each HN is $\alpha_H p_H$, where parameter α_H denotes the proportion of degrees of each HN for connections with other nodes in the same module. Hence, the value of S_1^b is equal to $2\alpha_H p_H$ (i.e., $S_1^b = 2\alpha_H p_H$) if $d_1 = H$.
- For design patterns with $d_2 = \text{BIG(block)}$ or $d_2 = \text{BIG(element)}$, the internal graph is a BIG-$(\theta_1, r_1, k_1, \lambda_1)$ with c_1 copies, where block nodes are all RNs and element nodes are all GNs or all HNs. Based on the discussion in Remark 2.6, such a BIG contains $\theta_1 k_1(c_1 + 1)$ element nodes and the inner-degree of each block node equals $k_1(c_1 + 1)$. Note that the inner-degree of each block node (RN) is $\alpha_R p_R$, i.e., $k_1(c_1 + 1) = \alpha_R p_R$, where parameter α_R denotes the proportion of degrees of each RN for connections with other nodes in the same module and parameter p_R denotes the degree of RNs. Hence, the number of element nodes in each module (i.e., the value of S_1^b) is equal to

  $$\theta_1 k_1(c_1 + 1) = \theta_1 \alpha_R p_R. \tag{A.13}$$

  Particularly, for design patterns with $d_2 = \text{BIG(element)}$, since RNs (block nodes) are not selected as dock nodes, the outer-degree of each RN is zero. This implies $\beta_R = 0$ and accordingly $\alpha_R = 1 - \beta_R = 1$. Hence, for design patterns with $d_2 = \text{BIG(element)}$, the value of S_1^b is equal to $\theta_1 \alpha_R p_R = \theta_1 p_R$.
- For design patterns with $d_2 = \text{TDG(block)}$ or $d_2 = \text{TDG(element)}$, TDG-$(n_1, \mu_1, \upsilon_1)$ with c_1 copies is employed as the internal graph, where block nodes are all RNs and element nodes are all GNs or all HNs. Based on the

discussion in Remark 2.8, such a TDG contains $n_1 \upsilon_1(c_1+1)$ element nodes and the inner-degree of each block node equals $n_1(c_1+1)$. Note that the inner-degree of each block node (RN) is $\alpha_R p_R$, i.e., $n_1(c_1+1) = \alpha_R p_R$, where parameter α_R denotes the proportion of degrees of each RN for connections with other nodes in the same module and parameter p_R denotes the degree of RNs. Hence, the number of element nodes in each module (i.e., the value of S_1^b) is equal to

$$n_1 \upsilon_1(c_1+1) = \upsilon_1 \alpha_R p_R. \tag{A.14}$$

Particularly, for design patterns with $d_2 =$ TDG(element), since RNs (block nodes) are not selected as dock nodes, the outer-degree of each RN is zero. This implies $\beta_R = 0$ and accordingly $\alpha_R = 1 - \beta_R = 1$. Hence, for design patterns with $d_2 =$ TDG(element), the value of S_1^b is equal to $\upsilon_1 \alpha_R p_R = \upsilon_1 p_R$.

In summary, the value of S_1^b is determined by the selected options of term d_1 and term d_2. Specifically,

- for design patterns with $d_2 =$ CG(any), we have $S_1^b = \alpha_G p_G + 1$ if $d_1 = G$ or $S_1^b = \alpha_H p_H + 1$ if $d_1 = H$;
- for design patterns with $d_2 =$ CBG(any), we have $S_1^b = 2\alpha_G p_G$ if $d_1 = G$ or $S_1^b = 2\alpha_H p_H$ if $d_1 = H$;
- for design patterns with $d_2 =$ BIG(block), we have $S_1^b = \theta_1 \alpha_R p_R$;
- for design patterns with $d_2 =$ BIG(element), we have $S_1^b = \theta_1 p_R$;
- for design patterns with $d_2 =$ TDG(block), we have $S_1^b = \upsilon_1 \alpha_R p_R$;
- for design patterns with $d_2 =$ TDG(element), we have $S_1^b = \upsilon_1 p_R$.

The proof is completed.

Based on the two propositions, the value of S_1 can be calculated as the product of the value of S_1^a and the value of S_1^b. Specifically,

- for design patterns with $d_2 =$ CG(any), we have $S_1 = \gamma_G p_G \cdot (\alpha_G p_G + 1)$ if $d_1 = G$ or $S_1 = (\gamma_H p_H + 1) \cdot (\alpha_H p_H + 1)$ if $d_1 = H$;
- for design patterns with $d_2 =$ CBG(any), we have $S_1 = \gamma_G p_G \cdot 2\alpha_G p_G$ if $d_1 = G$ or $S_1 = (\gamma_H p_H + 1) \cdot 2\alpha_H p_H$ if $d_1 = H$;
- for design patterns with $d_2 =$ BIG(block), we have $S_1 = \gamma_G p_G \cdot \theta_1 \alpha_R p_R$ if $d_1 = G$ or $S_1 = (\gamma_H p_H + 1) \cdot \theta_1 \alpha_R p_R$ if $d_1 = H$;
- for design patterns with $d_2 =$ BIG(element), we have $S_1 = \gamma_G p_G \cdot \theta_1 p_R$ if $d_1 = G$ or $S_1 = (\gamma_H p_H + 1) \cdot \theta_1 p_R$ if $d_1 = H$;
- for design patterns with $d_2 =$ TDG(block), we have $S_1 = \gamma_G p_G \cdot \upsilon_1 \alpha_R p_R$ if $d_1 = G$ or $S_1 = (\gamma_H p_H + 1) \cdot \upsilon_1 \alpha_R p_R$ if $d_1 = H$;
- for design patterns with $d_2 =$ TDG(element), we have $S_1 = \gamma_G p_G \cdot \upsilon_1 p_R$ if $d_1 = G$ or $S_1 = (\gamma_H p_H + 1) \cdot \upsilon_1 p_R$ if $d_1 = H$.

The proof is completed. ■

A.4.2 Proof of Lemma 4.2

Proof: We consider the number of modules (i.e., the value of S_2) for design patterns with $d_3 = P$ or $d_3 = B$, respectively.

- For design patterns with $d_3 = P$, the skeleton graph is fixed as a CG with m_2 modules, which implies $S_2 = m_2$.
- For design patterns with $d_3 = B$, the skeleton graph is fixed as a BIG-$(\theta_2, r_2, k_2, \lambda_2)$ with c_2 copies, where element nodes are modules and block nodes are RN clusters. By the properties of BIG with copies as skeleton graph (see Remark 2.10), there are $\theta_2 k_2(c_2 + 1)$ modules (element nodes) and the degree of each RN in RN clusters equals $k_2 t(c_2 + 1)$. Hence, we have $S_2 = \theta_2 k_2(c_2 + 1)$ and $k_2 t(c_2 + 1) = p_R$. By dividing θ_2 and multiplying θ_2 on both sides of equation $k_2 t(c_2 + 1) = p_R$, we have $\theta_2 k_2(c_2 + 1) = \theta_2 p_R/t$. Hence, the number of modules is $S_2 = \theta_2 k_2(c_2 + 1) = \theta_2 p_R/t$ for design patterns with $d_3 = B$.

The proof is completed. ■

A.4.3 Proof of Lemma 4.3

Proof: Note that the value of C_1 is equal to the sum of the following two terms (denoted by symbols C_1^a and C_1^b).

- The term C_1^a is the ratio of the number of GNs (or ϵ times the number of HNs) to the number of CNs and HNs.
- The term C_2 is the ratio of the number of RNs in modules to the number of CNs and HNs.

Hence, we introduce the following two propositions regarding the values of C_1^a and C_1^b, respectively.

Proposition 3 Let term C_1^a denote the ratio of the number of GNs (or ϵ times the number of HNs) to the number of CNs and HNs. Then the value of C_1^a is determined by the selected option of term d_1. Specifically,

- if $d_1 = G$, we have $C_1^a = 1/\gamma_G p_G$;
- if $d_1 = H$, we have $C_1^a = \epsilon/(\gamma_H p_H + 1)$.

Proof of Proposition 3. Note that term C_1^a denotes the ratio of the number of GNs (or ϵ times the number of HNs) to the number of CNs and HNs. Next, we consider the value of C_1^a for design patterns with $d_1 = G$ and $d_1 = H$, respectively.

- For design patterns with $d_1 = G$, term C_1^a denotes the ratio of the number of GNs to the number of CNs. Note that all CNs are directly connected to GNs. Hence, the value of C_1^a equals the ratio of 1 to the number of CNs directly connected to each GN. Recall that parameter γ_G denotes the proportion of degrees of GNs for connections with CNs. Hence, the value of C_1^a is equal to $1/\gamma_G p_G$ (i.e., $C_1^a = 1/\gamma_G p_G$) if $d_1 = G$.
- For design patterns with $d_1 = H$, term C_1^a denotes the ratio of ϵ times the number of HNs to the number of CNs and HNs. Note that all CNs are directly connected to HNs. Hence, the value of C_1^a is equal to the ratio of ϵ to the number of CNs directly connected to each HN plus one. Recall that parameter γ_H denotes the proportion of degrees for connections with CNs for each HN. Hence, the value of C_1^a is equal to $\epsilon/(\gamma_H p_H + 1)$ (i.e., $C_1^a = \epsilon/(\gamma_H p_H + 1)$) if $d_1 = H$.

In summary, the value of term C_1^a is equal to $1/\gamma_G p_G$ if $d_1 = G$ or $\epsilon/\gamma_H p_H + 1$ if $d_1 = H$. The proof of this proposition is completed.

Proposition 4 Let R_1 denote the ratio of the number of RNs in each module to the number of GNs and HNs in each module. Then the value of C_1^b (i.e., the ratio of the number of RNs in modules to the number of CNs and HNs) is determined by the selected option of term d_1. Specifically,

- if $d_1 = G$, we have $C_1^b = R_1/\gamma_G p_G$;
- if $d_1 = H$, we have $C_1^b = R_1/(\gamma_H p_H + 1)$.

The value of ratio R_1 is determined by the selected options of term d_1 and term d_2. Specifically,

- for design patterns with $d_2 =$ CG(any) or $d_2 =$ CBG(any), we have $R_1 = 0$;
- for design patterns with $d_2 =$ BIG(block) or $d_2 =$ TDG(block), we have $R_1 = \alpha_G p_G/\alpha_R p_R$ if $d_1 = G$ or $R_1 = \alpha_H p_H/\alpha_R p_R$ if $d_1 = H$;
- for design patterns with $d_2 =$ BIG(element) or $d_2 =$ TDG(element), we have $R_1 = \alpha_G p_G/p_R$ if $d_1 = G$ or $R_1 = \alpha_H p_H/p_R$ if $d_1 = H$.

Proof of Proposition 4. To prove this proposition, we first discuss the relationships between the values of C_1^b and R_1, followed by the consideration on the value of R_1.

- *Relationships between the values of C_1^b and R_1.* Based on the homogeneous assumptions, each module contains the same number of GNs (or HNs) and RNs, while each GN (or HN) attaches the same number of CNs. Hence, we express the value of C_1^b as follows:

$$
\begin{aligned}
C_1^b &= \frac{\text{\# of RNs in modules}}{\text{\# of CNs} + \text{\# of HNs}} \\
&= \frac{\text{\# of RNs in each module}}{\text{\# of CNs attached to each module} + \text{\# HNs in each module}} \\
&= \frac{\text{\# of RNs in each module}}{\text{\# of GNs or HNs in each module}} \cdot \frac{1}{\text{attachment number of each GN or HN}} \\
&= R_1 \cdot \frac{1}{\text{attachment number of each GN or HN}},
\end{aligned}
$$

where the attachment number of each GN is the number of its directly connected CNs and the attachment number of each HN is the number of its directly connected CNs plus one (itself).

In this way, we discuss the value of C_1^b for design patterns with $d_1 = G$ and $d_1 = H$, respectively.

- If $d_1 = G$, the attachment number of a GN is the number of its directly connected CNs. Recall that parameter γ_G denotes the proportion of degrees for connections with CNs and parameter p_G denotes the degree of GNs. Hence, the attachment number of each GN equals $\gamma_G p_G$. Therefore, the value of C_1^b is equal to $R_1/\gamma_G p_G$.
- If $d_1 = H$, the attachment number of an HN is the number of its directly connected CNs plus one (itself). Recall that parameter γ_H denotes the proportion of degrees for connections with CNs and parameter p_H denotes the degree of HNs. Hence, the attachment number of each HN equals $\gamma_H p_H + 1$. Therefore, the value of C_1^b is equal to $R_1/(\gamma_H p_H + 1)$.

• *The value of ratio R_1.* The value of R_1 is calculated for design patterns with different values of term d_2 as follows.

- For design patterns with $d_2 = \text{CG(any)}$ or $d_2 = \text{CBG(any)}$, all nodes in modules are GNs and there are no RNs in the modules. Hence, the value of R_1 (i.e., the number of RNs in each module to the number of GNs or HNs in each module) equals zero.
- For design patterns with $d_2 = \text{BIG(element)}$ or $d_2 = \text{TDG(block)}$, a BIG with copies is given as the internal graph, where block nodes are RNs and element nodes are GNs or HNs. Note that all BIGs with copies are biregular bipartite graphs. Hence, the value of R_1 (i.e., the number of RNs in each module to the number of GNs or HNs in each module) equals to the ratio of the inner-degree of each GN or HN to the inner-degree of each RN. Therefore, the value of R_1 is equal to $\alpha_G p_G/\alpha_R p_R$ if $d_1 = G$ or $\alpha_H p_H/\alpha_R p_R$ if $d_1 = H$. Particularly, for design patterns with $d_2 = \text{BIG(element)}$, we have $\alpha_R = 1$. Therefore, the value of R_1 is equal to $\alpha_G p_G/p_R$ if $d_1 = G$ or $\alpha_H p_H/p_R$ if $d_1 = H$.
- For design patterns with $d_2 = \text{TDG(element)}$ or $d_2 = \text{TDG(block)}$, a TDG with copies is given as the internal graph, where block nodes are RNs and

element nodes are GNs or HNs. Note that all TDGs with copies are biregular bipartite graphs. Hence, the value of R_1 (i.e., the number of RNs in each module to the number of GNs or HNs in each module) equals to the ratio of the inner-degree of each GN or HN to the inner-degree of each RN. Therefore, the value of R_1 is equal to $\alpha_G p_G/\alpha_R p_R$ if $d_1 = G$ or $\alpha_H p_H/\alpha_R p_R$ if $d_1 = H$. Particularly, for design patterns with $d_2 =$ TDG(element), we have $\alpha_R = 1$. Therefore, the value of R_1 is equal to $\alpha_G p_G/p_R$ if $d_1 = G$ or $\alpha_H p_H/p_R$ if $d_1 = H$.

The proof of this proposition is completed.

Based on these two propositions, the value of C_1 can be calculated as the sum of the values of C_1^a and C_1^b. Accordingly,

- if $d_1 = G$, we have $C_1 = (1 + R_1)/\gamma_G p_G$;
- if $d_1 = H$, we have $C_1 = (\epsilon + R_1)/(\gamma_H p_H + 1)$.

Meanwhile, the value of ratio R_1 is determined by the selected options of term d_1 and term d_2. Specifically,

- for design patterns with $d_2 =$ CG(any) or $d_2 =$ CBG(any), we have $R_1 = 0$;
- for design patterns with $d_2 =$ BIG(block) or $d_2 =$ TDG(block), we have $R_1 = \alpha_G p_G/\alpha_R p_R$ if $d_1 = G$ or $R_1 = \alpha_H p_H/\alpha_R p_R$ if $d_1 = H$;
- if $d_2 =$ BIG(element) or $d_2 =$ TDG(element), we have $R_1 = \alpha_G p_G/p_R$ if $d_1 = G$ or $R_1 = \alpha_H p_H/p_R$ if $d_1 = H$.

The proof is completed. ■

A.4.4 Proof of Lemma 4.4

Proof: To prove this lemma, we introduce the following two propositions, in which the relationships between the values of C_2 and R_2 are discussed and the value of R_2 is considered, respectively.

Proposition 5 The relationships between term C_2 and term R_2 is determined by the selected options of term d_1 and term d_3. Specifically,

- for design patterns with $d_3 = P$, we have $C_2 = 0$;
- for design patterns with $d_3 = B$, we have $C_2 = R_2/\gamma_G p_G p_R$ if $d_1 = G$ or $C_2 = R_2/(\gamma_H p_H + 1)p_R$ if $d_1 = H$.

Proof of Proposition 5. The relationships between term C_2 and term R_2 are considered under different options of term d_3, respectively.

- For design patterns with $d_3 = P$, there are no RN clusters in the topology. Since term C_2 is the ratio of the number of RNs in RN clusters to the number of CNs and HNs, we have $C_2 = 0$.
- For design patterns with $d_3 = B$, the bipartite mode is adopted for the interconnections between modules and RN clusters, which implies that the dock nodes in modules and RNs in RN clusters form a bipartite graph. Based on the homogeneous assumptions, such a bipartite graph is biregular because the degrees of all RNs in RN clusters are identical and the outer-degrees of all dock nodes are identical. Hence, we have

$$\begin{aligned}&(\text{\# of RNs in RN clusters}) \cdot (\text{the degree of each RN})\\ =\ &(\text{\# of dock nodes in modules}) \cdot (\text{the outer-degree of each dock node}).\end{aligned} \tag{A.15}$$

Based on the homogeneous assumptions, each module contains the same number of GNs or HNs and each RN cluster contains the same number of RNs. Hence, term C_2 can be expressed by

$$\begin{aligned}C_2 &= \frac{\text{\# of RNs in RN clusters}}{\text{\# of CNs and HNs attached to modules}}\\ &= \frac{\text{\# of RNs in RN clusters} \cdot p_R}{\text{\# of CNs and HNs attached to modules} \cdot p_R}\\ &= \frac{\text{\# of dock nodes in modules} \cdot \text{the outer-degree of each dock node}}{\text{\# of modules} \cdot \text{\# of CNs and HNs attached to each module} \cdot p_R}\\ &= \frac{\text{\# of dock nodes in each module} \cdot \text{the outer-degree of each dock node}}{\text{\# of CNs and HNs attached to each module} \cdot p_R}\\ &= \frac{\text{the total outer-degree of all dock nodes in each module}}{\text{\# of CNs and HNs attached to each module} \cdot p_R}.\end{aligned} \tag{A.16}$$

Meanwhile, by the definition of term R_2, we have

$$R_2 = \frac{\text{the total outer-degree of all dock nodes in each module}}{\text{\# of GNs and HNs in each module}}. \tag{A.17}$$

By combining equations (A.16) and (A.17), it follows that

$$C_2 = R_2 \cdot \frac{1}{p_R} \cdot \frac{\text{\# of GNs and HNs in each module}}{\text{\# of CNs and HNs attached to each module}}. \tag{A.18}$$

Regarding the ratio of the number of GNs and HNs in each module to the number of CNs and HNs attached to each module, it value is determined by the selected option for term d_1.

- If $d_1 = G$, there are no HNs and all CNs are attached to GNs. Hence, such a ratio turns into the ratio of the number of GNs in each module to the number of CNs attached to each module, which is equal to the ratio of 1 to the number of CNs attached to each GN. Note that parameter γ_G denotes the proportion of connections with CNs for each GN. Hence, such a ratio equals $1/\gamma_G p_G$.

- If $d_1 = H$, there are no GNs and all CNs are attached to HNs. Hence, such a ratio turns into the ratio of the number of HNs in each module to the number of CNs and HNs attached to each module, which is equal to the ratio of 1 to the number of CNs attached to each HN plus one (itself). Note that parameter γ_H denotes the proportion of connections with CNs for each HN. Hence, such a ratio equals $1/(\gamma_H p_H + 1)$.

As a result, for design patterns with $d_3 = B$, we have $C_2 = R_2/\gamma_G p_G p_R$ if $d_1 = G$ or $C_2 = R_2/(\gamma_H p_H + 1)p_R$ if $d_1 = H$.

In summary, for design patterns with $d_3 = P$, we have $C_2 = 0$; for design patterns with $d_3 = B$, we have $C_2 = R_2/\gamma_G p_G p_R$ if $d_1 = G$ or $C_2 = R_2/(\gamma_H p_H + 1)p_R$ if $d_1 = H$. The proof of this proposition is completed.

Proposition 6 The value of ratio R_2 is determined by the selected options of term d_1 and term d_2. Specifically,

- for design patterns with $d_2 =$ CG(any), $d_2 =$ CBG(any), $d_2 =$ BIG(element), or $d_2 =$ TDG(element), we have $R_2 = \beta_G p_G$ if $d_1 = G$ or $R_2 = \beta_H p_H$ if $d_1 = H$;
- for design patterns with $d_2 =$ BIG(block) or $d_2 =$ TDG(block), we have $R_2 = \beta_R \alpha_G p_G/\alpha_R$ if $d_1 = G$ or $R_2 = \beta_R \alpha_H p_H/\alpha_R$ if $d_1 = H$.

Proof of Proposition 6. Note that term R_2 denotes the ratio of the total outer-degree of all dock nodes in each module to the number of GNs and HNs in each module. Since the selection of dock nodes is related to the values in dimension d_2, we discuss the value of term R_2 for design patterns with distinct options of term d_2 as follows.

- Design patterns with $d_2 =$ CG(any), CBG(any), BIG(element), or TDG(element). For these design patterns, GNs or HNs are selected as dock nodes. Hence, term R_2 can be expressed as follows.

$$\begin{aligned} R_2 &= \frac{\text{outer-degree of all GNs or all HNs in each module}}{\text{\# of GNs or HNs in each module}} \\ &= \text{outer-degree of each GN or each HN} \end{aligned} \tag{A.19}$$

  Note that parameters β_G and β_H denote the proportions of outer-degrees for each GN and each HN, respectively. Hence, term R_2 is equal to $\beta_G p_G$ if $d_1 = G$ or $\beta_H p_H$ if $d_1 = H$.

- Design patterns with d_2 = BIG(block) or d_2 = TDG(block). For these design patterns, RNs are selected as dock nodes. Hence, term R_2 can be expressed as follows.

$$R_2 = \frac{\text{outer-degree of all RNs in each module}}{\text{\# of GNs or HNs in each module}} \tag{A.20}$$

Moreover, the internal graph is fixed as a biregular bipartite graph (BIG or TDG), where RNs are in one node partition and GNs or HNs are in the other node partition. Hence, we have

$$\begin{aligned}&(\text{\# of RNs in each module}) \cdot (\text{inner-degree of each RN})\\ &= (\text{\# of GNs or HNs in each module}) \cdot (\text{inner-degree of each GN or HN}).\end{aligned} \tag{A.21}$$

By substituting equation (A.21) into equation (A.20), it follows that

$$\begin{aligned}R_2 &= \frac{\text{outer-degree of all RNs in each module}}{\text{\# of GNs or HNs in each module}}\\ &= \frac{\text{outer-degree of all RNs in each module}}{\text{\# of RNs in each module}} \cdot \frac{\text{inner-degree of each GN or HN}}{\text{inner-degree of each RN}}\\ &= \text{outer-degree of each RN} \cdot \frac{\text{inner-degree of each GN or HN}}{\text{inner-degree of each RN}}\\ &= \text{inner-degree of each GN or HN} \cdot \frac{\text{outer-degree of each RN}}{\text{inner-degree of each RN}}\end{aligned} \tag{A.22}$$

Note that parameters α_R and β_R denote the proportions of inner-degree and outer-degree for each RN, respectively. Hence, the ratio of outer-degree of each RN to inner-degree of each RN is equal to β_R/α_R. Meanwhile, since parameter α_G (resp. α_H) denotes the proportion of inner-degree for each GN (resp. each HN), the inner-degree of each GN (resp. each HN) is equal to $\alpha_G p_G$ (resp. $\alpha_H p_H$). Hence, the value of R_2 is equal to $\beta_R \alpha_G p_G/\alpha_R$ if $d_1 = G$ or $\beta_R \alpha_H p_H/\alpha_R$ if $d_1 = H$.

In summary, for design patterns with d_2 = CG(any), d_2 = CBG(any), d_2 = BIG(element), or d_2 = TDG(element), we have $R_2 = \beta_G p_G$ if $d_1 = G$ or $R_2 = \beta_H p_H$ if $d_1 = H$; for design patterns with d_2 = BIG(block) or d_2 = TDG(block), we have $R_2 = \beta_R \alpha_G p_G/\alpha_R$ if $d_1 = G$ or $R_2 = \beta_R \alpha_H p_H/\alpha_R$ if $d_1 = H$. The proof of this proposition is completed.

Based on these two propositions, the proof of Lemma 4.4 is completed. ■

A.4.5 Proof of Lemma 4.5

Proof: By Definition 2.6, the latency L of a topology is defined as the maximum distance between any pair of nodes with computing function (i.e., CNs and HNs) in the topology. Then we consider the relationship between the values of L and l' under different values of term d_1.

- For design patterns with $d_1 = G$, there are no HNs in the topology. Hence, the latency is the maximum distance between any pair of CNs, while the value of l' is the maximum distance between two GNs. Note that each GN is directly connected to some CNs. Therefore, the maximum distance between any pair of CNs is equal to the maximum distance between any pair of GNs plus two (the distance between each GN and its directly connected CN). Accordingly, we have $L = l' + 2$.
- For design patterns with $d_1 = H$, there are no GNs in the topology. Nonetheless, it is not clear whether there are CNs or not. Next, we consider two cases where $\gamma_H = 0$ and $\gamma_H > 0$, respectively.
  - *Case 1:* $\gamma_H > 0$. In this case, each HN is directly connected to some CNs. Therefore, the maximum distance between any pair of CNs is equal to the maximum distance between any pair of HNs plus two (the distance between each GN and its directly connected CN). Accordingly, the latency is the maximum distance between any pair of CNs in the topology, while the value of l' is the maximum distance between two HNs. Hence, we have $L = l' + 2$.
  - *Case 2:* $\gamma_H = 0$. In this case, there are no CNs in the topology. Hence, the values of L and l' are both equal to the maximum distance between any pair of HNs in the topology. As a result, we have $L = l'$.

In summary, for design patterns with $d_1 = G$ or $d_1 = H$ and $\gamma_H > 0$, we have $L = l' + 2$; for design patterns with $d_1 = H$ and $\gamma_H = 0$, we have $L = l'$. The proof is completed. ■

A.4.6 Proof of Lemma 4.6

Proof: This proof is conducted by sequentially demonstrating the following propositions.

Proposition 7 For any topology under design patterns with $d_2 =$ CG(any) and $d_3 = P$, there exists a path with length less than or equal to three between each pair of GNs or HNs.

Proposition 8 For any topology under design patterns with $d_2 =$ CG(any) and $d_3 = P$, there exists a length-1 path (i.e., an edge) between each pair of GNs or HNs if and only if $x = N_D$.

Proposition 9 For any topology under design patterns with $d_2 = \text{CG(any)}$ and $d_3 = P$, there exists a path with length less than or equal to two between each pair of GNs or HNs if $tx = N_D$.

Proposition 10 There exists a topology under design patterns with $d_2 = \text{CG(any)}$ and $d_3 = P$, such that the value of l' is equal to 3.

If the described four propositions are established, then this lemma is proved. The reasons are given as follows. First, by Proposition 7, we prove that the value of l' can be 1, 2, or 3. Moreover, Propositions 8–10 show that 1, 2, and 3 are all feasible values of l'. Hence, it is proved that the feasible values of l' include 1, 2, and 3. Second, Proposition 8 proves that $l' = 1$ if and only if $x = N_D$. Third, Proposition 9 proves that $l' \leq 2$ if $tx = N_D$. By ruling out the case with $l' = 1$ and accordingly $x = N_D$, we can prove that $l' = 2$ if $tx = N_D$ and $t > 1$. The proofs of the four propositions described are very similar to the previous proofs of lemmas in this chapter and are therefore omitted here. ■

A.4.7 Proof of Lemma 4.7

Proof: This proof is conducted by sequentially demonstrating the following propositions.

Proposition 11 For any topology under design patterns with $d_2 = \text{CG(any)}$ and $d_3 = B$, there exists a path with length less than or equal to four between each pair of GNs or HNs.

Proposition 12 For any topology under design patterns with $d_2 = \text{CG(any)}$ and $d_3 = B$, there exists a pair of GNs or HNs at a distance of two.

Proposition 13 For any topology under design patterns with $d_2 = \text{CG(any)}$ and $d_3 = B$, there exists a path with length less than or equal to 3 between each pair of GNs or HNs if $tx = N_D$. Moreover, if we further have $x = 1$ or $c_2 = 0$, there exists a length-2 path between each pair of GNs or HNs.

Proposition 14 For any topology under design patterns with $d_2 = \text{CG(any)}$ and $d_3 = B$, there exists a length-3 path between each pair of GNs or HNs if $\theta_2 = 1$.

Proposition 15 There exists a topology under design patterns with $d_2 = \text{CG(any)}$ and $d_3 = B$, such that the value of l' is equal to 3.

Proposition 16 There exists a topology under design patterns with $d_2 = \text{CG(any)}$ and $d_3 = B$, such that the value of l' is equal to 4.

If these six propositions are established, then this lemma is proved. The reasons are given as follows. First, by Propositions 11 and 12, we prove that the value of l'

can be 2, 3, or 4. Moreover, Propositions 13–15 show that 2, 3, and 4 are all feasible values of l'. Hence, it is proved that the feasible values of l' include 2, 3, and 4. Second, Proposition 13 proves that $l' = 2$ if $x = 1$ or $tx = N_D$ and $c_2 = 0$. Third, Proposition 9 proves that $l' \leq 3$ if $tx = N_D$ or $\theta_2 = 1$. The proofs of the six propositions are very similar to the previous proofs of lemmas in this chapter and are therefore omitted here. ■

A.4.8 Proof of Lemma 4.8

Proof: This proof is conducted by sequentially demonstrating the following propositions.

Proposition 17 For any topology under design patterns with $d_2 = \text{CBG(any)}$ and $d_3 = P$, the distance between two GNs or HNs in the same node partition of CBG (internal graph) is two.

Proposition 18 For any topology under design patterns with $d_2 = \text{CBG(any)}$ and $d_3 = P$, there exists a path with length less than or equal to five between each pair of GNs or HNs.

Proposition 19 For any topology under design patterns with $d_2 = \text{CBG(any)}$ and $d_3 = P$, there exists a path with length less than or equal to four between each pair of GNs or HNs if $x = 1$ and $t \geq 2$. Moreover, if we further have $tx = N_D$ (hence $t = N_D$), then there exists a path with length less than or equal to three between each pair of GNs or HNs.

Proposition 20 For any topology under design patterns with $d_2 = \text{CBG(any)}$ and $d_3 = P$, there exists a path with length less than or equal to three between each pair of GNs or HNs if $x \geq 2$. Moreover, if we further have $tx = N_D$, then there exists a path with length less than or equal to two between each pair of GNs or HNs.

Proposition 21 There exists a topology under design patterns with $d_2 = \text{CBG(any)}$ and $d_3 = P$, such that the value of l' is equal to 3.

Proposition 22 There exists a topology under design patterns with $d_2 = \text{CBG(any)}$ and $d_3 = P$, such that the value of l' is equal to 4.

Proposition 23 There exists a topology under design patterns with $d_2 = \text{CBG(any)}$ and $d_3 = P$, such that the value of l' is equal to 5.

If these seven propositions are established, then this lemma is proved. The reasons are given as follows. First, by Propositions 17 and 18, we prove that the value of l' can be 2, 3, 4, or 5. Moreover, Propositions 19–23 show that 2, 3, 4, and 5 are all feasible values of l'. Hence, it is proved that the feasible values of l' include 2, 3, 4, and 5. Second, Proposition 19 proves that $l' \leq 4$ if $x = 1$ and $t \geq 2$ or $l' \leq 3$

if $x = 1$ and $t = N_D$. Third, Proposition 20 proves that $l' \leq 3$ if $x \geq 2$ or $l' \leq 2$ if $x \geq 2$ and $tx = N_D$. The proofs of the seven propositions are very similar to the previous proofs of lemmas in this chapter and are therefore omitted here. ■

A.4.9 Proof of Lemma 4.9

Proof: This proof is conducted by sequentially demonstrating the following propositions.

Proposition 24 For any topology under design patterns with $d_2 = \text{CBG(any)}$ and $d_3 = B$, there exists a path with length less than or equal to six between each pair of GNs or HNs.

Proposition 25 For any topology under design patterns with $d_2 = \text{CBG(any)}$ and $d_3 = B$, there exists a pair of GNs or HNs at a distance of two.

Proposition 26 For any topology under design patterns with $d_2 = \text{CBG(any)}$ and $d_3 = B$, there exists a path with length less than or equal to 3 between each pair of GNs or HNs if $tx = N_D$. Moreover, if we further have $x = 1$ or $c_2 = 0$, there exists a length-2 path between each pair of GNs or HNs.

Proposition 27 For any topology under design patterns with $d_2 = \text{CBG(any)}$ and $d_3 = B$, there exists a length-3 path between each pair of GNs or HNs if $\theta_2 = 1$.

Proposition 28 There exists a topology under design patterns with $d_2 = \text{CBG(any)}$ and $d_3 = B$, such that the value of l' is equal to 3.

Proposition 29 There exists a topology under design patterns with $d_2 = \text{CBG(any)}$ and $d_3 = B$, such that the value of l' is equal to 5.

Proposition 30 There exists a topology under design patterns with $d_2 = \text{CBG(any)}$ and $d_3 = B$, such that the value of l' is equal to 6.

If these seven propositions are established, then this lemma is proved. The reasons are given as follows. First, by Propositions 24 and 25, we prove that the value of l' can be 2, 3, 4, 5, or 6. Moreover, Propositions 26–30 show that 2, 3, 4, 5, and 6 are all feasible values of l'. Hence, it is proved that the feasible values of l' include 2, 3, 4, 5, and 6. Second, Proposition 26 proves that $l' = 2$ if $x = 1$ or $tx = N_D$ and $c_2 = 0$. Third, Proposition 27 proves that $l' \leq 3$ if $tx = N_D$ or $\theta_2 = 1$. The proofs of the seven propositions are very similar to the previous proofs of lemmas in this chapter and are therefore omitted here. ■

A.4.10 Proof of Lemma 4.10

Proof: This proof is conducted by sequentially demonstrating the following propositions.

Proposition 31 For any topology under design patterns with d_2 = BIG(block) and $d_3 = P$, there exists a path with length less than or equal to three between each pair of GNs or HNs.

Proposition 32 For any topology under design patterns with d_2 = BIG(block) and $d_3 = P$, the distance between two GNs or HNs in distinct modules is greater than two.

If these two propositions are established, then the maximum distance between two GNs or HNs is equal to three (i.e., $l' = 3$). Hence, this lemma is proved. The proofs of the two propositions are very similar to the previous proofs of lemmas in this chapter and are therefore omitted here. ■

A.4.11 Proof of Lemma 4.11

Proof: This proof is conducted by sequentially demonstrating the following three propositions.

Proposition 33 For design patterns with d_2 = BIG(block) and $d_3 = B$, the distance between two GNs or HNs is a positive even number.

Proposition 34 For design patterns with d_2 = BIG(block) and $d_3 = B$, the distance between two GNs or HNs in distinct modules is greater than two.

Proposition 35 For design patterns with d_2 = BIG(block) and $d_3 = B$, there is a path with length less than or equal to four between any pair of GNs or HNs.

If these three propositions are established, then the maximum distance between two GNs or HNs is equal to four (i.e., $l' = 4$). Hence, this lemma is proved. The proofs of the three propositions are very similar to the previous proofs of lemmas in this chapter and are therefore omitted here. ■

A.4.12 Proof of Lemma 4.12

Proof: This proof is conducted by sequentially demonstrating the following seven propositions.

Proposition 36 For design patterns with d_2 = BIG(element) and $d_3 = P$, the distance between two GNs or HNs in the same module is equal to two.

Proposition 37 For design patterns with d_2 = BIG(element) and $d_3 = P$, there is a path with length less than or equal to 5 between each pair of GNs or HNs.

Proposition 38 For design patterns with d_2 = BIG(element) and $d_3 = P$, there is a path with length less than or equal to 3 between each pair of GNs or HNs if $tx = N_D$. Moreover, if we further have $x = N_D$ (hence $t = 1$), then there is a length-2 path between each pair of GNs or HNs.

Proposition 39 For design patterns with d_2 = BIG(element) and $d_3 = P$, there is a path with length less than or equal to 4 between each pair of GNs or HNs if $m_2 = 3$.

Proposition 40 There exists a topology under design patterns with d_2 = BIG (element) and $d_3 = P$, such that the value of l' is equal to 3.

Proposition 41 There exists a topology under design patterns with d_2 = BIG (element) and $d_3 = P$, such that the value of l' is equal to 4.

Proposition 42 There exists a topology under design patterns with d_2 = BIG (element) and $d_3 = P$, such that the value of l' is equal to 5.

If the seven propositions are established, then this lemma is proved. The reasons are given as follows. First, by Propositions 36 and 37, we prove that the value of l' can be 2, 3, 4, or 5. Moreover, Propositions 38–42 show that 2, 3, 4, and 5 are all feasible values of l'. Hence, it is proved that the feasible values of l' include 2, 3, 4, and 5. Second, Proposition 38 proves that $l' \leq 3$ if $tx = N_D$. Moreover, if $x = N_D$, we have $l' = 2$. Third, Proposition 39 proves that $l' \leq 4$ if $m_2 = 3$. The proofs of the seven propositions are very similar to the previous proofs of lemmas in this chapter and are therefore omitted here. ■

A.4.13 Proof of Lemma 4.13

Proof: This proof is conducted by sequentially demonstrating the following six propositions.

Proposition 43 For design patterns with d_2 = BIG(element) and $d_3 = B$, the distance between two GNs or HNs is a positive even number.

Proposition 44 For design patterns with d_2 = BIG(element) and $d_3 = B$, there is a path with length less than or equal to six between any pair of GNs or HNs.

Proposition 45 For design patterns with d_2 = BIG(element) and $d_3 = B$, there is a path with length less than or equal to four between any pair of GNs or HNs if $tx = N_D$. Moreover, if we further have $x = 1$ (hence $t = N_D$) or $c_2 = 0$, there is a length-2 path between each pair of GNs or HNs.

Proposition 46 For design patterns with d_2 = BIG(element) and $d_3 = B$, there is a path with length less than or equal to four between any pair of GNs or HNs if $\theta_2 = 1$.

Proposition 47 For design patterns with d_2 = BIG(element) and $d_3 = B$, there exists a topology such that its value of l' is equal to four.

Proposition 48 For design patterns with d_2 = BIG(element) and $d_3 = B$, there exists a topology such that its value of l' is equal to six.

If these six propositions are established, then this lemma is proved. The reasons are given as follows. First, by Propositions 43 and 44, we prove that the value of l' can be 2, 4, or 6. Moreover, Propositions 45–48 show that 2, 4, and 6 are all feasible values of l'. Hence, it is proved that the feasible values of l' include 2, 4, and 6. Second, Proposition 45 proves $l' \leq 4$ if $tx = N_D$. Moreover, if we further have $x = 1$ (hence $t = N_D$) or $c_2 = 0$, then the value of l' is equal to two. Third, Proposition 46 proves $l' \leq 4$ if $\theta_2 = 1$. The proofs of the six propositions are very similar to the previous proofs of lemmas in this chapter and are therefore omitted here. ■

A.4.14 Proof of Lemma 4.14

Proof: This proof is conducted by sequentially demonstrating the following propositions.

Proposition 49 For any topology under design patterns with d_2 = TDG(block) and $d_3 = P$, there exists a path with length less than or equal to four between each pair of GNs or HNs.

Proposition 50 For any topology under design patterns with d_2 = TDG(block) and $d_3 = P$, the distance between two GNs or HNs in the same group of internal graph (TDG) is greater than three.

If the two propositions are established, then the maximum distance between two GNs or HNs is equal to four (i.e., $l' = 4$). Hence, this lemma is proved. The proofs of the two propositions are very similar to the previous proofs of lemmas in this chapter and are therefore omitted here. ■

A.4.15 Proof of Lemma 4.15

Proof: This proof is conducted by sequentially demonstrating the following four propositions.

Proposition 51 For design patterns with d_2 = TDG(block) and $d_3 = B$, the distance between two GNs or HNs is a positive even number.

Proposition 52 For design patterns with d_2 = TDG(block) and $d_3 = B$, the distance between two GNs or HNs in the same group of internal graph (TDG) is greater than two.

Proposition 53 For design patterns with d_2 = TDG(block) and $d_3 = B$, there is a path with length less than or equal to six between any pair of GNs or HNs. Moreover, if $c_2 = 0$, then there is a path with length less than or equal to four between any pair of GNs or HNs.

Proposition 54 For design patterns with d_2 = TDG(block) and $d_3 = B$, there exists a topology such that its value of l' is equal to six.

If these four propositions are established, then this lemma is proved. The reasons are given as follows. First, by Propositions 51 – 53, we prove that the value of l' can be 4 or 6. Moreover, Propositions 53 and 54 show that 4 and 6 are all feasible values of l'. Hence, it is proved that the feasible values of l' include 4 and 6. Second, Proposition 53 proves that $l' = 4$ if $c_2 = 0$. The proofs of the four propositions are very similar to the previous proofs of lemmas in this chapter and are therefore omitted here. ■

A.4.16 Proof of Lemma 4.16

Proof: This proof is conducted by sequentially demonstrating the following seven propositions.

Proposition 55 For design patterns with d_2 = TDG(element) and $d_3 = P$, the distance between two GNs or HNs in the same module is greater than 1.

Proposition 56 For design patterns with d_2 = TDG(element) and $d_3 = P$, there is a path with length less than or equal to 5 between each pair of GNs or HNs.

Proposition 57 For design patterns with d_2 = TDG(element) and $d_3 = P$, there is a path with length less than or equal to 4 between each pair of GNs or HNs if $tx = N_D$. Moreover, if we further have $x = N_D$ (hence $t = 1$), then there is a length-2 path between each pair of GNs or HNs.

Proposition 58 There exists a topology under design patterns with $d_3 = P$ and d_2 = TDG(element), such that the value of l' is equal to 3.

Proposition 59 There exists a topology under design patterns with $d_3 = P$ and d_2 = TDG(element), such that the value of l' is equal to 4.

Proposition 60 There exists a topology under design patterns with $d_3 = P$ and d_2 = TDG(element), such that the value of l' is equal to 5.

If these six propositions are established, then this lemma is proved. The reasons are given as follows. First, by Propositions 55 and 56, we prove that the value of l' can be 2, 3, 4, or 5. Moreover, Propositions 57–60 show that 2, 3, 4, and 5 are all feasible values of l'. Hence, it is proved that the feasible values of l' include 2, 3, 4, and 5. Second, Proposition 57 proves that $l' \leq 4$ if $tx = N_D$. Moreover, if $x = N_D$, we have $l' = 2$. The proofs of the six propositions are very similar to the previous proofs of lemmas in this chapter and are therefore omitted here. ■

A.4.17 Proof of Lemma 4.17

Proof: This proof is conducted by sequentially demonstrating the following six propositions.

Proposition 61 For design patterns with $d_2 = \text{TDG(element)}$ and $d_3 = B$, the distance between two GNs or HNs is a positive even number.

Proposition 62 For design patterns with $d_2 = \text{TDG(element)}$ and $d_3 = B$, there is a path with length less than or equal to six between any pair of GNs or HNs.

Proposition 63 For design patterns with $d_2 = \text{TDG(element)}$ and $d_3 = B$, there is a path with length less than or equal to four between any pair of GNs or HNs if $tx = N_D$ and $c_2 = 0$. Moreover, if we further have $x = p_R$ (hence $t = v_1$), there is a length-2 path between each pair of GNs or HNs.

Proposition 64 For design patterns with $d_2 = \text{TDG(element)}$ and $d_3 = B$, there is a path with length less than or equal to four between any pair of GNs or HNs if $\theta_2 = 1$ and $c_2 = 0$.

Proposition 65 For design patterns with $d_2 = \text{TDG(element)}$ and $d_3 = B$, there exists a topology such that its value of l' is equal to four.

Proposition 66 For design patterns with $d_2 = \text{TDG(element)}$ and $d_3 = B$, there exists a topology such that its value of l' is equal to six.

If these six propositions are established, then this lemma is proved. The reasons are given as follows. First, by Propositions 61 and 62, we prove that the value of l' can be 2, 4, or 6. Moreover, Propositions 63–66 show that 2, 4, and 6 are all feasible values of l'. Hence, it is proved that the feasible values of l' include 2, 4, and 6. Second, Proposition 63 proves $l' \leq 4$ if $tx = N_D$ and $c_2 = 0$. Moreover, if we further have $x = p_R$ (hence $t = v_1$), then the value of l' is equal to two. Third, Proposition 64 proves $l' \leq 4$ if $\theta_2 = 1$ and $c_2 = 0$. The proofs of the six propositions are very similar to the previous proofs of lemmas in this chapter and are therefore omitted here. ■

A.4.18 Proof of Lemma 4.18

Proof: By Definition 2.7, intra-module throughput T_1 is defined as a ratio A_1/B_1, where A_1 represents the minimum number of edge-disjoint paths between each pair of GNs or HNs in the same module and B_1 represents the minimum attachment number of each GN or HN. Correspondingly, we prove this lemma by considering the values of A_1 and B_1, respectively.

The value of $\mathbf{A_1}$. We consider the value of A_1 by sequentially demonstrating the following six propositions.

Proposition 67 Given two nodes in a graph, the number of edge-disjoint paths between them is less than or equal to the degree of each of the two nodes.

Proposition 68 Given a group of paths with even length, if their even-order node are different to each other, then these paths are edge-disjoint.

Proposition 69 For design patterns with $d_2 =$ CG(any), the number of edge-disjoint paths between any element nodes S and T is equal to the inner-degree of each GN or HN.

Proposition 70 For design patterns with $d_2 =$ CBG(any), the number of edge-disjoint paths between any element nodes S and T is equal to the inner-degree of each GN or HN.

Proposition 71 For design patterns with $d_2 =$ BIG(block) or $d_2 =$ BIG(element), the number of edge-disjoint paths between any element nodes S and T is equal to the inner-degree of each GN or HN.

Proposition 72 For design patterns with $d_2 =$ TDG(block) or $d_2 =$ TDG (element), the number of edge-disjoint paths between any element nodes S and T is equal to the inner-degree of each GN or HN.

If these six propositions are established, it can be shown that the value of A_1 is equal to the inner-degree of each GN or HN. Then it follows that the value of A_1 is equal to $\alpha_G p_G$ if $d_1 = G$ or $\alpha_H p_H$ if $d_1 = H$.

Proof of Proposition 67. Given two nodes S and T, the paths between them are edge-disjoint if and only if such two paths have no common edges. Let symbol d_S denote the degree of node S, then node S is incident to d_S edges. Note that each path between nodes S and T must traverse an edge incident to S. Suppose the number of edge-disjoint paths between S and T is greater than d_S, then at least two paths traverse the same edge incident to S, which is contradicted to the definition of edge-disjoint paths. Therefore, the number of edge-disjoint paths is less than or equal to the degree of node S. As a result, given two nodes in a graph, the number of edge-disjoint paths between them is less than or equal to the degree of each of the two nodes.

Proof of Proposition 68. Given a group of paths with even length, one of the endpoints of all edges in these paths must be a node with even order. Note that two edges are identical if and only if the two endpoints of such two edges are exactly same. Hence, if the even-order node of these paths are different to each other, then the edges of these paths must be different. According to the definition of edge-disjoint paths, these paths are edge-disjoint.

Proof of Proposition 69. For design patterns with $d_2 = \text{CG(any)}$, CG is fixed as the internal graph, where nodes are all GNs or all HNs. Let D denote the inner-degree of each GN or HN. Then there are $D + 1$ nodes in such a graph. Given any two nodes S and T in such a graph, there are D edge-disjoint paths between them, including one length-1 path and $D - 1$ length-2 paths. The length-1 path is the edge between nodes S and T. As for those $D - 1$ length-2 paths, each of them goes through a unique node other than nodes S and T. By Proposition 68, these length-2 paths are edge-disjoint. Meanwhile, the length-1 path contains edge $S - T$ only, which is also edge-disjoint with length-2 paths. Accordingly, there are D edge-disjoint paths between nodes S and T. In summary, for design patterns with $d_2 = \text{CG(any)}$, the number of edge-disjoint paths between nodes S and T is equal to the inner-degree of each GN or HN. The proof of this proposition is completed.

Proof of Proposition 70. For design patterns with $d_2 = \text{CBG(any)}$, CBG is fixed as the internal graph, where nodes are all GNs or all HNs. Let D denote the inner-degree of each GN or HN. Then there are two node partitions in such a graph, each of which contains D nodes. Next, we prove this proposition in two cases where nodes S and T are contained in the same node partition or not.

- *Case 1*: Nodes S and T are contained in the same node partition. In this case, there are D length-2 paths between nodes S and T, each of which goes through a unique node in the other node partition. By Proposition 68, these paths are edge-disjoint. As a result, there are D edge-disjoint paths between them.
- *Case 2*: Nodes S and T are contained in different node partitions. In this case, there are one length-1 path and $D - 1$ length-3 paths between nodes S and T. The length-1 path is the edge between them. As for those $D - 1$ length-3 paths, we label the other $D - 1$ nodes in the same node partition with node S (resp. node T) as $S_1, S_2, \cdots, S_{D-1}$ (resp. $T_1, T_2, \cdots, T_{D-1}$). In this way, the i-th length-3 path goes through nodes S_i and T_i with $i = 1, 2, 3, \cdots, D - 1$. Particularly, the i-th length-3 path is $S - T_i - S_i - T$. Since edges $S - T_i$, $T_i - S_i$

and $S_i - T$ are different for different values of i, these paths are edge-disjoint. Hence, there are D edge-disjoint paths between them.

Therefore, for design patterns with $d_2 =$ CBG(any), the number of edge-disjoint paths between nodes S and T is equal to the inner-degree of each GN or HN. The proof of this proposition is completed.

Proof of Proposition 71. When $d_2 =$ BIG(block) or $d_2 =$ BIG(element), BIG-$(\theta_1, r_1, k_1, \lambda_1)$ with c_1 copies is fixed as the internal graph, where block nodes are RNs and element nodes are all GNs or all HNs. For two element nodes S and T in such a graph, they can be duplicated from each other or not. Next, we prove this proposition in two cases where nodes S and T are duplicated from each other or not.

- *Case 1:* Nodes S and T are duplicated from each other. In this case, such two element nodes are adjacent to the same block nodes. According to the properties of BIG with copies (see Remark 2.6), the inner-degree of each element node is r_1. Hence, nodes S and T are adjacent to r_1 common block nodes. This implies that there exists r_1 length-2 paths between nodes S and T, each of which goes through a unique block node. By Proposition 68, these r_1 length-2 paths are edge-disjoint. Therefore, in this case, there are r_1 edge-disjoint paths between nodes S and T.
- *Case 2:* Nodes S and T are not duplicated from each other. In this case, such two element nodes are adjacent to λ_1 common block nodes according to the properties of BIG with copies (see Remark 2.6). Meanwhile, the inner-degree of each element node is r_1. This implies that each element node is adjacent to r_1 block nodes. Accordingly, nodes S and T are adjacent to $2r_1 - \lambda_1$ block nodes, which consists of the following three parts.
  1. The first part contains $r_1 - \lambda_1$ block nodes that are adjacent to node S but not adjacent to node T. Let symbol M_i denote the i-th block node in this part with $i = 1, 2, \cdots, r_1 - \lambda_1$.
  2. The second part contains $r_1 - \lambda_1$ block nodes that are adjacent to node T but not adjacent to node S. Let symbol N_j denote the j-th block node in this part with $j = 1, 2, \cdots, r_1 - \lambda_1$.
  3. The third part contains λ_1 block nodes that are adjacent to nodes S and T both. Let symbol O_k denote the k-th block node in this part with $k = 1, 2, \cdots, \lambda_1$.

  The subsequent proof relies on such a fact: through one-to-one assignment, there exists $r_1 - \lambda_1$ node pairs $< M_i, N_j >$, such that these pairs have no common nodes and each pair of nodes are directly connected through a length-2 path. Such a fact is demonstrated by the following three steps.

- *Step 1.* First, we construct a *logic* bipartite graph, where two node partitions are block nodes in the first part and second part, respectively. If there exists a length-2 path between a node in the first part and a node in the second part, then they are directly connected in the logic bipartite graph.
- *Step 2.* Second, we show that in the logic bipartite graph, given a subset X containing any x nodes in the first part, they are adjacent to at least x nodes in the second part.

  According to the described construction of logic bipartite graph, two nodes are adjacent in the logic bipartite graph if they are directly connected to some common element nodes in the original BIG. Note that in the original BIG, each block node is directly connected to k_1 element nodes. Hence, each node in the first part is directly connected to $k_1 - 1$ distinct element nodes other than node S. Consequently, any x nodes in the first part is directly connected to $x(k_1 - 1)$ element nodes other than node S. Note that such a number is counted with repetition because two nodes in the first part may be directly connected to some common element nodes. Nonetheless, once these x nodes is directly connected to some element node (say node v), then these x nodes are adjacent to all nodes in the second part who are directly connected to node v. Since the number of nodes in the second part that are directly connected to node v is greater than or equal to the number of nodes in subset X that are directly connected to node v, the nodes in subset X is adjacent to at least $x(k_1 - 1)$ nodes in the second part. Note that if one node in subset X and one node in the second part are directly connected to two or more common element nodes, such a number $x(k_1 - 1)$ counts two or more adjacent node but not one. Hence, such a number is also counted with repetition. Since each node in subset X is directly connected to at most $k_1 - 1$ common element nodes with node in the second part, the true number is greater than or equal to $x(k_1 - 1)/(k_1 - 1) = x$. Therefore, given any x nodes in the first part, they are adjacent to at least x nodes in the second part in the logic bipartite graph.
- *Step 3.* Third, we demonstrate the aforementioned fact by proving that the logic bipartite graph has a perfect matching. By the well-known Hall's theorem (see Theorem 5.2 in Bondy and Murty [1976]), a bipartite graph with two node partitions A and B has a perfect matching if and only if for any subset $X \subseteq A$, the number of adjacent nodes of nodes in X is greater than or equal to the number of nodes in X. Based on the conclusion in Step 2, the logic bipartite graph has a perfect matching. According to the definition of perfect matching, each node in the first part can be assigned to a unique node in the second part, such that they are directly connected by an edge in the logic graph (i.e., there exists a length-2 path in the original BIG with copies). As a result, the fact is proved.

Through these block nodes, we demonstrate that there exist r_1 edge-disjoint paths between nodes S and T, including λ_1 length-2 paths and $r_1 - \lambda_1$ length-4 paths. Next, we first show the existence of these paths, and then demonstrate that they are edge-disjoint.

- *Existence of λ_1 length-2 paths.* Regarding the λ_1 length-2 paths, each of which goes through a unique block node in the third part. In other words, the k-th length-2 path between nodes S and T is $S - O_k - T$, where $k = 1, 2, \cdots, \lambda_1$. The existence of these paths is established by the existence of nodes O_k with $k = 1, 2, \cdots, \lambda_1$.
- *Existence of $r_1 - \lambda_1$ length-4 paths.* As for the $r_1 - \lambda_1$ length-4 paths, each of which traverses a length-2 sub-path between a block node in the first part and a block node in the second part. Particularly, the i-th length-4 path between nodes S and T ($i = 1, 2, \cdots, r_1 - \lambda_1$) traverses three sub-paths, i.e., sub-path $S - M_i$, length-2 sub-path between M_i and N_j, and sub-path $N_j - T$. The existence of sub-paths $S - M_i$ and $N_j - T$ follows the existence of nodes M_i and N_j, while the existence of length-2 sub-path between M_i and N_j is established by this fact.
- *These paths are disjoint.* For these paths with length 2 or 4, the even-order nodes are block nodes either in the first part, second part, or third part. Particularly, the second node in the k-th length-2 path is O_k. The second node in the i-th length-4 path is M_i. By this fact, these length-4 paths have the distinct forth nodes. Since the even-order nodes in these paths are different, by Proposition 68, these paths are edge-disjoint.

As a result, in this case, there are r_1 edge-disjoint paths between nodes S and T, including λ_1 length-2 paths and $r_1 - \lambda_1$ length-4 paths.

Based on these discussions, there are r_1 edge-disjoint paths between nodes S and T. Note that the inner-degree of each GN or HN (element node) is equal to r_1 in BIG-$(\theta_1, r_1, k_1, \lambda_1)$ with copies (see Remark 2.6). Therefore, for design patterns with $d_2 =$ BIG(block) or $d_2 =$ BIG(element), the number of edge-disjoint paths between nodes S and T is equal to the inner-degree of each GN or HN. The proof of this proposition is completed.

Proof of Proposition 72. When $d_2 =$ TDG(block) or $d_2 =$ TDG(element), TDG-(n_1, μ_1, υ_1) with c_1 copies is fixed as the internal graph, where block nodes are RNs and element nodes are all GNs or all HNs. For two element nodes S and T in such a graph, they can be duplicated from each other if $c_1 \geq 1$ (i.e., the internal graph is TDG with copies). Next, we prove this proposition in two cases where $c_1 = 0$ and $c_1 \geq 1$, respectively.

- *Case 1:* $c_1 = 0$. In this case, the internal graph is TDG-(n_1, μ_1, υ_1) without copies. By Definition 1.24, element nodes are divided into n_1 groups with υ_1 element nodes, such that each pair of element nodes in distinct groups are adjacent to μ_1 common block nodes and each pair of element nodes in the same group are not adjacent to any common block nodes. Correspondingly, this case consists of two sub-cases about the relationships between nodes S and T. In the first sub-case, node S is contained in the same group with node T. In the second sub-case, node S is contained in a distinct group with node T. Next, we prove this proposition in such two sub-cases, respectively.
  - *Sub-case 1.* In this sub-case, node S is contained in the same group with node T. We consider another group of element nodes. Let symbol U_i denote the i-th element node in such a group with $i = 1, 2, \cdots, \upsilon_1$. By Definition 1.24, nodes S and U_i are adjacent to μ_1 common block nodes, while T and U_i are adjacent to another μ_1 common block nodes. Let symbol V_{ij} (resp. symbol W_{ij}) denote the j-th common block node adjacent to element nodes S (resp. T) and U_i, where $j = 1, 2, \cdots, \mu_1$. Then there exists $\mu_1\upsilon_1$ length-4 paths between nodes S and T. Particularly, the k-th length-4 path is $S - V_{ij} - U_i - W_{ij} - T$ with $k = \mu_1 \cdot (i-1) + j$. Note that block nodes V_{ij}s are adjacent to element node S, while block nodes W_{ij}s are adjacent to element node T. Hence, by Definition 1.24, block nodes V_{ij}s are different from block nodes W_{ij}s. Note that nodes V_{ij}s and W_{ij}s are the second node and the forth node of the $[\mu_1 \cdot (i-1) + j]$-th length-4 path. Consequently, by Proposition 68, these length-4 paths are edge-disjoint.
  - *Sub-case 2.* In this sub-case, node S is contained in distinct group with node T. We demonstrate that there exists $\mu_1\upsilon_1$ edge-disjoint paths between nodes S and T, including μ_1 length-2 paths and $\mu_1(\upsilon_1 - 1)$ length-4 paths. Next, we first show the existence of these paths, and then demonstrate they are edge-disjoint.
    - *Existence of μ_1 length-2 paths.* Regarding the μ_1 length-2 paths, each of which goes through a unique block node in the third part. Let R_k denote the k-th block node that are adjacent to nodes S and T both, where $k = 1, 2, cdots, \mu_1$. In this way, the k-th length-2 path between nodes S and T is $S - R_k - T$, where $k = 1, 2, \cdots, \mu_1$. The existence of these paths is established by the existence of nodes R_k with $k = 1, 2, \cdots, \mu_1$.
    - *Existence of $\mu_1(\upsilon_1 - 1)$ length-6 paths.* Note that each group of element nodes contains υ_1 nodes. Let symbols S_i and T_i denote the i-th element node in the same group with node S and node T, respectively, where $i = 1, 2, \cdots, \upsilon_1 - 1$. In this way, the $\mu_1(\upsilon_1 - 1)$ length-6 paths can be divided into $\upsilon_1 - 1$ groups, such that the paths in the i-th group goes through node S_i and node T_i. Particularly, the paths in the i-th group traverse three length-2 sub-paths, i.e., sub-path between S and T_i, sub-path between T_i and S_i, and sub-path between S_i and T. Note that

nodes S and S_i are contained in distinct groups with nodes T and T_i. By Definition 1.24, there exists μ_1 length-2 sub-paths between S and T_i, μ_1 length-2 sub-paths between T_i and S_i, and μ_1 length-2 sub-paths between S_i and T. As a result, there exists at least μ_1 length-6 paths between nodes S and T in the i-th group. Since there are $\upsilon_1 - 1$ groups of paths, there exists $\mu_1(\upsilon_1 - 1)$ length-6 paths between nodes S and T.

- *These paths are disjoint.* For these paths with length 2 or 4, the even-order nodes are common block nodes of either S and T, S and T_i, S_i and T, or S_i and T. Note that any two element nodes have no common block nodes. Hence, these even-order nodes are different from each other. By Proposition 68, these $\mu_1\upsilon_1$ paths are edge-disjoint.

As a result, in this sub-case, there are $\mu_1\upsilon_1$ edge-disjoint paths between nodes S and T, including μ_1 length-2 paths and $\upsilon_1(\mu_1 - 1)$ length-6 paths.

In summary, there are $\mu_1\upsilon_1$ edge-disjoint paths between nodes S, T in Case 1.

- *Case 2:* $c_1 \geq 1$. In this case, nodes S and T could be duplicated from each other or not. Next, we prove this proposition in such two sub-cases, respectively.
  - *Sub-case 1:* Nodes S and T are duplicated from each other. In this sub-case, nodes S and T are adjacent to the same block nodes. According to the properties of TDG with copies (see Remark 2.8), the inner-degree of each element node is $\mu_1\upsilon_1$. Hence, nodes S and T are adjacent to $\mu_1\upsilon_1$ common block nodes. This implies that there exists $\mu_1\upsilon_1$ length-2 paths between nodes S and T, each of which goes through a unique block node. By Proposition 68, these $\mu_1\upsilon_1$ length-2 paths are edge-disjoint. Therefore, in this sub-case, there are $\mu_1\upsilon_1$ edge-disjoint paths between nodes S and T.
  - *Sub-case 2:* Nodes S and T are not duplicated from each other. In this sub-case, let symbol T' denote the node satisfying the following two conditions.
    1. Nodes T' and T are duplicated from each other.
    2. Node T' and S are in the same copy of original element nodes.

  By discussions in Case 1, no matter nodes S and T' are contained in the same group or not, there exist $\mu_1\upsilon_1$ edge-disjoint paths between them. Meanwhile, since nodes T and T' are duplicated from each other, they have the same connection pattern with other nodes. Hence, there exist $\mu_1\upsilon_1$ edge-disjoint paths between nodes S and T.

In summary, there are $\mu_1\upsilon_1$ edge-disjoint paths between nodes S, T in Case 2.

Based on these discussions, there are $\mu_1\upsilon_1$ edge-disjoint paths between nodes S and T. Note that the inner-degree of each GN or HN (element node) is equal to $\mu_1\upsilon_1$ in TDG-(n_1, μ_1, υ_1) with copies (see Remark 2.6). Therefore, for design patterns with d_2 = TDG(block) or d_2 = TDG(element), the number of edge-disjoint paths between nodes S and T is equal to the inner-degree of each GN or HN. The proof of this proposition is completed.

Note that the inner-degree of each GN (resp. each HN) is equal to $\alpha_G p_G$ (resp. $\alpha_H p_H$). With these propositions, it follows that the value of A_1 (i.e., the number of edge-disjoint paths between any pair of GNs or HNs) is equal to $\alpha_G p_G$ if $p = G$ or $\alpha_H p_H$ if $p = H$.

The value of $\mathbf{B_1}$. The value of B_1 (i.e., the minimum attachment number of each GN or HN) is considered for design patterns with $d_1 = G$ and $d_1 = H$, respectively.

- For design patterns with $d_1 = G$, the value of B_1 represents the attachment number of each GN. Note that the attachment number of each GN is equal to its directly connected CNs. Since parameter γ_G denotes the proportions of degrees of GNs for the connections with CNs, the attachment number of each GN is equal to $\gamma_G p_G$, which implies $B_1 = \gamma_G p_G$.
- For design patterns with $d_1 = H$, the value of B_1 represents the attachment number of each HN. Note that the attachment number of each HN is equal to its directly connected CNs plus one. Since parameter γ_H denotes the proportions of degrees of HNs for the connections with CNs, the attachment number of each HN is equal to $\gamma_H p_H$, which implies $B_1 = \gamma_H p_H$.

In summary, the value of B_1 is equal to $\gamma_G p_G$ if $d_1 = G$ or $\gamma_H p_H$ if $d_1 = H$.

Based on these discussions, if $d_1 = G$, the intra-module throughput is equal to

$$T_1 = \frac{A_1}{B_1} = \frac{\alpha_G p_G}{\gamma_G p_G} = \frac{\alpha_G}{\gamma_G}. \tag{A.23}$$

Otherwise, the intra-module throughput is equal to

$$T_1 = \frac{A_1}{B_1} = \frac{\alpha_H p_H}{\gamma_H p_H + 1}. \tag{A.24}$$

The proof is completed. ■

A.4.19 Proof of Lemma 4.19

Proof: In this proof, we first demonstrate the relationship between the values of T_2 and R_3, and then calculate the values of ratio R_3.

- *Relationship between the values of T_2 and R_3.* By Definition 2.8, inter-module throughput T_2 is defined as a ratio A_2/B_2. The parameter A_2 represents the minimum number of edge-disjoint paths between any two modules, such that the edge-disjoint paths between two modules do not traverse through the nodes in other modules. The parameter B_2 denotes the maximum attachment number of modules. The relationship between ratio A_2/B_2 and ratio R_3 is considered for design patterns with $d_3 = P$ and $d_3 = B$, respectively.
  - For design patterns with $d_3 = P$, the skeleton graph is fixed as a CG with m_2 modules. Each pair of modules are connected via a super-edge as a t-parallelized all-to-all interconnection. Such a super-edge provides tx^2 edges

between each pair of modules (i.e., $A_2 = tx^2$) with occupying tx^2 outer-degrees of dock nodes in each module. Note that each module is directly connected to $m_2 - 1$ modules in the skeleton graph. Hence, the total outer-degree of dock nodes in each module is equal to $tx^2 \cdot (m_2 - 1)$. In this way, the value of inter-module throughput T_2 is equal to

$$T_2 = \frac{A_2}{B_2} = \frac{tx^2}{B_2} = \frac{tx^2 \cdot (m_2 - 1)}{B_2} \cdot \frac{1}{m_2 - 1} = R_3 \cdot \frac{1}{m_2 - 1}. \tag{A.25}$$

- For design patterns with $d_3 = B$, the skeleton graph is fixed as a BIG-$(\theta_2, r_2, k_2, \lambda_2)$ with c_2 copies. By the property of BIG with copies as the skeleton graph (see Remark 2.10), each pair of modules are interconnected through at least λ_2 length-2 paths over super-edges. Each length-2 path over super-edges provides $t\mu_2 x^2$ edge-disjoint paths over edges between dock nodes. Hence, the value of A_2 (i.e., the number of edge-disjoint paths over edges between any pair of modules) is equal to $\lambda_2 \cdot t\mu_2 x^2$. Meanwhile, each module is directly connected to r_2 RN clusters through r_2 super-edges between a module and an RN cluster. Each super-edge occupies $t\mu_2 x^2$ outer-degrees of dock nodes in each module. Hence, the total outer-degree of dock nodes in each module is equal to $r_2 \cdot t\mu_2 x^2$. Accordingly, we have

$$T_2 = \frac{A_2}{B_2} = \frac{\lambda_2 \cdot t\mu_2 x^2}{B_2} = \frac{r_2 \cdot t\mu_2 x^2}{B_2} \cdot \frac{\lambda_2}{r_2} = R_3 \cdot \frac{\lambda_2}{r_2}. \tag{A.26}$$

Based on these discussions, the value of inter-module throughput T_2 is equal to $R_3/(m_2 - 1)$ if $d_3 = P$ or $R_3 \cdot \lambda_2 / r_2$ if $d_3 = B$.

- *The value of ratio R_3.* The value of ratio R_3 is calculated for design patterns with different values of term d_2 as follows.
  - For design patterns with $d_1 = \text{CG(any)}$, $d_1 = \text{CBG(any)}$, $d_1 = \text{BIG(element)}$, or $d_1 = \text{TDG(element)}$, the GNs or HNs are selected as dock nodes. Hence, the ratio R_3 is expressed as follows:

$$\begin{aligned} R_3 &= \frac{\text{total outer-degree of dock nodes in each module}}{\text{attachment number of each module}} \\ &= \frac{\text{outer-degree of each GN/HN} \cdot \text{\# of GNs/HNs in each module}}{\text{attachment number of each GN/HN} \cdot \text{\# of GNs/HNs in each module}} \\ &= \frac{\text{outer-degree of each GN or HN}}{\text{attachment number of each GN or HN}}. \end{aligned}$$

  Based on this expression, we discuss the value of R_3 for design patterns with $d_1 = G$ and $d_1 = H$, respectively.
    - If $d_1 = G$, the attachment number of a GN is its directly connected CNs. Recall that parameters β_G and γ_G denote the proportions of degrees for outer-degrees and connections with CNs, respectively. Therefore, the value of ratio R_3 is equal to β_G / γ_G.

- If $d_1 = H$, the attachment number of an HN is its directly connected CNs plus one (itself). Recall that parameters β_H and γ_H denote the proportions of degrees for outer-degrees and connections with CNs, respectively. Therefore, the value of ratio R_3 is equal to $\beta_H p_H / \gamma_H + 1$.

- For design patterns with $d_2 =$ BIG(block) or $d_2 =$ TDG(block), RNs are selected as dock nodes. Meanwhile, the internal graph is fixed as a biregular bipartite graph (BIG or TDG), where RNs are in one node partition and GNs or HNs are in the other node partition. Hence, the ratio of the number of dock nodes (RNs) in each module to the number of GNs (or HNs) in each module is equal the the inner-degree of each GN (or HN) to the inner-degree of each RN. Accordingly, the ratio R_3 is expressed as follows:

$$R_3 = \frac{\text{total outer-degree of dock nodes in each module}}{\text{attachment number of each module}}$$
$$= \frac{\text{outer-degree of each RN} \cdot \text{\# of RNs in each module}}{\text{attachment number of each GN or HN} \cdot \text{\# of GNs or HNs in each module}}$$
$$= \frac{\text{outer-degree of each RN} \cdot \text{inner-degree of each GN or HN}}{\text{attachment number of each GN or HN} \cdot \text{inner-degree of each RN}}.$$

Based on this expression, we discuss the value of R_3 for design patterns with $d_1 = G$ and $d_1 = H$, respectively.

- If $d_1 = G$, the attachment number of a GN is its directly connected CNs. Recall that parameters α_G and γ_G denote the proportions of degrees of GNs for outer-degree and connections with CNs, respectively. Meanwhile, parameters α_R and β_R denote the proportions of degrees of RNs for inner-degree and outer-degree, respectively. Therefore, the value of ratio R_3 is equal to $(\beta_R \alpha_G)/(\alpha_R \gamma_G)$.
- If $d_1 = H$, the attachment number of an HN is its directly connected CNs plus one (itself). Recall that parameters β_H and γ_H denote the proportions of degrees of HNs for outer-degrees and connections with CNs, respectively. Meanwhile, parameters α_R and β_R denote the proportions of degrees of RNs for inner-degree and outer-degree, respectively. Therefore, the value of ratio R_3 is equal to $(\beta_R \alpha_H p_H)/[\alpha_R(\gamma_H p_H + 1)]$.

The proof is completed. ■

A.4.20 Proof of Lemma 4.20

Proof: This proof is conducted by calculating the value of fault-tolerance F for design patterns with $d_3 = P$ and $d_3 = B$, respectively.

- For design patterns with $d_3 = P$, the skeleton graph is fixed as a CG with m_2 modules. Let symbols M_1 and M_2 denote any pair of ordered modules who have the same connections with the RN clusters. Each pair of modules are connected via a super-edge as a t-parallelized all-to-all interconnection. By the property of t-parallelized all-to-all interconnection, each dock node in module M_1 is directly connected to all dock nodes in one ISN-(x, w, l) in module M_2. Meanwhile, for any GN in module M_2, there are w dock nodes in the ISN-(x, w, l) within distance l. Hence, given a dock node in module M_1, the failure of w dock nodes in M_2 would disconnect the short-distance communications between some pairs of GNs in such two modules through the given dock nodes in module M_1. By Definition 2.9, the fault-tolerance of each ordered module pairs is w. Accordingly, the fault-tolerance of the whole topology is w.

- For design patterns with $d_3 = B$, the skeleton graph is fixed as a BIG-$(\theta_2, r_2, k_2, \lambda_2)$ with c_2 copies. In such a BIG, there are $\theta_2 k_2(c_2 + 1)$ modules. Hence, the number of ordered module pairs N is equal to

$$N = 2 \cdot \binom{\theta_2 k_2(c_2 + 1)}{2} = \theta_2 k_2(c_2 + 1) \cdot [\theta_2 k_2(c_2 + 1) - 1]. \tag{A.27}$$

Meanwhile, these $\theta_2 k_2(c_2 + 1)$ modules are divided into $c_2 + 1$ groups, each of which corresponds to a copy of element nodes. In this way, the connections between each group of modules and RN clusters are exactly identical, which form a BIG-$(\theta_2, r_2, k_2, \lambda_2)$ with no copies. This implies that there are two cases for the fault-tolerance of an ordered module pair. The first case corresponds to the ordered module pairs who have the same connections with the RN clusters, while the second case corresponds to the ordered module pairs who have distinct connections with the RN clusters. Next, we consider the number of ordered module pairs and their fault-tolerances in such two cases, respectively.

  - *Case 1*: ordered module pairs have the same connections with RN clusters. Given two group of modules that correspond to two copies of element nodes, for each module in one group, there exists a module in the other group such that these two modules have the same connections with the RN clusters. Note that there are $c_2 + 1$ copies of element nodes (modules), each of which contains $\theta_2 k_2$ modules. Let symbol N_1 denote the number of ordered module pairs who have the same connections with the RN clusters, then we have

$$N_1 = \binom{c_2 + 1}{2} \cdot \theta_2 k_2 \cdot 2 = \theta_2 k_2 c_2(c_2 + 1). \tag{A.28}$$

Let symbols M_1 and M_2 denote any pair of ordered modules who have the same connections with RN clusters. By the properties of t-parallelized TD-(n, μ, x) interconnection, we have the following two conclusions. *First*, each RN cluster is directly connected to t ISNs-(x, w, l) in each module.

Second, the x dock nodes in each ISN-(x, w, l) are not directly connected to common RNs in the RN clusters. Therefore, given a dock node in module M_1, it is connected to t dock nodes belonging to t distinct ISNs-(x, w, l) in module M_2 via some RN in one RN cluster. Hence, when a GN in module M_1 selects a dock node in the same module to communicate with a GN in module M_2, the failure of t dock nodes in module M_2 would disconnect the short-distance communications from the GN in module M_1 to some GN in module M_2 via one RN cluster. By Definition 2.9, the fault-tolerance of each ordered module pair who has the same connections with the RN clusters is t.

- *Case 2*: ordered module pairs have distinct connections with RN clusters. Let N_2 denote the number of ordered module pairs who have distinct connections with the RN clusters, then we have

$$N_2 = N - N_1. \tag{A.29}$$

For any pair of ordered modules (denoted by symbols M_3 and M_4), given a dock node in module M_3, this dock node is connected to t ISNs-(x, w, l) in module M_4 through one RN cluster. Meanwhile, for any GN in module M_3, there are w dock nodes in each ISN-(x, w, l) within distance l. Hence, the failure of tw dock nodes of t ISNs-(x, w, l) in module M_4 would disconnect the short-distance communications from a GN in module M_3 to some GN in module M_4 through the given dock node in module M_3 via one RN cluster. By Definition 2.9, the fault-tolerance of these ordered module pairs is tw.

Based on these discussions, the fault-tolerance of the whole topology F is equal to

$$F = \frac{N_1 \cdot t + N_2 \cdot tw}{N} = t \cdot \frac{N_1 + (N - N_1) \cdot w}{N} = t \cdot \left[\left(1 - \frac{N_1}{N}\right) \cdot (w - 1) + 1 \right]. \tag{A.30}$$

Regarding the term $1 - N_1/N$, we substitute the values of N and N_1 in equations (A.27) and (A.28), leading to

$$
\begin{aligned}
1 - \frac{N_1}{N} &= 1 - \frac{\theta_2 k_2 c_2 (c_2 + 1)}{\theta_2 k_2 (c_2 + 1)[\theta_2 k_2 (c_2 + 1) - 1]} \\
&= 1 - \frac{c_2}{\theta_2 k_2 (c_2 + 1) - 1} \\
&= 1 - \frac{t c_2}{\theta_2 k_2 (c_2 + 1) t - t} \\
&= 1 - \frac{t c_2}{\theta_2 p_R - t} \\
&= \frac{\theta_2 p_R - t(c_2 + 1)}{\theta_2 p_R - t} \\
&= \frac{\theta_2 p_R - \frac{p_R}{k_2}}{\theta_2 p_R - t} \\
&= \frac{\theta_2 - \frac{1}{k_2}}{\theta_2 - \frac{t}{p_R}}.
\end{aligned}
\tag{A.31}
$$

By substituting the values of term $1 - N_1/N$ into equation (A.31), we have

$$
F = t \cdot \left[\frac{\theta_2 - \frac{1}{k_2}}{\theta_2 - \frac{t}{p_R}} \cdot (w - 1) + 1 \right]. \tag{A.32}
$$

The proof is completed. ■

References

AFL. AI Data Centers: Scaling Up and Scaling Out. Technical report. AFL, 2024. https://www.aflhyperscale.com/wp-content/uploads/2024/12/AI-Data-Centers-Scaling-Up-and-Scaling-Out-White-Paper.pdf.

Jung Ho Ahn, et al. HyperX: Topology, routing, and packaging of efficient large-scale networks. In *Proceedings of the Conference on High Performance Computing Networking, Storage and Analysis,* 2009.

Mohammad Al-Fares, Alexander Loukissas, and Amin Vahdat. A scalable commodity data center network architecture. In *Proceedings of ACM SIGCOMM*, 2008.

Alibaba Cloud. In-depth analysis of Alibaba Cloud Panjiu AL128 Supernode AI servers and their interconnect architecture. Alibaba Cloud Blog, 2025. https://www.alibabacloud.com/blog/in-depth-analysis-of-alibaba-cloud-panjiu-al128-supernode-ai-servers-and-their-interconnect-architecture_602665

Noga Alon. Eigenvalues and expanders. *Combinatorica*, 6(2):83–96, 1986.

Amazon Web Services. New Amazon EC2 P6e-GB200 ultraservers powered by NVIDIA Grace Blackwell GPUs for the highest AI performance. AWS Blog, 2025. https://aws.amazon.com/blogs/aws/new-amazon-ec2-p6e-gb200-ultraservers-powered-by-nvidia-grace-blackwell-gpus-for-the-highest-ai-performance/.

AMD. AMD Instinct™ MI300X Accelerator Data Sheet. Technical report, AMD, 2023. https://www.amd.com/content/dam/amd/en/documents/instinct-tech-docs/data-sheets/amd-instinct-mi300x-data-sheet.pdf.

AMD. AMD Instinct™ MI300X Platform Data Sheet. Technical report, AMD, 2024. https://www.amd.com/content/dam/amd/en/documents/instinct-tech-docs/data-sheets/amd-instinct-mi300x-platform-data-sheet.pdf.

AMD. AMD Q4 2024 earnings call transcript. AMD Investor Relations, 2025a. https://ir.amd.com/financial-information/quarterly-results.

AMD. LLM inference on AMD instinct accelerators. AMD ROCm Blog, 2025b. https://rocm.blogs.amd.com/artificial-intelligence/LLM_Inference/README.html

Wei An, et al. Fire-flyer AI-HPC: A cost-effective software-hardware co-design for deep learning. In *Proceedings of IEEE SC*, 2024.

Alexey Andreyev. Introducing data center fabric, the next generation Facebook Data Center Network, 2014. https://engineering.fb.com/2014/11/14/production-engineering/introducing-data-center-fabric-the-next-generation-facebook-data-center-network/.

Anthropic. Claude opus. Anthropic, 2025. https://www.anthropic.com/claude/opus.

Oluwatosin Oladayo Aramide. Ultra ethernet vs. infiniband for AI/ML clusters: A comparative study of performance, cost and ecosystem viability. *Open Access Research Journal of Science and Technology*, 12(02):169–179, 2024.

Sadoon Azizi, Naser Hashemi, and Ahmad Khonsari. A flexible and high-performance data center network topology. *The Journal of Supercomputing*, 73(4):1484–1503, 2017.

Baidu. Baidu successfully lights up Kunlun core third-generation 10,000-card cluster and further 30,000-card cluster. Baidu News, 2025. https://home.baidu.com/home/index/news_detail/id/18018.

Baidu Cloud. Baidu Kunlun Supernode 32 XPU release. Technical Article, Baidu Cloud, 2025. https://cloud.baidu.com/article/3546962.

Travis Beals, et al. Towards a future space-based, highly scalable AI infrastructure system design. *arXiv preprint arXiv:2511.19468*, 2025.

John Adrian Bondy and Uppaluri Siva Ramachandra Murty. *Graph Theory with Applications*. London, Macmillan, 1976.

Raj Chandra Bose. A note on the resolvability of balanced incomplete block designs. *The Indian Journal of Statistics*, 6:105–110, 1942.

Andries E. Brouwer and Willem H. Haemers. *Spectra of Graphs*. Springer, 2011.

Yijia Chang, et al. Systematic topology design for large-scale networks: A unified framework. In *Proceedings of IEEE INFOCOM*, 2020.

Tao Chen, Xiaofeng Gao, and Guihai Chen. The features, hardware, and architectures of data center networks: A survey. *Journal of Parallel & Distributed Computing*, 96:45–74, 2016.

Shou-Yi Cheng and Jen-Hui Chuang. Varietal hypercube-A new interconnection network topology for large scale multicomputer. In *Proceedings of IEEE ICPADS*, 1994.

Mung Chiang, et al. Layering as optimization decomposition: A mathematical theory of network architectures. *Proceedings of the IEEE*, 95(1):255–312, 2007.

China Daily. Baidu unveils new Kunlun chips. China Daily, 2025. https://www.chinadaily.com.cn/a/202511/14/WS69166bfba310d6866eb29629.html.

China Mobile. China Mobile Next-generation Network Infrastructure for Cloud AI White Paper. Technical report. Future Network Development Conference, 2025. https://www.gfnds.com/uploads/20250819/31d0bc6afa27c9f478bcd4b0f0081e7e.pdf.

Cisco Corporation. *Cisco Annual Internet Report (2018–2023) White Paper*. Cisco Corporation, 2020.

Charles Colbourn and Jeffrey Dinitz. *Handbook of Combinatorial Designs*. CRC press, 2006.

Andrew R. Curtis, Srinivasan Keshav, and Alejandro Lopez-Ortiz. Legup: Using heterogeneity to reduce the cost of data center network upgrades. In *Proceedings of ACM CoNEXT*, 2010.

Andrew R. Curtis, et al. Rewire: An optimization-based framework for unstructured data center network design. In *Proceedings of IEEE INFOCOM*, 2012.

Nicolaas Govert De Bruijn and Paul Erdös. On a combinatorial problem. *Proceedings of the Section of Sciences of the Koninklijke Nederlandse Akademie van Wetenschappen te Amsterdam*, 51(10):1277–1279, 1948.

Daniele De Sensi, et al. Exploring GPU-to-GPU communication: Insights into supercomputer interconnects. In *Proceedings of IEEE SC*, 2024.

DeepSeek-AI, et al. Deepseek-R1: Incentivizing reasoning capability in LLMs via reinforcement learning. *arXiv:2501.12948*, 2025. https://arxiv.org/abs/2501.12948.

Reinhard Diestel. *Graph Theory*. Springer, 2005.

Eric Ding, Chuhan Ouyang, and Rachee Singh. Photonic rails in ML datacenters. In *Proceedings of the 24th ACM Workshop on Hot Topics in Networks*, pages 149–159, 2025.

Michael Dinitz, Michael Schapira, and Gal Shahaf. Approximate Moore graphs are good expanders. *Journal of Combinatorial Theory, Series B*, 141:240–263, 2020.

Dyxnet. 2025 AI Network Technology White Paper. Technical report, Dyxnet Group, 2025. https://www.dyxnet.com/hk/wp-content/uploads/sites/2/2025/08/2025AI%E7%BD%91%E7%BB%9C%E6%8A%80%E6%9C%AF%E7%99%BD%E7%9A%AE%E4%B9%A6.pdf.

Epoch AI. Trends in machine learning training compute (2025 update). Epoch AI Blog, 2025. https://epochai.org/blog/compute-trends-2025.

European Space Agency. ASCEND: Advanced Space Cloud for European Net Zero Emission and Data Sovereignty. Technical report, European Space Agency, 2025. https://ascend-horizon.eu/.

Nathan Farrington and Alexey Andreyev. Facebook's data center network architecture. In *Proceedings of IEEE OIC*, 2013.

Keqin Feng. Spectra of hypergraphs and applications. *Journal of Number Theory*, 60(1):1–22, 1996.

Joel Friedman. A proof of Alon's second eigenvalue conjecture. In *Proceedings of the ACM STOC*, pages 720–724, 2003.

Adithya Gangidi, et al. RDMA over ethernet for distributed training at meta scale. In *Proceedings of ACM SIGCOMM*, 2024.

Talia Gershon, et al. The infrastructure powering IBM's GEN AI model development. *arXiv preprint arXiv:2407.05467*, 2024.

Chris Godsil and Gordon Royle. *Algebraic Graph Theory*. Springer, 2001.

Google Cloud. TPU7x (Ironwood). Technical report, Google Cloud, 2025. https://cloud.google.com/tpu/docs/tpu7x.

T. G. Griffin, F. B. Shepherd, and G. Wilfong. The stable paths problem and interdomain routing. *IEEE/ACM Transactions on Networking*, 10(2):232–243, 2002.

Chuanxiong Guo, et al. Dcell: A scalable and fault-tolerant network structure for data centers. In *Proceedings of ACM SIGCOMM*, 2008.

Chuanxiong Guo, et al. BCube: A high performance, server-centric network architecture for modular data centers. In *Proceedings of ACM SIGCOMM*, 2009.

Deke Guo, et al. BCN: Expansible network structures for data centers using hierarchical compound graphs. In *Proceedings of IEEE INFOCOM*, 2011.

Z. Guo and Y. Yang. On nonblocking multicast fat-tree data center networks with server redundancy. *IEEE Transactions on Computers*, 64(4):1058–1073, 2015.

László Gyarmati and Tuan Anh Trinh. Scafida: A scale-free network inspired data center architecture. *SIGCOMM Computer Communication Review*, 73(4):1484–1503, 2010.

Jacques Hadamard. Resolution d'une question relative aux determinants. *Bulletin des Sciences Mathematiques*, 2:240–246, 1893.

Tom Høholdt and Heeralal Janwa. Eigenvalues and expansion of bipartite graphs. *Designs, Codes and Cryptography*, 65(3):259–273, 2012.

Shlomo Hoory, Nathan Linial, and Avi Wigderson. Expander graphs and their applications. *Bulletin of the American Mathematical Society*, 43(4):439–561, 2006.

IEEE Standards Association. IEEE Std 802.3-2022—IEEE Standard for Ethernet. Technical report, IEEE, 2022. https://standards.ieee.org/ieee/802.3/10422/.

InfiniBand Trade Association. InfiniBand Architecture Specification. Technical report, InfiniBand Trade Association, 2023. https://www.infinibandta.org/ibta-specification/.

Intel. Intel® Gaudi® 3 AI Accelerator HLB-325 Baseboard Product Brief. Technical report, Intel, 2024a. https://cdrdv2-public.intel.com/817489/gaudi-3-ai-accelerator-hlb-325-baseboard-product-brief.pdf.

Intel. Intel unleashes enterprise AI with Gaudi 3 AI open systems. Intel Newsroom Press Release, 2024b. https://www.intc.com/news-events/press-releases/detail/1689/intel-unleashes-enterprise-ai-with-gaudi-3-ai-open-systems.

Intel. Intel® Gaudi® 3 AI Accelerator Technical Paper Generative AI Compute AI for Enterprise. Technical report, Intel, 2025. https://cdrdv2-public.intel.com/817486/gaudi-3-ai-accelerator-white-paper.pdf.

Peter Jones. Ethernet for AI networking. Cisco Live Presentation CISCOU-2038, 2025. https://www.ciscolive.com/c/dam/r/ciscolive/emea/docs/2025/pdf/CISCOU-2038.pdf.

John Kim, William Dally, and Dennis Abts. Flattened butterfly: A cost-efficient topology for high-radix networks. In *Proceedings of ISCA*, 2007.

John Kim, et al. Technology-driven, highly-scalable dragonfly topology. In *Proceedings of ISCA*, 2008.

Thomas P. Kirkman. Note on an unanswered prize question. *Cambridge and Dublin Mathematical Journal*, 5:255–262, 1850.

J. Lee, et al. Reverse-engineering MAC: A non-cooperative game model. *IEEE Journal on Selected Areas in Communications*, 25(6):1135–1147, 2007.

Ang Li, et al. Evaluating modern GPU interconnect: PCIe, NVLink, NV-SLI, NVSwitch and GPUDirect. *IEEE Transactions on Parallel and Distributed Systems*, 31(1):94–110, 2019.

Dan Li, et al. FiConn: Using backup port for server interconnection in data centers. In *Proceedings of IEEE INFOCOM*, 2009.

Heng Liao, et al. UB-mesh: A hierarchically localized nD-fullmesh datacenter network architecture. *arXiv:2503.20377*, 2025. https://arxiv.org/abs/2503.20377.

Vincent Liu, et al. F10: A fault-tolerant engineered network. In *Proceedings of USENIX NSDI*, 2013.

Vincent Liu, et al. Subways: A case for redundant, inexpensive data center edge links. In *Proceedings of ACM CoNEXT*, 2015.

Joseph T. Lizier, et al. Analytic relationship of relative synchronizability to network structure and motifs. *Proceedings of the National Academy of Sciences*, 120(37):e2303332120, 2023.

Harris F. MacNeish. Euler squares. *Annals of Mathematics*, 23(3):221–227, 1922.

Qingkai Meng, et al. Astral: A datacenter infrastructure for large language model training at scale. In *Proceedings of ACM SIGCOMM*, 2025.

Meta. Llama 3.3 model card and prompt formats. Meta, 2024. https://llama.meta.com/docs/model-cards-and-prompt-formats/llama3_3/.

Meta AI. Introducing llama 3.1: Our most capable models to date. Meta AI Blog, 2024. https://ai.meta.com/blog/meta-llama-3-1/.

Meta Engineering Team. Roce networks for distributed AI training at scale. Engineering at Meta Blog, 2024. https://engineering.fb.com/2024/08/05/data-center-engineering/roce-network-distributed-ai-training-at-scale/.

Microsoft. Microsoft azure delivers the first large-scale cluster with NVIDIA GB300 NVL72 for openAI workloads. Microsoft Azure Blog, 2025. https://azure.microsoft.com/en-us/blog/microsoft-azure-delivers-the-first-large-scale-cluster-with-nvidia-gb300-nvl72-for-openai-workloads/.

Jayaram Mudigonda, Praveen Yalagandula, and Jeffrey C. Mogul. Taming the flying cable monster: A topology design and optimization framework for data-center networks. In *Proceedings of USENIX ATC*, 2011.

NADDOD. How to deploy NVIDIA DGX spark. NADDOD Medium Blog, 2025. https://naddod.medium.com/how-to-deploy-nvidia-dgx-spark-7aa4d8151346.

NVIDIA. NVIDIA DGX-1 with Tesla V100 System Architecture. Technical report, NVIDIA, 2017. https://images.nvidia.com/content/pdf/dgx1-v100-system-architecture-whitepaper.pdf.

NVIDIA. NVIDIA A100 40GB PCIe GPU Accelerator Product Brief. Technical report, NVIDIA, 2020a. https://www.nvidia.com/content/dam/en-zz/Solutions/Data-Center/a100/pdf/A100-PCIE-Prduct-Brief.pdf.

NVIDIA. NVIDIA Quantum QM8700 HDR InfiniBand Switch Product Brief. Technical report, NVIDIA, 2020b. https://network.nvidia.com/sites/default/files/doc-2020/pb-qm8700.pdf.

NVIDIA. NVIDIA NVSwitch Technical Overview. Technical report, NVIDIA, 2021. https://images.nvidia.com/content/pdf/nvswitch-technical-overview.pdf.

NVIDIA. ConnectX-6 InfiniBand Adapter Cards Datasheet. Technical report, NVIDIA, 2022. https://www.macnica.co.jp/business/semiconductor/manufacturers/connectx-6-infiniband-datasheet.pdf.

NVIDIA. NVIDIA DGX SuperPOD: Next Generation Scalable Infrastructure for AI Leadership Reference Architecture Featuring NVIDIA DGX H100 Systems. NVIDIA Technical Whitepaper, 2023a. https://docs.nvidia.com/https:/docs.nvidia.com/dgx-superpod-reference-architecture-dgx-h100.pdf.

NVIDIA. NVIDIA DGX A100 System User Guide—Network Configuration. Technical report, NVIDIA, 2023b. https://docs.nvidia.com/dgx/dgxa100-user-guide/network-config.html.

NVIDIA. NVIDIA hopper GPUs expand reach as demand for AI grows. NVIDIA Newsroom, 2023c. https://nvidianews.nvidia.com/news/nvidia-hopper-gpus-expand-reach-as-demand-for-ai-grows.

NVIDIA. Doubling all2all performance with NVIDIA collective communication library 2.12. NVIDIA Developer Blog, 2024a. https://developer.nvidia.com/blog/doubling-all2all-performance-with-nvidia-collective-communication-library-2-12/.

NVIDIA. NVIDIA Blackwell Architecture Technical Brief. Technical report, NVIDIA Corporation, 2024b. https://resources.nvidia.com/en-us-blackwell-architecture/blackwell-architecture-technical-brief.

NVIDIA. NVIDIA DGX H200 Datasheet. Technical report, NVIDIA, 2024c. https://resources.nvidia.com/en-us-dgx-systems/dgx-h200-datasheet.

NVIDIA. NVIDIA DGX B200 System User Guide. Technical report, NVIDIA, 2025a. https://docs.nvidia.com/dgx/dgxb200-user-guide/dgxb200-user-guide.pdf.

NVIDIA. NVIDIA DGX B300 System User Guide. Technical report, NVIDIA, 2025b. https://docs.nvidia.com/dgx/dgxb300-user-guide/dgxb300-user-guide.pdf.

NVIDIA. NVIDIA DGX GB300 Datasheet. Technical report, NVIDIA, 2025c. https://resources.nvidia.com/en-us-dgx-systems/dgx-gb300-datasheet.

NVIDIA. NVIDIA DGX SuperPOD: Next Generation Scalable Infrastructure for AI Leadership Reference Architecture Featuring NVIDIA DGX B200. Technical report, NVIDIA, 2025d. https://docs.nvidia.com/dgx-superpod/reference-architecture-scalable-infrastructure-b200/latest/_downloads/4b3d3302fa4fc6854b595ccff50b9932/RA11334001-DSPB200-ReferenceArch.pdf.

NVIDIA. NVIDIA SuperPOD with DGX B300 Systems, NVIDIA Quantum-X800 InfiniBand Switching and AC Power Reference Architecture. Technical report, NVIDIA, 2025e. https://docs.nvidia.com/dgx-superpod/reference-architecture/scalable-infrastructure-b300-xdr/latest/_downloads/f2d6423ee942e3f144511f5df58e3080/RA11339001-DSPB300-XDR-ReferenceArch.pdf.

NVIDIA. NVIDIA DGX SuperPOD: Next Generation Scalable Infrastructure for AI Factories Reference Architecture Featuring NVIDIA DGX GB300 Systems. Technical report, NVIDIA, 2025f. https://docs.nvidia.com/pdf/dgx-spod-gb300-ra.pdf.

NVIDIA. Introducing NVIDIA Dynamo, a low-latency distributed inference framework for scaling reasoning AI models. NVIDIA Developer Blog, 2025g. https://developer.nvidia.com/blog/introducing-nvidia-dynamo-a-low-latency-distributed-inference-framework-for-scaling-reasoning-ai-models/.

NVIDIA. NVIDIA GB200 NVL72 delivers trillion-parameter LLM training and real-time inference. NVIDIA Developer Blog, 2025h. https://developer.nvidia.com/blog/nvidia-gb200-nvl72-delivers-trillion-parameter-llm-training-and-real-time-inference/.

NVIDIA. NVIDIA GB200 NVL72. NVIDIA Data Center Product Page, 2025i. https://www.nvidia.com/en-us/data-center/gb200-nvl72/.

NVIDIA. GPU-Accelerated AI Inference Whitepaper. NVIDIA Technical Whitepaper, 2025j. https://www.nvidia.com/en-us/lp/ai/inference-whitepaper/.

NVIDIA. NVIDIA announces financial results for fourth quarter and fiscal 2025. NVIDIA Newsroom Press Release, 2025k. https://nvidianews.nvidia.com/news/nvidia-announces-financial-results-for-fourth-quarter-and-fiscal-2025.

NVIDIA. Demystifying AI inference deployments for trillion parameter large language models. NVIDIA Developer Blog, 2025l. https://developer.nvidia.com/blog/demystifying-ai-inference-deployments-for-trillion-parameter-large-language-models/.

NVIDIA. NVIDIA DGX H100/H200 System User Guide. Technical report, NVIDIA, 2025m. https://docs.nvidia.com/dgx/dgxh100-user-guide/index.html.

NVIDIA. NVIDIA Fabric Manager User Guide. Technical report, NVIDIA, 2025n. https://docs.nvidia.com/datacenter/tesla/fabric-manager-user-guide/index.html.

NVIDIA. NVIDIA NVLink and NVLink switch, 2025o. https://www.nvidia.com/en-us/data-center/nvlink/.

S. R. Ohring, et al. On generalized fat trees. In *Proceedings of International Parallel Processing Symposium*, pages 37–44, 1995.

Open Data Center Committee (ODCC). Towards the light: Eth-x ultra-full optical interconnect supernode technology and cost analysis. ODCC News Information, 2025. https://www.odcc.org.cn/news/p-1970680806776844289.html.

Oracle. Oracle offers first zettascale cloud computing cluster. Oracle News, 2024a. https://www.oracle.com/news/announcement/ocw24-oracle-offers-first-zettascale-cloud-computing-cluster-2024-09-11/.

Oracle. Oracle and NVIDIA to deliver sovereign AI worldwide. Oracle News, 2024b. https://www.oracle.com/news/announcement/oracle-and-nvidia-to-deliver-sovereign-ai-worldwide-2024-03-18/.

Afshin Oroojlooy and Davood Hajinezhad. A review of cooperative multi-agent deep reinforcement learning. *Applied Intelligence*, 53(11):13677–13722, 2023.

Jianping Pan, Jinwei Zhao, and Lin Cai. Measuring a low-earth-orbit satellite network. In *Proceedings of IEEE PIMRC*, 2023.

Giannis Patronas, et al. Optical switching for data centers and advanced computing systems. *Journal of Optical Communications and Networking*, 17(1):A87–A95, 2024.

Larry Peterson and Bruce Davie. *Computer Networks: A Systems Approach*. Elsevier, 2007.

Lorenzo Pichetti, et al. Benchmarking ethernet interconnect for HPC/AI workloads. In *Proceedings of ISCA*, 2024.

Kun Qian, et al. Alibaba HPN: A data center network for large language model training. In *Proceedings of ACM SIGCOMM*, 2024.

Zhenghang Ren, et al. Enabling efficient {GPU} communication over multiple {NICs} with {FuseLink}. In *Proceedings of OSDI*, 2025.

Brandon Schlinker, et al. Condor: Better topologies through declarative design. In *Proceedings of ACM SIGCOMM*, 2015.

Steve Scott, et al. The blackwidow high-radix clos network. In *Proceedings of ISCA*, 2006.

SemiAnalysis. AWS trainium3 deep dive: A potential game-changer? SemiAnalysis Newsletter, 2025. https://newsletter.semianalysis.com/p/aws-trainium3-deep-dive-a-potential.

Russell Shackelford, et al. Computing curricula 2005: The overview report. *ACM SIGCSE Bulletin*, 38(1):456–457, 2006.

Ziyu Shao, et al. Intra-data-center traffic engineering with ensemble routing. In *Proceedings of IEEE INFOCOM*, 2013.

Amit Sharma. AI accelerators for large language model in-ference: Architecture analysis and scaling strategies. *arXiv preprint arXiv:2506.00008*, 2025.

Li Shen, et al. On efficient training of large-scale deep learning models. *ACM Computing Surveys*, 57(3):1–36, 2024.

Ji-Yong Shin, Bernard Wong, and Emin Gün Sirer. Small-world datacenters. In *Proceedings of ACM SOCC*, pages 1–13, 2011.

Alexander Shpiner, et al. Dragonfly+: Low cost topology for scaling datacenters. In *Proceedings of IEEE HiPINEB*, 2017.

Shakhrul Iman Siam, et al. Artificial intelligence of things: A survey. *ACM Transactions on Sensor Networks*, 21(1):1–75, 2025.

Arjun Singh, et al. Jupiter rising: A decade of clos topologies and centralized control in google's datacenter network. *Proceedings of ACM SIGCOMM*, 45(4):183–197, 2015.

Ankit Singla, et al. Jellyfish: Networking data centers randomly. In *Proceedings of USENIX NSDI*, 2012.

Amirhossein Sojoodi, et al. Accelerating intra-node GPU communication: A performance model for multi-path transfers. In *Proceedings of IEEE SC*, 2025.

Stanford Institute for Human-Centered Artificial Intelligence (HAI). Artificial Intelligence Index Report 2025. Technical report, Stanford University, 2025. https://aiindex.stanford.edu/report/2025/.

Douglas R. Stinson. *Combinatorial Designs: Constructions and Analysis*. Springer, 2007.

Yantao Sun, et al. Diamond: An improved fat-tree architecture for large-scale data centers. *Journal of Communications*, 9(1):91–98, 2014.

Chunqiang Tang. Meta's hyperscale infrastructure: Overview and insights. *Communications of the ACM*, 68(2):52–63, 2025.

Roman Trobec, et al. Interconnection networks in petascale computer systems: A survey. *ACM Computing Surveys*, 49(3):44:1–44:24, 2016.

Didem Unat, et al. The landscape of GPU-centric communication. *arXiv preprint arXiv:2409.09874*, 2024.

Asaf Valadarsky, et al. Xpander: Towards optimal-performance datacenters. In *Proceedings of ACM CoNEXT*, pages 205–219, 2016.

E. R. van Dam and E. Spence. Combinatorial designs with two singular values II. Partial geometric designs. *Linear Algebra and its Applications*, 396:303–316, 2005.

Walter Wallis and John George. *Introduction to Combinatorics*. CRC Press, 2016.

Meg Walraed-Sullivan, Amin Vahdat, and Keith Marzullo. Aspen trees: Balancing data center fault tolerance, scalability and cost. In *Proceedings of ACM CoNEXT*, pages 85–96, 2013.

Mowei Wang, et al. Neural network meets DCN: Traffic-driven topology adaptation with deep learning. In *Proceedings of ACM SIGMETRICS*, 2018.

Weiyang Wang, et al. Rail-only: A low-cost high-performance network for training LLMs with trillion parameters. In *Proceedings of IEEE HOTI*, 2024.

D. X. Wei, et al. FAST TCP: Motivation, architecture, algorithms, performance. *IEEE/ACM Transactions on Networking*, 14(6):1246–1259, 2006.

Wikipedia. IEEE 802.3. Wikipedia, 2025. https://en.wikipedia.org/wiki/IEEE_802.3.

Wenfeng Xia, et al. A survey on data center networking: Infrastructure and operations. *IEEE Communications Surveys & Tutorials*, 19(1):640–656, 2017.

Ao Xiao, et al. xDeepServe: Model-as-a-service on Huawei CloudMatrix384. *arXiv:2508.02520*, 2025. https://arxiv.org/abs/2508.02520.

Jiajun Xu, et al. On-device language models: A comprehensive review. *arXiv preprint arXiv:2409.00088*, 2024. https://arxiv.org/abs/2409.00088.

Zihan Yan, et al. From ATOP to ZCube: Automated topology optimization pipeline and a highly cost-effective network topology for large model training. In *Proceedings of ACM SIGCOMM*, 2025.

Qiang Yang, et al. Federated machine learning: Concept and applications. *ACM Transactions on Intelligent Systems and Technology*, 10(2):12:1–12:19, 2012.

Zhisheng Ye, et al. Deep learning workload scheduling in GPU datacenters: A survey. *ACM Computing Surveys*, 56(6):1–38, 2024.

Y. Yu and C. Qian. Space shuffle: A scalable, flexible, and high-performance data center network. *IEEE Transactions on Parallel and Distributed Systems*, 27(11):3351–3365, 2016.

Eitan Zahavi, Isaac Keslassy, and Avinoam Kolodny. Quasi fat trees for HPC clouds and their fault-resilient closed-form routing. In *Proceedings of IEEE HOTI*, 2014.

Chenggang Zhao, et al. Insights into deepseek-V3: Scaling challenges and reflections on hardware for AI architectures. In *Proceedings of ISCA*, 2025.

Lie Zhu. Some recent developments on BIBDS and related designs. *Discrete Mathematics*, 123(1-3):189–214, 1993.

Pengfei Zuo, et al. Serving large language models on Huawei CloudMatrix384. *arXiv:2506.12708,* 2025. https://arxiv.org/abs/2506.12708.

Index

Printed in the USA/Agawam, MA
August 12, 2026

909020.010